SILICON-ON-INSULATOR TECHNOLOGY: MATERIALS TO VLSI

3rd Edition

SILICON-ON-INSULATOR TECHNOLOGY: MATERIALS TO VLSI

3rd Edition

by

Jean-Pierre Colinge
University of California

Kluwer Academic Publishers
Boston/Dordrecht/London

Distributors for North, Central and South America:
Kluwer Academic Publishers
101 Philip Drive
Assinippi Park
Norwell, Massachusetts 02061 USA
Telephone (781) 871-6600
Fax (781) 871-6528
E-Mail <kluwer@wkap.com>

Distributors for all other countries:
Kluwer Academic Publishers Group
Post Office Box 322
3300 AH Dordrecht, THE NETHERLANDS
Telephone 31 78 6576 000
Fax 31 78 6576 474
E-Mail <orderdept@wkap.nl>

 Electronic Services <http://www.wkap.nl>

Library of Congress Cataloging-in-Publication

Title: Silicon-on-Insulator Technology:
Materials to VLSI, 3rd Edition
Author (s): Jean-Pierre Colinge
ISBN: 1-4020-7773-4

Printed in the United States of America

Contents

Preface

The first edition of this book was published in 1991. In the preface one can read *"At first SOI technology was only considered as a possible replacement for SOS in some niche applications. It has, however, been discovered since then that thin-film SOI MOSFETs have excellent scaling properties which make them extremely attractive for deep-submicron ULSI applications"*. Six years later one could read in the preface of the second edition: *"SOI technology has boomed since the first edition... SOI chips are commercially available and SOI wafer manufacturers have gone public. SOI has finally made it out of the academic world and is now a big concern for every major semiconductor company."* All this was true, but six years later, at the time of this third edition, these sentences sound so innocent and unaware of a revolution yet to come. The end of the 20th century turned out to be a milestone for the semiconductor industry, as high-quality SOI wafers suddenly became available in large quantities. From then on, it took only a few years to witness the use of SOI microprocessors in personal computers and SOI audio amplifiers in car stereo systems. Chances are the watch you are wearing around your wrist has an SOI chip.

This book retraces the evolution of SOI materials, devices and circuits over a period of roughly twenty years. Twenty years of progress, research and development during which SOI material fabrication techniques have been born and abandoned, devices have been invented and forgotten, sometimes to be re-discovered and re-named a few years later, and during which SOI circuits have little by little proven they could outperform bulk CMOS in every possible way.

> «La vie ne fait pas de cadeau. Et nom de dieu c'est triste Orly le dimanche. Avec ou sans Bécaud.» - Jacques Brel, *Orly*, 1977

Acknowledgements

This book is dedicated to those pioneers who kept on believing that developing exotic devices and theories for the most unreliable, defective, rare and expensive form of silicon substrates could some day lead to practical, or even mainstream applications. It is also dedicated to those individuals who managed to render SOI wafers reliable, defect-free, abundant and relatively inexpensive, and to those who demonstrated that a hundred million SOI transistors could do the same job as a hundred million bulk transistors, only faster and with lower energy consumption.

In particular, I want to dedicate this book to Akira, Alberto, Alexei, André, Anne, Bernard, Bohdan, Carlos, Cor, Daniel, Danielle, Denis, Dimitris, Duy-Phach, Edval, Eric, Fernand, Herman, Gracie, Guido, Guy, Igor, Isabelle, Jean-Marc, Jean-Paul, Jean-Pierre, Jae-Woo, Jian, João Antonio, Jong-Tae, Kuntjoro, Laurent, Luis, Marcelo, Maurice, Michel, Minghui, Pascale, Paul, Peter, Pierrot, Renaud, Shang-Yi, Sorin, Stefan, Tamara, Ted, Ulf, Valeriya, Vincent, Vitaly, Weize, Xavier, and Xiaohui: working with you has been a pleasure and a privilege. And of course, I have special thanks to Cindy who edited the manuscript of this book.

To my Parents, Wife and Children

Chapter 1

INTRODUCTION

The idea of realizing semiconductor devices in a thin silicon film that is mechanically supported by an insulating substrate has been around for several decades. The first description of the insulated-gate field-effect transistor (IGFET), which evolved into the modern silicon metal-oxide-semiconductor field-effect transistor (MOSFET), is found in the historical patent of Lilienfield dating from 1926 [1]. This patent depicts a three-terminal device where the source-to-drain current is controlled by a field effect from a gate, dielectrically insulated from the rest of the device. The piece of semiconductor that constituted the active part of the device was a thin semiconductor film deposited on an insulator. In a sense, it can thus be said that the first MOSFET was a Semiconductor-on-Insulator (SOI) device. The technology of that time was unfortunately unable to produce a successfully operating Lilienfield device. IGFET technology was then forgotten for a while, completely overshadowed by the enormous success of the bipolar transistor discovered in 1947 [2].

It was only years later, in 1960, that Kahng and Atalla realized the first working MOSFET [3], when technology had reached a level of advancement sufficient for the fabrication of good quality gate oxides. The advent of the monolithic integrated circuits gave MOSFET technology an increasingly important role in the world of microelectronics. CMOS technology is currently the driving technology of the whole microelectronics industry.

CMOS integrated circuits are almost exclusively fabricated on bulk silicon substrates, for two well-known reasons: the availability of electronic-grade material produced either by the Czochralski of by the float-zone technique, and because a good-quality oxide can be readily grown on silicon, a process which is not possible on germanium or on compound semiconductors. Yet, modern MOSFETs made in bulk silicon are far from the ideal structure described by Lilienfield. Bulk MOSFETs are made in silicon wafers having a thickness of approximately 800 micrometers, but only the first micrometer at the top of the wafer is used for transistor fabrication. Interactions between the devices and the substrate give rise to a range of unwanted parasitic effects.

One of these parasitics is the capacitance between diffused source and drain and the substrate. This capacitance increases with substrate doping, and becomes larger in modern submicron devices where the doping concentration in the substrate is higher than in previous MOS technologies. Source and drain capacitance is not limited to the obvious capacitance of the depletion regions associated with the junctions, but it also includes the capacitance between the junction and the heavily-doped channel stop located underneath the field oxide. In addition, latchup, which consists of the unwanted triggering of a PNPN thyristor structure inherently present in all bulk CMOS structures, becomes a severe problem in devices with small dimensions.

Of course, some "tricks" have been found to reduce these parasitic components. The area of the source and drain junctions can be minimized by creating local interconnections and placing contacts over the field area, and occurrence of latchup can be reduced by using epitaxial substrates or deep trench isolation. These different techniques, however, necessitate sophisticated processing, which impacts both the cost and the yield of manufacturing.

If an SOI substrate is used, quasi-ideal MOS devices can be fabricated. The SOI MOSFET contains the traditional three terminals (a source, a drain, and a gate which controls a channel in which current flows from source to drain). However, the full dielectric isolation of the devices prevents the occurrence of most of the parasitic effects experienced in bulk silicon devices. To illustrate this, Figure 1-1 schematically represents the cross-sections of both a bulk CMOS inverter and an SOI CMOS inverter. Most parasitic effects in bulk MOS devices find their origin in the interactions between the device and the substrate. Latchup in bulk devices finds its origin in the parasitic PNPN structure of the CMOS inverter represented in Figure 1-1. The latchup path can be symbolized by two bipolar transistors formed by the substrate, the well and the source and drain junctions. Latchup can be triggered by different mechanisms such as node voltage overshoots, displacement current, junction avalanching and photocurrents. For latchup to occur the current gain of the loop formed by the two bipolar transistors be larger than unity $(\beta_F > 1)$[4]. In an SOI CMOS inverter the silicon film containing the active devices is thin enough for the junctions to reach through to the buried insulator. A latchup path is ruled out because there is no current path to the substrate. In addition the lateral PNPN structures contain heavily doped bases (the N+ and P+ drains) that virtually reduce the gain of the bipolar devices to zero.

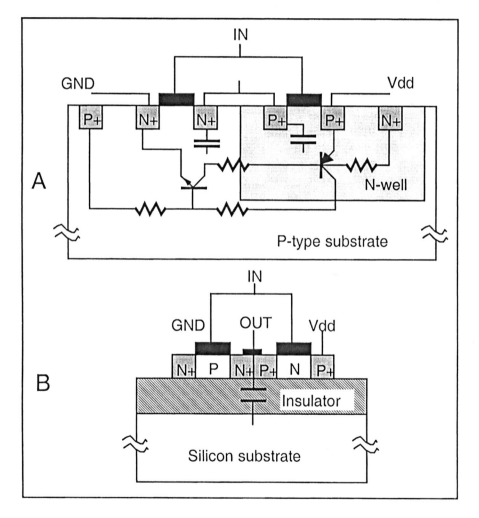

Figure 1-1. A: Cross section showing the latchup path in a bulk CMOS inverter. B: Cross section of an SOI CMOS inverter. The drain parasitic capacitances are also presented.

Bulk circuits utilize reverse-biased junctions to isolate devices from one another. Consider the drain of the n-channel transistor of Figure 1-1A. The drain is always positively or zero biased with respect to the substrate (the drain voltage can range between GND and +V_{DD}). The depletion capacitance of the drain junction reaches a maximum value when the drain voltage is zero and it increases with increased substrate doping concentration. Modern submicron circuits use high substrate doping concentrations, which contributes to an increase in junction capacitance. In addition, an added parasitic capacitance occurs between the junctions and the channel stop implant underneath the field oxide. In SOI circuits the maximum capacitance between the junctions and the substrate is the capacitance of the buried insulator (which tends towards zero if thick insulators are used, such as in SOS technology). This capacitance is proportional to the dielectric constant of the capacitance material. Silicon dioxide, used as buried insulator, has a dielectric constant ($\varepsilon_{ox}=3.9\ \varepsilon_0$) which is three times smaller than that of silicon ($\varepsilon_{si}=11.7\ \varepsilon_0$), therefore, a junction located on a buried oxide layer gives rise to a parasitic capacitance approximately three times smaller than that of a bulk junction. Since SOI buried insulator thickness does scale with device scaling, parasitic capacitances do not necessarily increase as technology progresses. In addition, a lightly doped, p-type silicon wafer can be utilized as mechanical support. In that case, a depletion layer can be created beneath the insulator, which further contributes to a reduction in junction-to-substrate capacitance.

Silicon-on-Insulator CMOS technology is also attractive because it involves fewer processing steps than bulk CMOS technology and because it suppresses some yield hazard factors present in bulk CMOS. To illustrate this we can take the example of realizing a shallow junction and making contact to it (Figure 1-2). Fabricating a shallow junction is not an obvious task in bulk CMOS. If a thin (100 nm) SOI substrate is used, on the other hand, the depth of the junction will automatically be equal to the thickness of the silicon film. Electrical contact to a shallow junction can be made using a metal (*e.g.:* tungsten), an alloy (*e.g.:* Al:Si or Ti:W) or a metal silicide (*e.g.:* $TiSi_2$). In bulk silicon devices, unwanted reactions can take place between the silicon and the metal or the silicide, such that the metal "punches through" the junction (Figure 1-2.A). This effect is well known in the case of aluminum (aluminum spiking), but can also occur with other metal or silicide systems. Such a junction punch-through gives rise to uncontrolled leakage currents. If the devices are realized in thin SOI material, the N^+ or P^+ source and drain diffusions extend to the buried insulator (reach-through junctions). In that case, there is no metallurgical junction underneath the metal-silicon contact area, and hence no leakage will be produced if some uncontrolled metal-silicon reaction occurs (Figure 1-2.B).

The absence of latch-up, the reduced parasitic source and drain capacitances, and the ease of making shallow junctions are merely three obvious examples of the advantages presented by SOI technology over bulk. There are many other properties allowing SOI devices and circuits to exhibit performances superior to those of their bulk counterparts (radiation hardness, high-temperature operation, improved transconductance and sharper subthreshold slope).

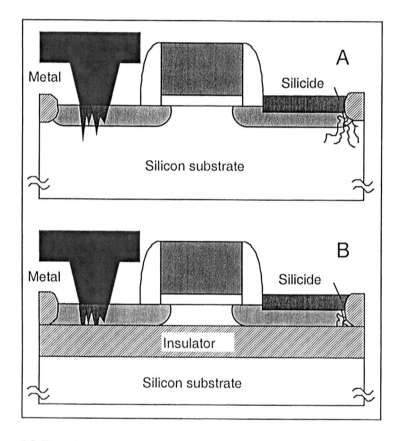

Figure 1-2. Formation contact or silicide on shallow junctions in the case of bulk silicon (A) and thin-film SOI (B).

It is a fact that silicon-on-Insulator integrated circuits are a commercial success. Chances are you have an SOI microprocessor in your computer, an SOI chip in your wristwatch or SOI power transistors in your car radio or even your fluorescent light bulbs. One of the reasons for this success, which began to manifest itself around year 2000, is the availability of large quantities of high-quality SOI wafers. Figure 1-3 shows an estimate of the worldwide SOI wafer production capability, in 8-inch equivalent. Chapters 2 and 3 are dedicated to the description of techniques used to produce SOI wafers and methods used to characterize these wafers.

Chapter 4 describes SOI CMOS technology. This includes fundamentals of processing and transistor design. Chapter 5 is dedicated to the most common SOI device: the single-gate MOS transistor. The physics of partially depleted and fully depleted inversion-mode devices, as well as that of accumulation-mode MOSFETs, is analyzed in that chapter. The evolution and diversity of the SOI MOSFET *genus* is shown in Figure 1-4. As one can see, there are many more device types than just partially and fully depleted single-gate transistors. Chapter 6 deals with these other SOI MOSFETs devices, such as the hybrid bipolar-MOS (DTMOS, MTCMOS) devices, and double, triple and quadruple-gate MOSFETs. Other devices, such as power transistors, are described in Chapter 6 as well.

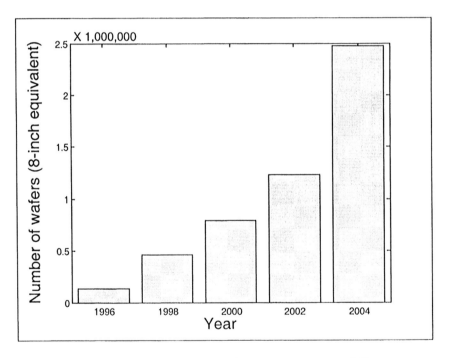

Figure 1-3. Worldwide SOI wafer production capability. [5,6,7]

SOI technology was confined to niche markets for many years. During the 1980's and 90's SOI was mostly used for radiation-hardened and high-temperature applications. Thus the behavior of SOI devices in harsh environment conditions is described in Chapter 7. Finally, Chapter 8 describes integrated circuits using SOI technology. SOI offers tremendous advantages for both mainstream CMOS circuits such as microprocessors, and niche markets such as high-temperature, smart power or radiation-hardened circuits.

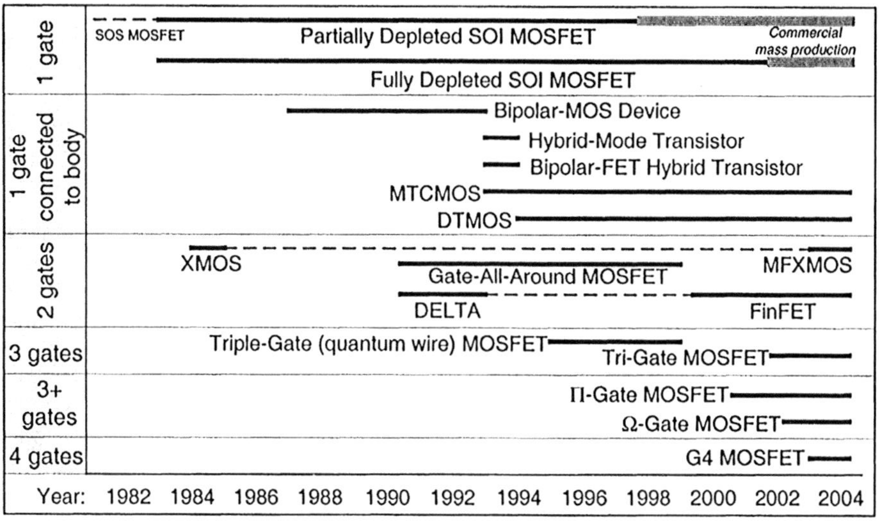

1 gate	SOS MOSFET		Partially Depleted SOI MOSFET							*Commercial mass production*		
			Fully Depleted SOI MOSFET									
1 gate connected to body					Bipolar-MOS Device							
					Hybrid-Mode Transistor							
					Bipolar-FET Hybrid Transistor							
			MTCMOS									
			DTMOS									
2 gates		XMOS									MFXMOS	
				Gate-All-Around MOSFET								
				DELTA							FinFET	
3 gates			Triple-Gate (quantum wire) MOSFET					Tri-Gate MOSFET				
3+ gates								Π-Gate MOSFET				
								Ω-Gate MOSFET				
4 gates										G4 MOSFET		
Year:	1982	1984	1986	1988	1990	1992	1994	1996	1998	2000	2002	2004

Figure 1-4. Evolution of SOI MOSFETs

REFERENCES

[1] J.E. Lilienfield, U.S. patents 1,745,175 (filed 1926, issued 1930), 1,877,140 (filed 1928, issued 1932), and 1,900,018 (filed 1928, issued 1933)

[2] See for example: W. Shockley, "The path to the conception of the junction transistor", IEEE Transactions on Electron Devices, Vol. 23, no. 7, p. 597, July 1976

[3] See for example: D. Kahng, "A historical perspective on the development of MOS transistors and related devices", IEEE Transactions on Electron Devices, Vol. 23, no. 7, p. 655, July 1976

[4] R.R. Troutman, *Latchup in CMOS Technology*, Kluwer Academic Publishers, 1986

[5] A.J. Auberton-Hervé, "SOI: materials to systems", Technical Digest of IEDM, p. 3, 1996

[6] Extrapolated from data in: C. Maleville, "300 mm Unibond® ramp-up", presented at the SOI Workshop, September 2002, Sainte-Maxime (France)

[7] L.P. Allen, W. Skinner, A. Cate, "Status and technology of SOI substrate material", IEEE International SOI Conference Proceedings, p. 5, 2001

Chapter 2

SOI MATERIALS

2.1 INTRODUCTION

Researchers have spent over 20 years developing reliable techniques to produce SOI wafers. The challenge of fabricating SOI material is to produce a thin film of single-crystal silicon sitting on top of an insulator (usually SiO_2) across an entire silicon wafer. Furthermore, the thin silicon crystal must sit on top of a high-quality amorphous silicon dioxide layer with no mechanical stress or electrically active defects. The thickness of the silicon layer typically is in the range of 100 nm, while the diameter of the wafer can reach 30 cm. There can thus be a 3,000,000:1 ratio between the width (diameter) and the thickness of an SOI crystal.

Many techniques have been developed for producing a film of single-crystal silicon on an insulator. Some of them are based on the epitaxial growth of silicon on either a silicon wafer covered with an insulator (homoepitaxial techniques) or on a crystalline insulator (heteroepitaxial techniques). Other techniques are based on the crystallization of a thin polysilicon layer by melt and regrowth (laser recrystallization, e-beam recrystallization and zone-melting recrystallization). Silicon-on-insulator material can also be produced from a bulk silicon wafer by isolating a thin silicon layer from the substrate through the formation and oxidation of porous silicon (FIPOS) or through the ion beam synthesis of a buried insulator layer (SIMOX, SIMNI and SIMON). SOI material can also be obtained by thinning a silicon wafer bonded to an insulator and a mechanical substrate (BESOI). More recently, layer transfer techniques such as the Smart-Cut® and Eltran® processes have enabled the "peeling-off" of a thin silicon layer from a wafer and its transfer onto an oxidized wafer. Strained silicon-on-insulator and strained SiGe-on-insulator layers can also be produced. Table 2-1 presents the most important SOI materials and the techniques involved in their fabrication.

Table 2-1. SOI Materials and associated fabrication techniques [1]

Generic Process	Fabrication Technique (acronym)	Process Variations
Silicon heteroepitaxy	Silicon-on-Sapphire (SOS)	SPEAR DSPE UTSi
	Graphoepitaxy Silicon on Cubic Zirconia (SOZ) Silicon on Spinel Silicon on CaF$_2$	
Thick polysilicon deposition	Dielectric Isolation (DI)	
Polysilicon melting and recrystallization	Laser Recrystallization	Selective Annealing
	Electron Beam Recrystallization	
	Zone-Melting Recrystallization (ZMR)	LEGO
Silicon homoepitaxy	Epitaxial Lateral Overgrowth (ELO)	CLSEG PACE
	Lateral Solid-Phase Epitaxy (LSPE)	MILC
Formation of porous silicon	Full Isolation by Porous Silicon (FIPOS)	
Ion beam synthesis of a buried insulator	Separation by Implanted Oxygen (SIMOX)	ITOX SIMOX MLD
	Separation by Implanted Nitrogen (SIMNI)	
	Separation by Implanted O+N (SIMON)	
	Synthesis of silicon carbide (SiCOI)	
Wafer bonding	Wafer Bonding and Etch Back (BESOI)	PACE
Layer transfer	H$^+$-Induced Splitting (Smart-Cut®)	UNIBOND® NanoCleave®
	Porous Silicon Splitting	Eltran®
SiGe epitaxy	Strained silicon-on-insulator (SSOI)	
	Strained SiGe-on-insulator (SiGeOI)	
Diamond layer formation	Silicon on Diamond (SOD)	
Preferential etching	Silicon on Nothing (SON)	

2.2 HETEROEPITAXIAL TECHNIQUES

Heteroepitaxial Silicon-on-Insulator films are obtained by epitaxially growing a silicon layer on a single-crystal insulator. Reasonably good epitaxial growth is possible on insulating materials when the lattice parameters are sufficiently close to those of single-crystal silicon. Substrates can either be single-crystal bulk material, such as (0 1 -1 2) Al_2O_3 (sapphire) or thin insulating films grown on a silicon substrate (epitaxial CaF_2). Heteroepitaxial growth of a silicon film can never produce a defect-free material by itself if the lattice parameters of the insulator do not perfectly match those of silicon. Table 2.2 illustrates the differing materials properties. Furthermore, the silicon film will never be stress-free if the thermal expansion coefficients of the silicon and the insulating substrate are not equal.

Table 2-2. Important parameters of materials used in heteroepitaxial SOI materials fabrication. [2]

Material	Crystal Structure	Dielectric Constant	Lattice Parameter (nm)	Thermal Expansion Coefficient (K^{-1})
Si	Diamond	11.7	0.5430	3.8 x 10^{-6}
Sapphire (0 1 –1 2)	Rhombohedral	9.3	0.4759	9.2 x 10^{-6}
Cubic Zirconia	Cubic	20	0.5206	11.4 x 10^{-6}
Spinel	Cubic	8.4	0.808	8.1 x 10^{-6}
CaF$_2$	Cubic	6.8	0.5464	26.5 x 10^{-6}

Heteroepitaxial silicon-on-insulator films are grown using silane or dichlorosilane at temperatures around 1000°C. All the insulating substrates have thermal expansion coefficients that are 2 to 3 times higher than that of silicon. Therefore, thermal mismatch is the single most important factor determining the physical and electrical properties of heteroepitaxial silicon films grown on bulk insulators. Indeed, the silicon films have a thickness which is typically 1000 times smaller than that of the insulating substrate. While the films are basically stress-free at growth temperature, the important thermal coefficient mismatch results in a compressive stress in the silicon film which reaches $\cong -7 \times 10^9$ dynes/cm^2 at the surface of a 0.5 μm silicon-on-sapphire (SOS) film. An even higher value is reached at the Si-sapphire interface. Such stresses may equal or exceed the yield stress of silicon, resulting in relaxation in the silicon film via generation of crystallographic defects such as microtwins, stacking faults and dislocations. It is worth noting that silicon crystals can be grown on an amorphous material such as

SiO$_2$ if periodically spaced grooves are patterned in it. Using such a technique, called "graphoepitaxy", thin silicon crystals of limited size have been successfully grown.[34]

2.2.1 Silicon-on-Sapphire (SOS)

Silicon-on-Sapphire (SOS) material was first introduced 1964.[5] It is obtained by epitaxial growth of silicon on a (1 1 −1 2)-oriented crystalline alumina (α-Al$_2$O$_3$ also called sapphire) wafer. The sapphire crystals are produced using Czochralski growth. The sapphire boule is sliced into wafers that are then subjected to mechanical and chemical polishing. The sapphire wafers receive a final hydrogen etching at 1150°C in an epitaxial reactor, and a silicon film is deposited using the pyrolysis of silane at temperatures between 900 and 1000°C. The lattice constant of silicon and (1 1 −1 2) sapphire are 0.543 and 0.475 nm, respectively, The thermal expansion for silicon and sapphire are 3.8x10^{-6} and 9.2x10^{-6} K^{-1}, respectively. Due to the lattice mismatch between sapphire and silicon the defect density in the silicon film is quite high, especially in very thin films. As the film thickness increases, however, the defect density appears to decrease as a simple power law function of the distance from the Si-Sapphire interface. The main defects present in the as-grown SOS films are stacking faults and (micro)twins. Typical defect densities near the Si-Sapphire interface reach values as high as 10^6 planar faults/cm and 10^9 line defects/cm^2. These account for low values of resistivity, mobility, and lifetime near the interface. Because the epitaxial silicon is deposited at high temperature and because the thermal expansion coefficients of silicon and sapphire are different the silicon film is under compressive stress at room temperature.

The electron mobility observed in SOS devices is lower than bulk mobility. This is a result of both the high defect density found in as-grown SOS films and the compressive stress measured in the silicon film. Indeed, in the case of (100) SOS, the compressive stress causes the kx and ky ellipsoids to become more populated with electrons than the kz ellipsoid (which is normal to the silicon surface). As a result, the effective mass of electrons becomes larger than in bulk silicon [6] and a relatively low channel electron mobility is observed in SOS n-channel MOSFETs ($\cong$250-350 cm^2/V.s). On the other hand, the effective mass of holes is smaller than in bulk silicon, due to the same compressive stress. The hole surface mobility, which could in principle be higher than in bulk because of the compressive stress, is, however, affected by the presence of defects, such that the value of surface mobility for holes in SOS p-channel MOSFETs is comparable to that in bulk silicon. A beneficial consequence of the very low electron mobility at the Si-Sapphire interface is the reduction of the back-channel leakage current in n-channel devices. The minority carrier lifetime found in as-grown SOS films is a fraction of a nanosecond. As a result, relatively high junction leakage currents ($\cong 1$ pA/μm) are observed.[7]

Several techniques have been developed to reduce both the defect density and the stress in the SOS films. The Solid-Phase Epitaxy and Regrowth (SPEAR) and the Double Solid-Phase Epitaxy (DSPE) techniques are other more successful methods for improving the crystal quality of SOS films.[8,9,10] These techniques employ the following steps. First, silicon implantation is used to amorphize the silicon film, with the exception of a thin superficial layer, where the original defect density is lowest. Then a thermal annealing step is used to induce solid-phase regrowth of the amorphized silicon, the top silicon layer acting as a seed. A second silicon implant is then used to amorphize the top of the silicon layer, which is subsequently recrystallized in a solid-phase regrowth step using the bottom of the film as a seed. In the SPEAR process, an additional epitaxy step is performed after solid-phase regrowth. Using such techniques, substantial improvement of the defect density is obtained. Noise in MOS devices is reduced, and the minority carrier lifetime is increased by two to three orders of magnitude.[11,12].

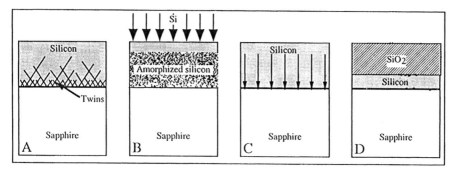

Figure 2-1. UTSi SOS process. A: Growth of a relatively thick epitaxial silicon film; B: Amorphization using silicon ion implantation; C: Solid-phase regrowth downward from the defect-free surface; D: Thinning of the silicon film by thermal oxidation.

The most recent technique used to produce high-quality SOS is the UTSi (Ultra-Thin Silicon) process: a relatively thick film of silicon is grown on sapphire and, as in the SPEAR process, silicon ion implantation is used to amorphize the silicon film below the most superficial layer, which is relatively defect-free (Figure 2-1). Low-temperature annealing is then used to regrow the defect-free silicon downward from the surface through a solid-phase epitaxy mechanism. The silicon film is then thinned to the desired thickness (100 nm) by thermal oxidation and oxide strip. This process delivers relatively defect-free and stress free SOS material in which devices with a high effective mobility can be fabricated.[13,14]

2.2.2 Other heteroepitaxial SOI materials

Silicon can be epitaxially grown on a number of other insulating materials such as cubic boron phosphide (BP) and rhombohedral $B_{13}P_2$, AlN [15], cubic β-SiC [16], BeO [17] and NaCl. [18] The best results, however, have been obtained by growing silicon on cubic zirconia, spinel, and calcium fluoride.

2.2.2.1 Silicon-on-Zirconia (SOZ)

Yttrium-stabilized cubic zirconia $[(Y_2O_3)_m \cdot (ZrO_2)_{1-m}]$ can also be used as an alternative dielectric substrate for silicon epitaxy.[19] Indeed, zirconia is an oxygen conductor at high temperature. This means that, while being an excellent insulator at room temperature ($\rho > 10^{13}$ Ω.cm), cubic zirconia is permeable to oxygen at high temperature. This unique property has been used to grow an SiO_2 layer at the silicon-zirconia interface (*i.e.* to oxidize the most defective part of the silicon film) by the transport of oxygen through a 500 μm-thick zirconia substrate.[20,21] The growth of a 160 nm-thick film at the interface necessitates only 100 min at 925°C in pyrogenic steam.

2.2.2.2 Silicon-on-Spinel

Spinel $[(MgO)_m \cdot (Al_2O_3)_{1-m}]$ can be used as a bulk insulator material or can be grown epitaxially on a silicon substrate at a temperature between 900 and 1000°C.[22] Stress-free 0.6 μm silicon-on-spinel films have been grown, but the properties of MOSFETs made in this material are inferior to those of devices made in SOS films due to higher defect density.[23] Better device performance is obtained when much thicker 3-40 μm silicon-on-spinel films are used. Again as with SOZ, oxygen can diffuse at high temperature through thin spinel films. Using this property, a Si/spinel/SiO$_2$/Si structure has been produced.[24]

2.2.2.3 Silicon on Calcium Fluoride

Like spinel, calcium fluoride (CaF_2) can be grown epitaxially on silicon. Fluoride mixtures can also be formed and their lattice parameters can be matched to those of most semiconductors. For example, $(CaF_2)_{0.55} \cdot (CdF_2)_{0.45}$ has the same lattice parameters as silicon at room temperature, and $(CaF_2)_{0.42} \cdot (SrF_2)_{0.58}$ is matched to germanium.

Unfortunately, the thermal expansion coefficient of these fluorides is quite different from that of silicon, and lattice match cannot be maintained over any appreciable temperature range. Silicon can, in turn, be grown on CaF_2 using MBE or e-gun evaporation at a temperature of approximately 800°C.[25,26] As in the case of Si films on epitaxial spinel, $Si/CaF_2/Si$ films are essentially stress-free, which can be readily understood by noticing that the mechanical support is a silicon wafer, which, of course, has the same thermal expansion coefficient as the top silicon film. MOSFETs have been made in $Si/CaF_2/Si$ material and exhibit surface electron and hole mobility of 570 and 240 $cm^2/V.s$, respectively.[27]

2.3 DIELECTRIC ISOLATION (DI)

The dielectric isolation process (DI) is used to create relatively thick (several micrometers) silicon islands on SiO_2. The fabrication process is illustrated by Figure 2-2.

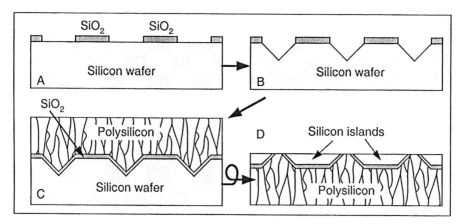

Figure 2-2. DI process A: Oxide mask patterning; B: Anisotropic etch using KOH or TMAH; C: Oxidation and deposition of a thick polysilicon layer; D: Silicon wafer polishing.

Silicon dioxide islands are patterned on a (100) silicon wafer and used as a hard mask for anisotropic etching of the silicon using potassium hydroxide (KOH) or tetramethyl ammonium hydroxide (TMAH). After the etching of V-shaped groves in the silicon, oxide is grown and a thick (500 µm) polysilicon layer is deposited. Mechanical polishing is then used to etch the silicon wafer until the oxide layer is reached and silicon-on-insulator island are formed. The DI process is used to fabricate thick-film SOI devices such as power transistors, bipolar transistors, and radiation-hard components. [28,29,30] A major disadvantage of the DI process is the stress introduced by the deposition of the very thick polysilicon layer, which makes it difficult to apply the technique to large-diameter wafers.

2.4 POLYSILICON MELTING AND RECRYSTALLIZATION

MOS transistors can be fabricated in a layer of polysilicon deposited on an oxidized silicon wafer [31,32], but the presence of grain boundaries brings about low surface mobility values ($\cong 10$ cm^2/V.s) and high threshold voltages (several volts). Grain boundaries contain silicon dangling bonds giving rise to a high density of interface states (several 10^{12} cm^{-2}V^{-1}) which must be filled with channel carriers before threshold voltage is reached. Above threshold, once the traps are filled, the grain boundaries generate potential barriers which have to be overcome by the channel carriers flowing from source to drain. This gives rise to the low values of mobility observed in polysilicon devices.[33] Mobility can be improved and more practical threshold voltage values can be reached if the silicon dangling bonds in the grain boundaries are passivated. This can be performed by exposing the wafers to a hydrogen plasma, during which the fast diffusing atomic hydrogen can penetrate the grain boundaries and passivate the dangling bonds.[34] This treatment can improve the drive current of polysilicon devices by a factor of 10, due to both an increase of mobility and a reduction of threshold voltage. Hydrogen passivation also significantly improves the leakage current of the devices.[35] Such devices can find application as active polysilicon loads, and 64k SRAMs with p-channel polysilicon loads have been demonstrated.[36] High-performance IC applications, however, require much better device properties, and grain boundaries must be eliminated from the silicon film. This is the goal of the polysilicon melting and recrystallization techniques described next. In these techniques a polysilicon film is deposited on an oxidized silicon wafer and melted using a focused laser beam, an electron beam, a heated carbon strip or a halogen lamp. Quenching conditions are then controlled such that the polysilicon film is converted into relatively large silicon crystals.

2.4.1 Laser recrystallization

Experiments of laser recrystallization of polycrystalline silicon have been carried out with pulsed lasers (ruby lasers and Nd:YAG lasers) [37] at a time where pulsed laser annealing was popular for activating dopants.[38] The technique, however, was rapidly abandoned for the fabrication of device-worthy SOI layers because of its lack of controllability and difficulty of growing large silicon grains. Continuous-wave (cw) lasers such as CO_2, Ar and YAG:Nd lasers have proven to be much more effective for producing

SOI films. Silicon is transparent at the 10.6 μm wavelength produced by CO_2 lasers. Therefore, silicon films cannot be directly heated by such lasers; SiO_2, on the other hand, absorbs the 10.6 μm wavelength with a penetration depth of $\cong 10$ μm. Polysilicon films deposited on SiO_2 (quartz or an oxidized silicon wafer) or covered by an SiO_2 cap can thus be melted through "indirect" heating produced by CO_2 laser irradiation.[39] CO_2 lasers have the advantage of a high output power (> 100 W), such that wide elliptical beams can be produced. The polysilicon layer has to be surrounded by relatively thick SiO_2 layers (to efficiently absorb the 10.6 μm wavelength) limits the use of CO_2 lasers for SOI materials fabrication. CO_2 lasers are also impractical for 3D applications, where heating of only the top silicon layer is desired. Continuous-wave (CW) YAG:Nd lasers can output high power beams (300 W) at a wavelength of 1.06 μm. Silicon is transparent at this wavelength, but if the wafer is preheated to a temperature of 1200-1300°C, free carriers are generated in the silicon and the 1.06 μm wavelength can be absorbed.[40] CW Ar lasers, on the other hand, emit two main spectral lines at 488.0 and 514.5 nm (blue and green), and can reach an output power of 25 W when operated in the multiline mode. These wavelengths are well absorbed by silicon. In addition to this, the reflectivity of silicon increases abruptly once melting is reached. This effect is very convenient since it acts as negative feedback on the power absorption and prevents the silicon from overheating above melting point. The laser beam is focused on the sample by means of an achromatic lens (or a combination of lenses) into a circular or, more often, an elliptical spot. Scanning of the beam is achieved through the motion of galvanometer-driven mirrors. The size of the molten zone and the texture of the recrystallized silicon depend on parameters such as laser power, laser intensity profile, substrate preheating temperature (the wafer is held on a heated vacuum chuck), and scanning speed.[41] Typical recrystallization conditions of a 500 nm-thick LPCVD polysilicon film deposited on a 1 μm thermal oxide grown on a silicon wafer are: spot size of 50-150 μm (defined as the laser spot diameter at $1/e^2$ intensity, TEM_{00} mode), power of 10-15 watts, scanning speed of 5-50 cm/sec, and substrate heating at 300-600°C.

Silicon films recrystallized on an amorphous SiO_2 substrate have a random crystal orientation. X-ray diffraction studies of polysilicon recrystallized with a Gaussian laser profile indicate the presence of crystallites having (111), (220), (311), (400), (331), (110), and (100) orientations.[42] This is clearly unacceptable for device fabrication, since different crystal orientations will result in different gate oxide growth rates. In addition, in between the crystallites, grain boundaries exist which act to reduce carrier mobility. Ideally, one wants a uniform (100) orientation for all crystallites. From there comes the idea of opening a window (seeding area) in the insulator to allow contact between the silicon substrate and the polysilicon layer. Upon melting and recrystallization, lateral epitaxy can take place and the recrystallized silicon will have a uniform (100) orientation, as shown in Figure 2-3.[43] Unfortunately, the (100) crystal orientation can only be formed up to 100 μm away from the seeding window, at which distance defects appear which cause loss of the (100) orientation. Growth of longer SOI crystals can be obtained if the trailing edge of the molten zone is concave. This can be achieved by masking part of the beam.[44] In this approach, however, the available laser power is reduced. More efficient beam shaping can be achieved by merging different laser modes (doughnut-shaped beam) [45] or by recombining a split laser beam.[46]

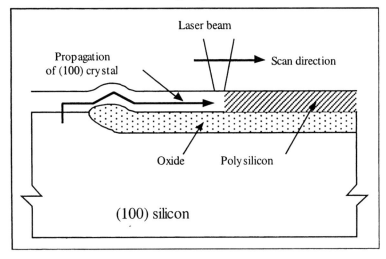

Figure 2-3: Principle of the lateral seeding process.

In order to obtain not only a single large crystal, but a large single-crystal area, the laser beam must be raster-scanned on the wafer with some overlap between the scans. Unfortunately, small random crystallites arise at the edges of the large crystals, which precludes the formation of large single-crystal areas, and grain boundaries are formed between the single-crystal stripes. The location of these grain boundaries depends on the scanning parameters and the stability of the beam. In other words, from a macroscopic point of view, the location of the boundaries is quasi-random, and the yield of large circuits made in this material will be zero. A solution to this problem is to use stripes of an anti reflecting (AR) material (SiO_2 and/or Si_3N_4) to obtain the photolithographically-controlled shaping of the molten zone and produce a succession of concave solidification fronts at the trailing edge of the molten zone.[47,48,49]

This technique is called "selective annealing" because more energy is selectively deposited on the silicon covered by AR material. It permits the growth of large adjacent crystals with straight grain boundaries, the location of which is controlled by a lithography step (Figure 2-4). Although this technique imposes constraints to circuit design, it allows for placing film defects outside the active area of transistors. The technique can be used with a laser scan parallel, slanted or perpendicular to the anti reflection stripes (AR stripes). Using this technique, chip-wide (several mm × several mm) defect-free, (100)-oriented single-crystal areas have been produced.[50]

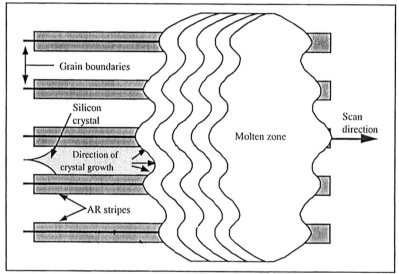

Figure 2-4: Recrystallization using anti reflection stripes (selective annealing).

2.4.2 E-beam recrystallization

The recrystallization of a polysilicon film on an insulator using an electron beam (e-beam) is in many respects very similar to the recrystallization using a continuous-wave (cw) laser. Similar seeding techniques are used, and an SiO_2 (or Si_3N_4) encapsulation layer is used to prevent the melted silicon from de-wetting.[51] The use of an e-beam for recrystallizing SOI layers has some potential advantages over laser recrystallization since the scanning of the beam can be controlled by electrostatic deflection, which is far more flexible than the galvanometric deflection of mirrors. Indeed, the oscillation frequency of a laser beam scanned using galvanometer-driven mirrors is limited to a few hundred hertz, while e-beam scan frequencies of 50 MHz have been utilized.[52] The absorption of the energy deposited by the electron beam is almost the same in most materials, such that the energy absorption in a sample is quite independent of crystalline state and optical reflectivity of the different materials composing it. This improves the uniformity of the recrystallization of silicon deposited over an uneven substrate, but precludes the use of a patterned anti reflection coating. Structures with tungsten stripes have, however, been proposed to achieve differential absorption.

2.4.3 Zone-melting recrystallization

One of the main limitations of laser recrystallization is the small molten zone produced by the focused beam, which results in a long processing time needed to recrystallize a whole wafer. Recrystallization of a polysilicon film on an insulator can also be carried out using incoherent light (visible or near IR) sources. In this case, a narrow (a few millimeters) but long molten zone can be created on the wafer. A molten zone length of the size of an entire wafer diameter can readily be obtained. As a result, full recrystallization of a wafer can be carried out in a single pass. Such a recrystallization technique is generally referred to as Zone-Melting Recrystallization (ZMR) because of the analogy between this technique and the float-zone refining process used to produce silicon ingots. An excellent review of the ZMR mechanisms can be found in a book by E.I. Givargizov.[53] The first method which successfully achieved recrystallization of large-area samples makes use of a heated graphite strip which is scanned across the sample to be recrystallized. The set-up is called "graphite strip heater" (Figure 2-5). A heated graphite susceptor is used to raise the temperature of the entire sample up to within a few hundred degrees below the melting temperature of silicon. Additional heating is locally produced at the surface of the wafer using a heated graphite strip located a few millimeters above the sample and scanned across it.[54, 55]

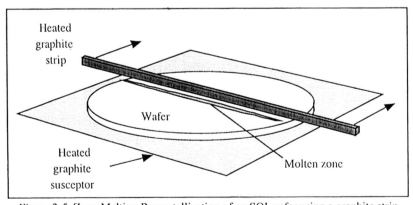

Figure 2-5: Zone-Melting Recrystallization of an SOI wafer using a graphite strip heater.

A typical sample is made of a silicon wafer on which a 1 μm-thick oxide is grown and a 0.5 μm-thick layer of LPCVD amorphous or polycrystalline silicon is then deposited. The whole structure is capped with a 2 μm-thick layer of deposited SiO_2 covered by a thin Si_3N_4 layer. The capping layer helps minimize mass transport and protects the molten silicon from contaminants (such as carbon from the strip heater). Recrystallization is carried out in a vacuum or an inert gas ambient in order to keep the graphite elements from burning.

Both the graphite susceptor and the graphite strip can be replaced by lamps to achieve ZMR of SOI wafers. A lamp recrystallization system is composed of a bank of halogen lamps which is used to heat the wafer from the back to a high temperature (1100°C or above), and a top halogen or mercury lamp whose light is focused on the sample by means of an elliptical reflector (Figure 2-6).[56,57,58,59] An unpolished quartz plate may be inserted between the lamp bank and the wafer in order to diffuse the light and homogenize the energy deposition at the back of the wafer. As in the case of strip heater recrystallization, a narrow, wafer-long molten zone is created and scanned across the wafer with a speed on the order of 0.1-1 mm/sec.

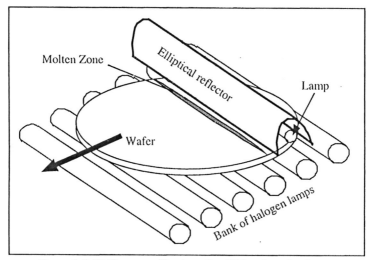

Figure 2-6: ZMR recrystallization of an SOI wafer using lamps.

ZMR can also be carried out using an elongated laser spot; a linear molten zone can be created using a high-power (300 W) continuous-wave YAG:Nd laser (wavelength=1.06 μm). The circular laser beam can be transformed into a linear beam using 90°-crossed cylindrical lenses. The produced linear spot, which is focused on the sample, can exceed 10 cm in length. In order to get free carrier absorption from the silicon, the substrate has to be heated to a temperature of 1200-1300°C by a bank of halogen lamps. The laser beam melts a 0.5-1 mm-wide zone of the polysilicon layer sitting on top of an oxidized silicon substrate. Crystallization is performed in a single pass of the molten zone across the wafer. The scanning speed can vary from 0.1 to 2 mm/s. The lateral temperature gradient is controlled by slightly focusing or defocusing the beam.[60]

It is also possible to fabricate *thick* silicon films (10-100 μm) by recrystallization over an insulator using the LEGO technique (Lateral Epitaxial Growth over Oxide).[61] This technique makes use of a bank of stationary halogen lamps which heat the front side of a wafer. A thick polysilicon film is deposited on an oxidized silicon wafer where windows have been opened in the oxide for seeding purposes. As the temperature of the wafer is raised, the top polysilicon melts across the whole wafer. The temperature is then carefully ramped down, and crystal growth proceeds epitaxially from the seeding windows over the oxide until the entire silicon film is recrystallized. Recrystallization of thick silicon films can also be carried out on non-planar substrates using a scanning lamp apparatus.[62]

2.5 HOMOEPITAXIAL TECHNIQUES

Silicon-on-insulator can be produced by homoepitaxial growth of silicon on silicon, provided that the crystal growth can extend laterally on an insulator (SiO_2, typically). This can be achieved either using a classical epitaxy reactor or by lateral solid-phase crystallization of a deposited amorphous silicon layer.

2.5.1 Epitaxial lateral overgrowth

The Epitaxial Lateral Overgrowth technique (ELO) consists of the epitaxial growth of silicon from seeding windows over SiO_2 islands or devices capped with an insulator. It can be performed in an atmospheric or in a reduced-pressure epitaxial reactor [63]. The principle of ELO is illustrated in Figure 2-7. Typical sample preparation for ELO involves patterning windows in an oxide layer grown on a (100) silicon wafer. The edges of the windows are oriented along the <010> direction. After cleaning, the wafer is loaded into an epitaxial reactor and submitted to a high-temperature hydrogen bake to remove the native oxide from the seeding windows. Epitaxial growth is performed using *e.g.* an SiH_2Cl_2 + H_2 + HCl gas mixture. Unfortunately, nucleation of small silicon crystals with random orientation occurs on the oxide. These crystallites can be removed by an *in-situ* HCl etch step. Once the small nuclei are removed, a new epitaxial growth step is performed, followed by an etch step, and so on, until the oxide is covered by epitaxial silicon. The epitaxial growth proceeds from the seeding windows both vertically and laterally, and the silicon crystal is limited by <100> and <101> facets (Figure 2-7.A). When two growth fronts, seeded from opposite sides of the oxide, join together, a continuous silicon-on-insulator film is formed, which contains a low-angle subgrain boundary where the two growth fronts meet. Because of the presence of <101> facets on the growing crystals, a groove is observed over the center of the SOI area (Figure 2-7.B). This groove, however, eventually disappears if additional epitaxial growth is performed (Figure 2-7C).[64] Three-dimensional stacked CMOS inverters have been realized by lateral overgrowth of silicon over MOS devices.[65] The major disadvantage of the ELO technique is the nearly 1:1 lateral-to-vertical growth ratio, which means that a 10 μm-thick film must be grown to cover 20 μm-wide oxide patterns (10 μm from each side). Furthermore, 10 additional micrometers must be grown in order to get a planar surface. Thinner SOI films can, however, be obtained by polishing the wafers after the growth of a thick ELO film.[66] The ELO technique has been used to fabricate three-dimensional and double-gate devices.[67,68]

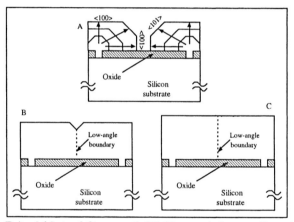

Figure 2-7: Epitaxial Lateral Overgrowth (ELO): growth from seeding windows (A), coalescence of adjacent crystals (B), self-planarization of the surface (C).

A variation of the ELO technique, called "tunnel epitaxy", "confined lateral selective epitaxy" (CLSEG) or "pattern-constrained epitaxy" (PACE), has been reported by several groups.[69,70,71,72] In this technique, a "tunnel" of SiO_2 is created, which forces the epitaxial silicon to propagate laterally (Figure 2-8). With this method, a 7:1 lateral-to-vertical growth ratio has been obtained.

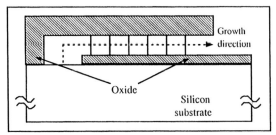

Figure 2-8: Principle of tunnel epitaxy.

2.5.2 Lateral solid-phase epitaxy

Lateral Solid-Phase Epitaxy (LSPE) is based on the lateral epitaxial growth of crystalline silicon through the controlled crystallization of amorphous silicon (α-Si).[73,74,75,76,77,78] A seed is needed to provide the crystalline information necessary for the growth. The thin amorphous silicon film can either be deposited or obtained by amorphizing a polysilicon film by means of a silicon ion implantation step. LSPE is performed at relatively low temperature (575-600°C) in order to obtain regrowth while minimizing random nucleation in the amorphous silicon film. The LSPE technique has been used to fabricate double-gate SOI MOSFETs.[79]

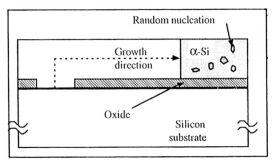

Figure 2-9: Principle of lateral solid-phase epitaxy.

The lateral epitaxy rate is on the order of 0.1 nm/s in undoped α-Si and 0.7 nm/s in heavily (3×10^{20}cm^{-3}) phosphorous-doped α-Si. The distance over which lateral epitaxy can be performed over an oxide layer is in the order of 8 µm for undoped material, and 40 µm if a heavy phosphorus doping is used, further lateral extension of LSPE being limited by random nucleation in the α-Si film (Figure 2-9). An increase of the lateral extension of LSPE in undoped silicon can be obtained by alternating stripes of P-doped and undoped material, the stripes being perpendicular to the growth front (this can be performed through a mask step and ion implantation).[80] The growth in the doped material "stimulates" the growth front in the undoped material.

Solid-phase growth of crystals from amorphous silicon can also be stimulated by the presence or contact of a metal such as palladium, aluminum or nickel.[81,82,83] This technique, called "metal-induced lateral crystallization" (MILC) has been used to fabricate thin-film transistors (TFTs),[84] gate-all-around transistors [85,86], and three-dimensional CMOS SOI integrated circuits.[87] In a typical MILC process used for making TFTs, amorphous silicon is deposited on SiO_2 and capped with a low-temperature deposited oxide layer. After gate formation, contact holes are opened in the source/drain areas and nickel is deposited. Upon annealing at a temperature close to 400°C $NiSi_2$ forms in the contact holes. Annealing at 500°C is then performed during which MILC occurs, as part of the original $NiSi_2$ moves through the amorphous silicon, leaving behind a trail of long, needle-shaped silicon crystals. The passage of $NiSi_2$ through the silicon leaves behind approximately 0.02 atomic percent of nickel in the crystallized silicon. MILC recrystallization of over 40 µm has been demonstrated. The typical growth rate ranges between 0.25 and 1 µm/hour.[88,89,90,91]

2.6 FIPOS

The process of Full Isolation by Porous Oxidized Silicon (FIPOS) was invented in 1981.[92,93] It relies on the conversion of a layer of silicon into porous silicon and on the subsequent oxidation of this porous layer. The oxidation rate of porous silicon being orders of magnitude higher than that of monolithic silicon, a full porous silicon buried layer can be oxidized while barely growing a thin oxide on silicon islands on top of it.

The original FIPOS process is described in Figure 2-10. It relies on the fact that p-type silicon can readily be converted into porous silicon by electrochemical dissolution of p-type silicon in HF (the sample is immersed into an HF solution and a potential drop is applied between the sample and a platinum electrode dipped into the electrolyte). The conversion rate of n-type silicon is much lower. Porous silicon formation proceeds as follows: at first, an Si_3N_4 film is patterned over a p-type silicon wafer, and boron is implanted to control the density of the porous silicon surface layer. The N⁻ material is formed by conversion of the P⁻ silicon into N⁻ silicon by proton (H⁺ ion) implantation. The p-type silicon is then converted into porous silicon by anodization in a hydrogen fluoride solution. Optimal conversion yields a 56% porosity (porosity is controlled by the HF concentration in the electrolyte, the applied potential, and the current density at the sample surface during anodization). In this way, the volume of the buried oxide formed by oxidation of the porous layer is equal to that of the porous silicon, and stress in the films can be minimized.

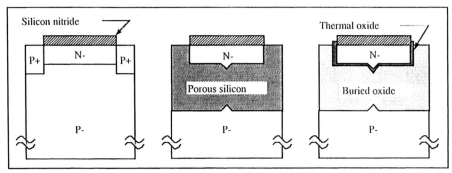

Figure 2-10: The original FIPOS process. From left to right: formation of N⁻ islands and P⁺ current paths, formation of porous silicon, and oxidation of the porous silicon.

Porous silicon contains an intricate network of pores. As a result, the surface area of silicon exposed to the ambient is extremely high, and porous silicon oxidizes very rapidly. This allows the grow of a thick buried oxide (oxidation depth is comparable to the width of the N⁻ silicon islands) while growing only a thin oxide at the edges of the N⁻ silicon islands in which the devices will be made. The thermally grown oxide provides the silicon islands with a high-quality bottom interface.

The original FIPOS technique produces high-quality SOI islands, and has been used to produce devices with good electrical characteristics [94,95] but it has several limitations. Indeed, the formation of a thick oxidized porous silicon layer is needed to isolate even small islands. Such a thick oxide can induce wafer warpage, especially if the islands are unevenly distributed across the wafer. A second limitation is the formation of a little cusp of unanodized silicon at the bottom-center of the silicon islands, where the two (left and right) anodization fronts meet during porous silicon formation (Figure 2-10). This limits the use of such a technique for thin-film SOI applications, where thickness uniformity of the silicon islands is of crucial importance. Several variations of the process have been proposed, such as the formation of a buried P⁺ layer (over a P⁻ substrate) located below N⁻ silicon islands, the whole structure being produced *e.g.* by epitaxy.[96] This technique solves the problem of having to produce a very thick buried layer to isolate wide islands, since it permits an island width-to-porous silicon layer thickness ratio larger than 50.

A different approach for the fabrication of FIPOS structures is based on the preferential anodization of the N⁺ layer of an N⁻/N⁺/N⁻ structure (Figure 2-11).[97,98,99] With this method, the thickness of both the silicon islands and the porous silicon layer are uniform and easily controlled by the N⁺ doping profile (*e.g.* obtained by antimony implant on a lightly-doped, n-type wafer, and subsequent epitaxy of an N⁻ superficial layer). Another advantage of the N⁻/N⁺/N⁻ approach is the automatic endpoint on the island isolation. As soon as all of the N⁺ layer is converted into porous silicon the anodization current drops, and anodization stops due to a change in anodization potential threshold between the N⁺ layer and the N⁻ silicon in the substrate and the islands. Such an automatic end of reaction control is not available in the N⁻/P⁺/P⁻ approach where anodization of the P⁻ substrate occurs as soon as the P⁺-layer has been converted into porous silicon. After oxidation of the porous silicon layer [100], a dense buried oxide layer is obtained.

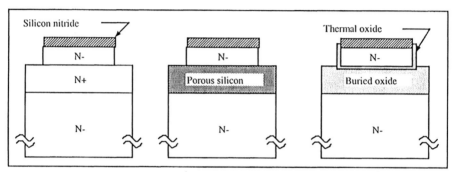

Figure 2-11: FIPOS formation (N⁻/N⁺/N⁻ technique). From left to right: N⁻/N⁺/N⁻ structure, formation of porous silicon, and oxidation of the porous silicon.

It is worth noticing that porous silicon is a single-crystal material, in spite of the fact that it contains many voids. A blanket porous silicon layer can, therefore, be created on a wafer, and epitaxial single-crystal silicon can be grown on it using MBE (molecular-beam epitaxy) or low-temperature PECVD (plasma-enhanced chemical vapor deposition).[101] Other semiconductors such as GaAs can also be grown on porous silicon.[102] The growth of epitaxial silicon on porous silicon is a key processing step of the Eltran® process which will be described in section 9 of this chapter.

2.7   ION BEAM SYNTHESIS OF A BURIED INSULATOR

SOI material can be obtained by implanting ions of oxygen or nitrogen into silicon and annealing the structure to form a buried insulator layer, commonly referred to as the BOX (buried oxide). SIMOX is the most successful material based on this technique.

2.7.1 Separation by implanted oxygen (SIMOX)

The SIMOX technique was invented by K. Izumi, M. Doken and H. Ariyoshi of NTT in 1978.[103] In this technique a high dose of oxygen ions is implanted in a silicon wafer followed by high-temperature annealing step to form a buried oxide layer (Figure 2-12). Ion implantation is traditionally used in the semiconductor industry to introduce dopant atoms, and doses higher than a few 10^{15} cm^{-2} are rarely employed. In the SIMOX technique implanted oxygen atoms are used to synthesize a new material, namely silicon dioxide. As a result a very high dose of oxygen ions (typically 1.8×10^{18} cm^{-2} at 200 keV in the "standard" SIMOX process) must be implanted to form the buried oxide (BOX) layer.

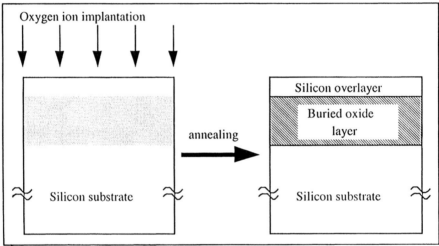

Figure 2-12: The principle of SIMOX: a high dose of oxygen is implanted into silicon, followed by an annealing step. The result is a buried layer of silicon dioxide below a thin, single-crystal silicon overlayer.

2.7.1.1 "Standard" SIMOX

Stoichiometric SiO_2 contains 4.4×10^{22} oxygen atoms/cm^3. Therefore, the implantation of 4.4×10^{17} atoms/cm^2 should be sufficient to produce a 100 nm-thick buried oxide layer. Unfortunately, due to the statistical nature of ion implantation, the oxygen profile in silicon does not have a box shape, but rather a skewed Gaussian profile. The implanted atoms spread over more than 100 nm, such that SiO_2 stoichiometry is not reached (Figure 2-13). If the wafer is annealed after implanting an oxygen dose that is too low, oxide precipitates form at a depth equal to the depth of maximum oxygen concentration, but no continuous layer of SiO_2 is produced. Experiments show that a dose of 1.4×10^{18} cm^{-2} must be implanted (at an energy of 200 keV) in order to create a continuous buried oxide layer. The most commonly used dose is 1.8×10^{18} cm^{-2}, which produces a 400 nm-thick buried oxide layer upon annealing. Figure 2-13 illustrates the evolution of the profile of oxygen atoms implanted into silicon with an energy of 200 keV. At low doses, a Gaussian oxygen profile is obtained. When the dose reaches 1.4×10^{18} cm^{-2}, stoichiometric SiO$_2$ is formed (66 at.% of oxygen for 33 at.% of silicon), and further implantation does not increase the peak oxygen concentration, but rather broadens the overall profile (*i.e.* the buried oxide layer becomes thicker). This is possible because the diffusivity of oxygen in SiO2 is high enough for the oxygen to readily diffuse to the Si-SiO2 interface where oxidation occurs. The dose at which the buried oxide starts to form ($\cong 1.4 \times 10^{18}$ cm^{-2}) is called the "critical dose".

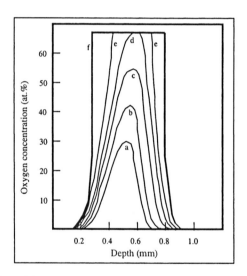

Figure 2-13: Evolution of the oxygen concentration profile with the implanted dose for an implantation energy of 200 keV: a) 4×10^{17} cm^{-2}, b) 6×10^{17} cm^{-2}, c) 10^{18} cm^{-2}, d) 1.2×10^{18} cm^{-2}, e) 1.8×10^{18} cm^{-2}, and f) 2.4×10^{18} cm^{-2}.

The temperature at which the implantation is performed is also an important parameter that influences the quality of the silicon overlayer. Indeed, the oxygen implantation step does amorphize the silicon which is located above the projected range. If the temperature of the silicon wafer during implantation is too low, the silicon overlayer becomes completely amorphized, and forms polycrystalline silicon upon subsequent annealing, an undesirable effect. When the implantation is carried out at higher temperatures (above 500°C) the amorphization damage anneals out during the implantation process ("self annealing"), and the single-crystal nature of the top silicon layer is maintained. The silicon overlayer, however, is highly defective and the ion implantation step must be followed by a high-temperature anneal step to improve the quality of both the BOX and the silicon layer. Since every single implanted oxygen atom must traverse the top silicon layer (the future silicon-on-insulator layer) a large number of defects is created. The typical defect density of early SIMOX layers was in excess of 10^9 defects per square centimeter. Constant improvement in the SIMOX process was brought about by many research groups worldwide. In summary, these improvements consist of maintaining the wafer at a temperature where most defects would self anneal during implantation, and performing a subsequent thermal treatment at high temperature (1350°C) in an appropriate ambient (argon + 2% oxygen). This allows for the stabilization and densification of the BOX, as well as for the removal of oxide precipitates and other defects in the top silicon layer.

During annealing, both dissolution of the oxide precipitates and precipitation of the dissolved oxide take place in the oxygen-rich silicon layers. In order to minimize the total surface energy of the SiO_2 precipitates, thus creating a more stable system, small precipitates dissolve into silicon, and large precipitates grow from the dissolved oxygen. At any given temperature (and for a given concentration of oxygen in the silicon), there exists a critical precipitate radius, r_c, below which a precipitate will disappear, and above which it will be stable.

It can be shown that $r_c = -\dfrac{2\sigma}{\Delta H_v}\dfrac{T_E}{T_E\text{-}T}$, where T is the temperature, T_E is the temperature of equilibrium between the two phases (solid precipitate or dissolved oxide), ΔH_v is the volume enthalpy of formation of the SiO_2 phase, and σ is the surface energy of the precipitates. The critical radius increases with temperature and becomes essentially infinite at very high temperatures (*i.e.*, when $T{\to}T_E$), such that the only stable precipitate is the BOX itself, with has an infinite radius of curvature. This phenomenon of growth of the large precipitates at the expense of smaller ones is known as "Ostwald ripening".

The use of a nitrogen ambient during annealing can induce the formation of silicon oxynitride around the oxide precipitates and inhibit their dissolution into the silicon matrix. Therefore, the use of an inert gas, such as argon, is preferred to nitrogen for the high-temperature annealing step used in the SIMOX formation process.[104,105] Figure 2-14 shows TEM cross sections of SIMOX samples fabricated in 1985 and 1998. The improvement in top silicon layer quality can readily be seen. The small silicon inclusions at the bottom of the BOX in Figure 2-14B are characteristic of the standard SIMOX material.[106,107]

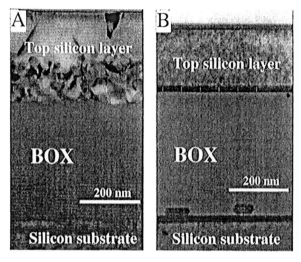

Figure 2-14: A: SIMOX fabricated in 1985, 10^9 defects/cm^2; B: SIMOX produced in 1998, less than 1000 defects/cm^2 (Courtesy S. Krause, M. Anc and P. Roitman).[108]

2.7.1.2   Low-dose SIMOX

In 1990 Nakashima and Izumi proposed reducing the implanted oxygen dose to drastically reduce the dislocation density in the silicon overlayer film. They found that the dislocation density drops significantly as the dose is reduced below 1.4×10^{18} cm^{-2} with an implantation energy of 180 keV (Figure 2-15).[109] Beside the reduction of defect density in the silicon layer there are other motivations for reducing the oxygen dose used to produce SIMOX material. At first, the total-dose radiation hardness of thin buried oxides is expected to be better than that of thicker ones. Secondly, direct fabrication of thin buried oxides is attractive for thin-film device applications. Finally, the production cost of a SIMOX wafer is proportional to the implanted dose. A potential additional benefit from this technique is the reduction of contamination of the wafers by impurities (carbon, heavy metals), which is proportional to the implanted oxygen dose.

Low-dose SIMOX is obtained by implanting O$^+$ ions within a narrow dose window of approximately 4×10^{17} atoms/cm^{-2}, called the "Izumi window".[110] Under these conditions implantation followed by a 6 hour anneal at 1320°C forms a continuous BOX having a thickness of 80-nm. Figure 2-16 schematically represents the structure of the buried oxide versus dose around the process window for an implant energy of 120 keV. [111] At a doses of 3×10^{17} cm^{-2} isolated oxide precipitates are formed. For a dose of 5×10^{17} cm^{-2} silicon precipitates form in the BOX. Only doses within a very narrow process window around 4×10^{17} cm^{-2} produce a continuous, precipitate-free BOX.

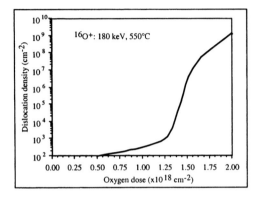

Figure 2-15: Evolution of dislocation density in the silicon overlayer with implanted oxygen dose.[112]

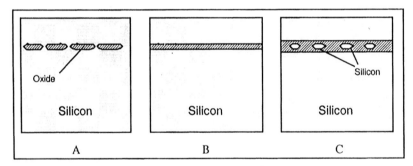

Figure 2-16: Evolution of the buried oxide structure for a dose of A: 3×10^{17}; B: 4×10^{17}; C: 5×10^{17} O$^+$ cm^{-2} and an energy of 120 keV.[113]

Figure 2-17 shows TEM cross sections of 80 and 150 nm-thick buried oxides produced by low-dose implantation. The choice of the implantation energy is a critical parameter. At 190 keV, which is close to the standard SIMOX energy, some SiO$_2$ islands are found in the silicon overlayer. Indeed, when the peak of implant defect generation and the projected range of the oxygen ions are distinct, two precipitation sites can occur and oxide precipitates can form at both the oxygen projected range and the defect projected range. A reduction of the implant energy to 120 keV is sufficient to merge the two precipitation sites and obtain a single and continuous BOX (Figure 2-18). The use of a lower implantation dose significantly reduces the defect density. Dislocation densities on the order of 300 cm^{-2} are found in low-dose SIMOX material.

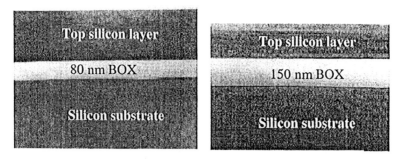

Figure 2-17: 80-nm (left) and 150-nm (right) buried oxides produced by low-dose oxygen implantation. (Courtesy C. Maleville)

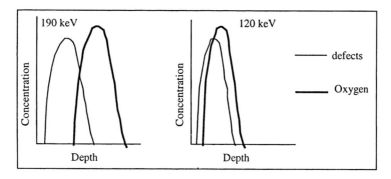

Figure 2-18: Defect and oxygen ion concentration produced by 190 and 120 keV implants.

From an industrial point of view, the use of low-dose implantation is obviously advantageous, since the throughput of the oxygen implanter is inversely proportional to the implanted dose. The lowest-dose SIMOX material reported uses a 2×10^{17} O$^+$/cm^2 implant which produced a 56 nm-thick BOX.[114] An empirical relationship between implant dose and energy for the production of thin buried oxide layers has been experimentally established and it has been shown that, for doses ranging between 2 and 6×10^{17} O$^+$ cm^{-2}, continuous BOX layers can be formed using an implant energy equal to $E \cong 30D$, where D is expressed in 10^{17} cm^{-2} and E is expressed in keV.[115]

2.7.1.3 ITOX

It is possible to increase the thickness of the BOX produced by low-dose oxygen implantation. Indeed, high-temperature (1350°C) oxidation of a low-dose SIMOX wafer causes an increase of the buried oxide thickness (Figure 2-19). This phenomenon is called high-temperature internal oxidation

(ITOX). As long as the thermal oxide grown on the silicon overlayer is thinner than 500 nm, there exists a linear relationship between the thickness of this oxide layer (t_{OX}) and the thickness increase of the buried oxide (Δt_{BOX}): $\Delta t_{BOX} = 0.06\, t_{OX}$. The internal oxide grows at the expense of the bottom of the silicon overlayer. High-temperature internal oxidation has been shown to significantly improve the roughness of interface between the silicon overlayer and the BOX and to densify the oxide itself.

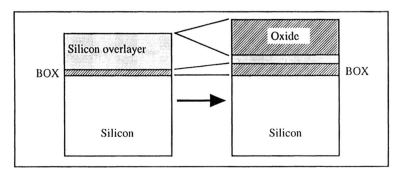

Figure 2-19: Principle of internal thermal oxidation (ITOX).[116,117,118]

2.7.1.4   SIMOX MLD

In low-dose SIMOX the formation of a continuous buried oxide layer is possible if the peak oxygen concentration is located in highly defective silicon. This observation is the basis for the modified low-dose (MLD) SIMOX process.[119] The MLD process overcomes the problem of oxide continuity encountered during low-dose SIMOX processing. To promote the formation of an ultrathin buried oxide during post-implantation annealing, the implantation process is modified to produce a microstructure that promotes coalescence of the oxygen into a continuous layer. This is accomplished by performing a two-step implant. In the first step, oxygen is implanted at a dose of 3×10^{17} cm^{-2} and an energy of 150 keV at a temperature of 525°C. These dose and temperature values ensure minimal generation of defects in the silicon overlayer. Unfortunately, the defects generated by the implant are not located at the peak of the oxygen concentration. To introduce additional defects a second implantation is carried out at room temperature. The dose (10^{15} cm^{-2}) is chosen to selectively amorphize the region near the oxygen peak to form a highly defective layer. This layer provides a template or guide upon which the oxide forms during subsequent annealing. Buried oxides prepared in this way are shown to be continuous and without silicon inclusions.[120] It is worth noting that additional implant of oxygen after the formation of can be used to improve the stoichiometry of the oxide and improve its electrical properties. It has been shown that the additional implantation of oxygen doses ranging between 10^{15} and 10^{17} cm^{-2} in the BOX of an already formed SIMOX structure (either "standard" or low-dose) increases the density of the oxide and reduces the trapping of charges in the oxide when the material is exposed to ionizing radiations.[121] The thickness uniformity of MLD SIMOX wafers is better than 2-3 nm (6σ) and the SOI RMS surface roughness is 0.12 nm (1x1 μm^2 AFM scan). The throughput of a high-current oxygen implanter (beam current=80 mA) is approximatively 30,000 300-mm MLD SIMOX wafers/year.[122]

2.7.1.5 Related techniques

It is possible to form a buried oxide layer without actually implanting oxygen. As we have seen earlier the simultaneous presence of defects and oxygen in silicon can result in the formation of a continuous buried oxide layer. In 2001, A. Ogura reported the formation of a continuous BOX obtained by implantation of light ions (H$^+$ or He$^+$) and subsequent annealing in an oxygen-containing ambient. The implantation of hydrogen or helium ions creates a defective layer near the projected range of the ions. Few defects, however, are created in the rest of the silicon, including in the future silicon overlayer, since H$^+$ and He$^+$ are light ions. Reported implant conditions are H$^+$, 5x10^{16} cm^{-2}, 45 keV, and He$^+$, 1-5x10^{17} cm^{-2}, 45 keV, all implanted at room temperature. Upon annealing in an argon/oxygen ambient at temperatures ranging from 1200 to 1350°C, oxygen diffuses through the top of the silicon layer and an internal oxidation process takes place which forms a buried oxide layer where the defects were presents (Figure 2-20). This technique makes it possible to produce a SIMOX-like SOI structure without the need for oxygen implantation and with less damage to the silicon overlayer. It is worth noting that oxygen implantation can be used to create the defective layer, as an alternative to hydrogen or helium.[123]

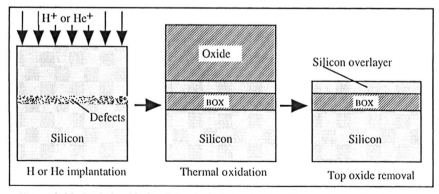

Figure 2-20: Buried oxide formation by light ion implantation and annealing in oxygen atmosphere.[124]

2.7.1.6 Material quality

The silicon film and buried oxide thicknesses are quite uniform in SOI wafers produced by the SIMOX technique. However, in early SIMOX wafers silicon and BOX thickness variations were observed. The thickness of the layers depends on the dose of oxygen introduced in the substrate. If the implantation is not uniform local variations of the thickness can occur. Figure 2-21 shows the silicon and BOX thickness measured on a SIMOX wafer that received a non-unifom implantation. The higher the dose, the thicker the BOX and the thinner the silicon overlayer.

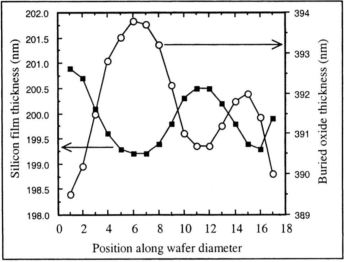

Figure 2-21: Silicon overlayer thickness and BOX thickness of an early "standard" SIMOX wafer, measured at 17 positions along a diameter of the wafer.[125]

2.7.1.6.1 Silicon overlayer

The dislocation density in SIMOX has been reduced from 10^9 cm^{-2} in early material to a few 10^3 cm^{-2} in modern material. [126] This improvement is attributed to the better control of the implant conditions (cleanliness, stability of the beam, optimization of scanning techniques, and uniformity of wafer heating), and, more recently, to the introduction of multiple implant techniques and low-dose SIMOX formation.

Early SIMOX material contained high concentrations of light metals such as aluminum and heavy metals such as iron, chromium and copper. These where sputtered from the implanter walls or the rotating drum which held the wafers. The contaminants were subsequently thermally driven into the silicon during the thermal annealing step. Such metallic impurities increase the leakage currents of junctions made in SIMOX material and degrade the radiation hardness of SIMOX MOSFETs. Great efforts have been made to reduce the metal contamination sources, and silicon or silicon dioxide shields have been placed in the implanters such that the beam never "sees" any metallic part. The design of high-temperature furnaces used to anneal the SIMOX material has also been optimized in order to minimize the diffusion of metallic contaminants from the heating elements into the furnace tubes.

One defect is typical of SIMOX silicon layers. It is called "HF defect" since it is revealed when a SIMOX wafer is dipped for a certain time into HF. Normally, HF does not etch silicon unless some oxidizing agent is present. When HF defects are present, HF etches small holes in the silicon overlayer and reaches the buried oxide, where it etches a cavity which can easily be observed by optical microscopy. HF defects are due to the presence of metallic silicide and silicate compounds, such as $CaSi_2$, $CaSiO_3$, Ca_2SiO_4, $FeSi$, $FeSiO_3$, etc. which react with HF and allow for the formation of micro-holes in the silicon overlayer.[127]

2.7.1.6.2 Buried oxide

The electrical characteristics of the SIMOX buried oxide are inferior to those of thermally-grown SiO_2.[128] The most important difference between the SIMOX buried oxide (BOX) and thermally grown oxide is that the former usually contains excess silicon in some form.[129] In this sense the SIMOX BOX is similar to deposited SiO_2 with deliberately introduced excess silicon. The electrical conduction in the BOX has two components: a "bulk" component, which is area-dependent, and a localized component, which is due to defects, and which is not area dependent.[130]

♦ *Bulk conduction*: The buried oxide of SIMOX structures exhibit a well-defined and reproducible bulk conduction which, in contrast to defect conduction, is area dependent. [131] This bulk conduction is quasi ohmic and time dependent up to an applied voltage of 80 V for a 400 nm-thick BOX. The temperature dependence of the bulk conduction has an activation energy of 0.3 eV and can be associated with trapping defects. At higher applied voltages the current-voltage characteristics are super-linear and resemble those observed for deposited SiO_2 films that contain excess silicon in the form of Si clusters. The high-field conduction depends very little on temperature, indicating that the conduction is controlled by tunneling between silicon clusters or O_3Si-SiO_3 bonds (E' centers).

The average size of the clusters can be estimated to be approximately 0.5 nm (*i.e.* three silicon atoms), and their density to be 2×10^{19} cm^{-3}, which means that the concentration of excess silicon is on the order of 6×10^{19} cm^{-3}. The bulk conduction depends on polarity. This effect is probably associated with the asymmetry in the distribution of the excess silicon in the BOX. Supplemental oxygen implantation eliminates the trapping-controlled conduction at low field and renders the high-field conduction somewhat similar to that characteristic of thermally grown oxide, without being identical to it (Figure 2-22). Supplemental oxygen implantation and annealing can be used to match the density of the BOX to that of a thermal oxide.[132] This process has been demonstrated to improve the total-dose hardness of SIMOX buried oxides.[133,134]

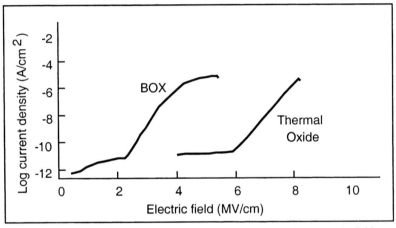

Figure 2-22: Oxide "bulk" current density as a function of the electric field.

♦ *Defect conduction*: The BOX layers exhibit localized, defect-induced conduction superimposed on the background (bulk) conduction. Some of this conduction can be associated with the presence of vertical silicon filaments (silicon pipes, pinholes) which run across the buried oxide from the silicon wafer to the silicon overlayer.[135] These silicon pipes find their origin in the micro-masking effect caused by particles on the surface of the wafer during the oxygen implantation step. They present a high conductivity of ohmic nature until the current density becomes high enough, at which point the filament "blows" like a fuse, creating an open circuit. The conductivity then returns back to the background "bulk" conductivity.

Another type of "defect" conductivity can be attributed to silicon filaments that do not extend all the way from the silicon substrate to the silicon overlayer. These defect give rise to a very low current, until the applied voltage reaches several volts. The current then rises suddenly, probably due to a field-emission mechanism. A blow-out mechanism similar to that found in continuous silicon pipes occurs at higher applied voltages (Figure 2-23).

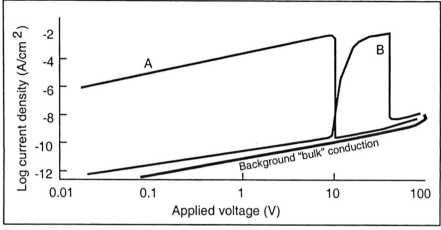

Figure 2-23: "Defect" current density in the BOX as a function of the applied (ramped) voltage ($t_{BOX} \cong 400$ nm). A: Current due to silicon pipes; B: Field-emission current.

The region situated near the perimeter of an SOI wafer, called the "edge exclusion" region is of poorer quality than the rest of the wafer and usually cannot be used. This is true not only for SIMOX, but for all SOI materials. The edge exclusion region is a few millimeters wide. Some of the characteristics of modern SIMOX material are listed in Table 2-3.

Table 2-3. SIMOX material properties (in 2003) [1,136]

Parameter	Standard SIMOX	SIMOX MLD
Wafer diameter	up to 200 mm	up to 300 mm
Silicon film thickness	210 nm	20 to 145 nm
Silicon film thickness uniformity	±2.5 %	±2 nm
Buried oxide (BOX) thickness	375 nm	135 or 145 nm
Buried oxide (BOX) thickness uniformity	±10 nm	±5 nm
Surface roughness (RMS)	0.7 nm	<0.15 nm
Dislocation density	< 1000 cm^{-2}	< 1000 cm^{-2}
HF defect density	< 0.5 cm^{-2}	< 0.1 cm^{-2}
BOX pipe (pinhole) density	< 0.1 cm^{-2}	< 0.1 cm^{-2}
Metallic contamination	< 5x10^{10} cm^{-2}	< 3x10^{10} cm^{-2}
BOX dielectric breakdown	>5 MV/cm	>7 MV/cm

2.7.2 Separation by implanted nitrogen (SIMNI)

Just as buried oxide can be synthesized by oxygen ion implantation, a buried silicon nitride layer (Si_3N_4) can be obtained by implanting nitrogen into silicon. The critical nitrogen dose for forming a buried nitride layer is 1.1×10^{18} cm^{-2} at 200 keV [137], but lower doses can be employed if the implant energy is lower. The obtained buried nitride and silicon overlayer thickness are 190 nm and 215 nm, respectively, in the case of a 7.5×10^{17} N$^+$ ions cm^{-2} implantation at 160 keV followed by a 1200°C anneal.[138] The most significant difference between oxygen and nitrogen inmplantation is that the peak of the nitrogen distribution does not saturate once stoichiometry is attained. This is due to the low diffusion coefficient of nitrogen in Si_3N_4 (10^{-28} cm^2 s^{-1} at a temperature of 500°C compared with 10^{-17} cm^2 s^{-1} for O_2 in SiO_2). The result of this low diffusion coefficient is that unreacted, "free" nitrogen is found in the buried layer if supercritical doses are implanted, and, as a result, nitrogen bubbles form. Nitride buried layers are polycrystalline, as opposed to buried oxide layers which are amorphous, and the grain boundaries between the Si_3N_4 crystallites can cause leakage currents between devices made in the silicon overlayer and the underlying silicon substrate. Furthermore, Si-Si_3N_4 interfaces are known to have a higher density of surface states than Si-SiO_2 interfaces. MOS device fabrication has, however, been demonstrated in SIMNI material.[139]

2.7.3 Separation by implanted oxygen and nitrogen (SIMON)

Implantation of both oxygen and nitrogen ions into silicon has been achieved by several research groups.[140] These were attempting to combine the advantages of both SIMNI (formation of a buried layer by implantation of a relatively low dose and low defect generation) with those of SIMOX (formation of an amorphous rather than polycrystalline buried layer with good Si-dielectric interfaces). Buried oxynitride layers may also present better radiation hardness performances than pure SIMOX material. Buried oxynitride layers have been formed by implantation of different doses of both nitrogen and oxygen into silicon. Different implant schemes have been proposed (oxygen can be implanted before nitrogen, or vice-versa). The kinetics of synthesis of oxynitrides by ion implantation is more complicated than that of pure SIMOX or pure SIMNI materials. In some instances, gas (nitrogen) bubbles can be formed within the buried layer. It is, however, possible to synthesize buried oxynitride layers that are stable and remain amorphous after annealing at 1200°C. The resistivity of the buried oxynitride layers can reach up to 10^{15} Ω.cm, which is comparable to that of oxynitride layers formed by other means.

2.7.4 Separation by implanted Carbon

Using a high-dose carbon implantation in silicon it is possible to form a buried β-SiC layer. To obtain a flat carbon concentration as a function of depth it may be necessary to implant multiple doses of carbon at different doses. For instance, using the following implantation sequence: 2.6×10^{17} cm^{-2} at 100 keV, 3.3×10^{17} cm^{-2} at 120 keV, 4.7×10^{17} cm^{-2} at 150 keV, and 10^{18} cm^{-2} at 195 keV (implants performed at 500°C), and annealing the wafer in a nitrogen ambient for 6 hours at 1150°C, a 300 nm-thick buried SiC layer can be produced under a 250 nm-thick silicon overlayer. Combining this technique with wafer bonding, silicon carbide-on-insulator (SiCOI) can be fabricated as well.[141]

2.8 WAFER BONDING AND ETCH BACK (BESOI)

The expression "wafer bonding" refers to the phenomenon whereby mirror-polished, flat and clean wafers of almost any material, when brought into contact, are locally attracted to each other by Van der Waals forces and adhere or "bond" to each other. Bonds formed at room temperature are usually relatively weak. Therefore, for many applications the room temperature bonded wafers must undergo a heat treatment to strengthen the bonds across the interface. After wafer bonding one of the wafers is subsequently polished or etched down to a thickness suitable for SOI applications. The other wafer serves as a mechanical substrate, and is called handle wafer (Figure 2-24).

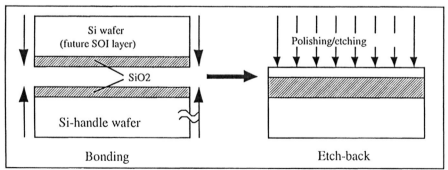

Figure 2-24: Bonding of two oxidized silicon wafers (left), and polishing/etching back of one of the wafers.

2.8.1   Hydrophilic wafer bonding

When two flat, hydrophilic surfaces such as oxidized silicon wafers are placed against one another, bonding naturally occurs, even at room temperature. The contacting force is caused by the attraction of hydroxyl groups (OH)⁻ adsorbed on the two surfaces. This attraction propagates from a first site of contact across the whole wafer in the form of a "bonding wave" with a speed of several cm/s. Figure 2-25 shows the propagation of a bonding wave between two oxidized, 100-mm silicon wafers. The presence of particles between the wafers creates unbonded areas called "voids". Voids with a diameter of several millimeters are readily created by particles 1 μm or less in size. Because of the usually organic nature of the particulates the extrinsic voids cannot be eliminated by high-temperature annealing. The only way of obtaining void-free bonding is, therefore, to clean the wafers with ultra-pure chemicals and water and to carry out all bonding operations in an ultra-clean environment.

The presence of voids is easily revealed by infrared, ultrasonic imaging, or X-ray tomography. The presence of voids is, of course, undesirable, since they can lead to delamination of devices located in imperfectly bonded areas during device processing. Wafer bonding must be carried out in a very clean environment to avoid the presence of particulates between the bonded wafers. A micro-clean room bonding apparatus has been developed, which involves bonding the wafers in a particle-free enclosure. Deionized filtered water is first flushed between the two closely-spaced wafers. The wafers are then dried by spinning under an infrared lamp and finally brought in contact.[142, 143]

Directly after room temperature bonding the adhesion between the two wafers is determined by Van der Waals interactions or hydrogen bridge bonds and one or two orders of magnitude lower than typical for covalent bonding. The typical bonding energy due to the Van der Waals is on the order of 50-100 mJ/m^2.[*] For most practical applications a higher bond energy is required which may be accomplished by an appropriate heating step which frequently, for commercial SOI production, is performed at temperatures as high as 1100°C. Stengl *et al.* developed a mechanistic model for direct wafer bonding for oxidized wafers describing the bonding chemistry at different temperatures [144]. When silica surfaces are hydrated, water molecules cluster on the oxidized wafer surface.

---

[*] Some authors express the bonding energy in ergs/cm^2. Conversion is 1 erg/cm^2 = 1 mJ/m^2.

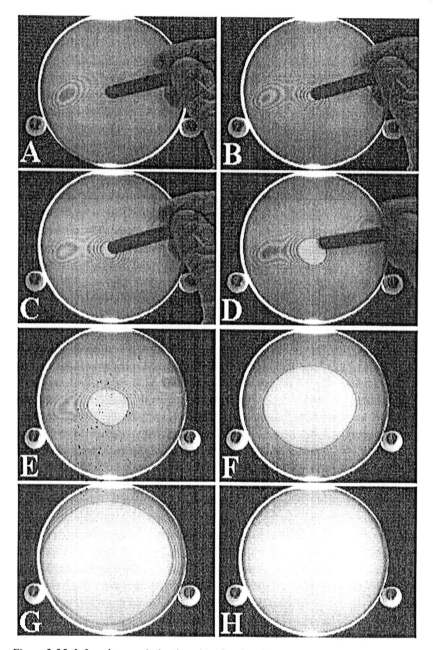

Figure 2-25: Infrared transmission imaging showing the propagation of a bonding wave between two oxidized 100-mm silicon wafers (from A to H). Time between different pictures is approximately 0.5 second. (Courtesy K. Hobart)

When two such surfaces are brought into contact, hydrogen bonding occurs via the adsorbed water, as shown in Figure 2-26A. At temperatures above 200°C the adsorbed water separates from the SiOH group and forms a tetramer water cluster. For temperatures greater than 700°C the water clusters decompose and diffuse away leaving Si-O-Si bonds (Figure 2-26). Maszara's experimental findings agree with Stengl's model, where reaction bonding proceeds by two different reactions [145]. He confirms the mechanism by measuring surface energies of bonded wafers using the crack propagation method.

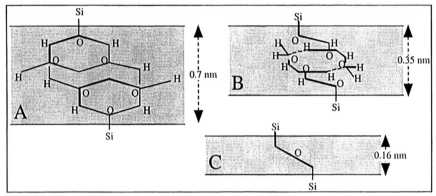

Figure 2-26: Stengl's proposed model for silicon wafer bonding at different temperatures. A: Room temperature, SiOH:(OH₂)₂:(OH₂)₂:HOSi; B: T = 200°C, SiOH:HOSi + (H₂O)₄ C: T > 700°C, SiOSi + H₂O.[146]

Bond strength for room temperature contact bonded wafers varies between 60-85 mJ/m², which is consistent with the surface energy of silica bonded through hydrogen bonding. In addition, the surface energy increases at the transition temperature of 300°C. This is the temperature where hydrogen bonds begin to convert to Si-O-Si bonds. Surface energy is constant for bonded wafers annealed in the region of 600°C to 1100°C for anneal times between 10 seconds to 6 hours and reaches values in excess of 1000 mJ/m². Maszara concludes that the bonding process does not involve mass transport in that temperature range. Rather the bond strength is limited in that regime by the amount of contacted area of the bonded wafers, which is a function of how well the wafers can elastically deform at a specific temperature. The kinetics of the deformation is so fast that the bond strength appears to be a function of temperature only. For temperatures greater than 1100°C, the bond energy does increase with time of anneal, but this is due to the viscous flow of the oxides at these high temperatures.[147] It is worth noting that high bonding energies (>1000 mJ/m²) can be obtained by activating the surface of the wafers using a plasma (*e.g:.* oxygen plasma) prior to bonding and then annealing the wafer pair at low temperature (*e.g.*: 400°C).[148,149] To be complete, one should mention that the bonding of silicon wafers to sodium-containing glass substrates can be facilitated by the application of an electric field during the annealing step. This process is called "anodic bonding".[150]

2.8.2 Etch back

After bonding of the wafers has been carried out, the top wafer has to be thinned down from a thickness of ...600 μm... to a few micrometers or less in order to be useful for SOI device applications. The Reader can find a detailed list of etch-back techniques in the book by Tong and Gösele.[151] Two basic thinning approaches can be used: grinding followed by chemico-mechanical polishing, and grinding followed by selective etch-back. The grinding operation is a rather crude but rapid step that is used to remove all but the last several micrometers of the (top) bonded wafer. The thinning method using chemico-mechanical polishing is cheap, but its use is, so far, limited to the fabrication of rather thick SOI films because of the absence of an etch stop. Much more accurate are the techniques using, after initial grinding, a chemical etch-back procedure with etch stop(s). The etch stop is usually obtained by creating doping concentration gradients at the surface (*i.e.* right next to the oxide layer used for bonding) of the top wafer. For instance, in the double etch-stop technique, a lightly doped wafer is used, and a P^{++} layer is created at its surface by ion implantation. Then, a low-doped epitaxial layer is grown onto it. This epitaxial layer will be the SOI layer at the end of the process. After bonding and grinding, two chemical etch steps are used. First, a potassium hydroxide solution [152] is used to selectively etch the substrate and to stop on the P^{++} layer. Then a 1:3:8 $HF:HNO_3:CH_3COOH$ etch is used to remove the P^{++} layer. The combined selectivity of the two etching solutions is better than 10,000:1. The final thickness uniformity of the SOI layer depends on the uniformity of the silicon thickness grown epitaxially, as well as on the uniformity of the P^{++} layer formation, but thickness standard deviations better than 12 nm can be obtained. Other etch stop techniques can be used as well (SiGe layer, carbon or nitrogen-doped silicon layer, etc.). Precision polishing of the top silicon film can also be achieved using a computer-controlled scanning plasma electrode. This technique, called plasma-assisted chemical etching (PACE) is used to planarize the silicon film and can produce 100 nm-thick SOI material with an rms thickness variation better than 2 nm ($\sigma=1...1.6$ nm).[153,154,155,156] A more recent polishing technique, called the magnetorheological finishing (MRF) technique can reach a silicon film thickness uniformity of 0.8 nm.[157] It makes use of a polishing fluidic slurry whose viscosity can be locally controlled by the application of a magnetic field. The rate of material removal during MRF polishing is controlled by a computer to achieve high-accuracy results.

2.9   LAYER TRANSFER TECHNIQUES

In the layer transfer techniques a thin superficial layer is peeled off from a silicon wafer and transferred to an oxidized silicon handle wafer. Splitting the superficial layer from a wafer is achieved either by gas (usually hydrogen) implantation and anneal and/or by applying pressure along a weakened crystal plane underneath the surface of a wafer.

2.9.1   Smart-Cut®

The Smart-Cut® process combines ion implantation technology and wafer bonding to transfer a thin surface layer from a wafer onto another wafer or an insulating substrate.[158,159] It consists of the three steps: implantation of gas ions (usually hydrogen), bonding to a stiffener, and thermal annealing (Figure 2-27): SOI wafers fabricated using the Smart-Cut® process are called UNIBOND® wafers.

◊ Ion implantation of hydrogen ions into an oxidized silicon wafer (called the "seed wafer"). The implanted dose is on the order of 5×10^{16} cm^{-2}. At this stage hydrogen-decorated defects (microbubbles) are formed at a depth equal to the implantation range (R_p). The wafer is preferably capped with thermally grown SiO_2 prior to implantation to protect the silicon top surface during implantation. This oxide layer will be used for hydrophilic bonding to a separate oxidized wafer.

◊ Hydrophilic bonding of the seed wafer to another oxidized silicon wafer, called the to "handle wafer", is performed.

◊ A two-phase heat treatment of the bonded wafers is then carried out. During the first phase, which takes places at a temperature around 500°C, a crystalline rearrangement and coalescence of the hydrogen-decorated defects into larger structures occurs in the hydrogen-implanted region of the seed wafer. Hydrogen pressure builds up in the growing cavities and eventually the seed wafer splits in two parts: a thin layer of monocrystalline silicon which remains bonded onto the handle wafer, and the remainder of the seed wafer which can be recycled for later use. The basic mechanism of the wafer splitting upon hydrogen implantation and thermal treatment is similar to surface flaking and blistering of materials exposed to helium or proton bombardment. During annealing, the average size of the microcavities increases. This size increase takes place along a (100) direction (*i.e.* parallel to the wafer surface) and an interaction between cavities is observed, which eventually results into the propagation of a crack across the whole wafer. This crack is quite parallel to the bonding wafer. The second heat treatment takes place at a higher temperature (1100°C) and is aimed at strengthening the bond between the handle wafer and the SOI film.

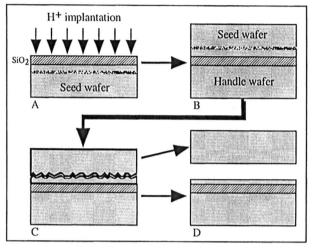

Figure 2-27: The Smart-Cut® process. A: Hydrogen implantation; B: Wafer bonding; C: Splitting of wafer A; D: Polishing of both wafers. Wafer A is recycled as a future handle wafer.

Finally, chemo-mechanical polishing is performed on the SOI film to give it the desired mirror-like surface. Indeed, this layer exhibits some micro-roughness after the splitting of wafer wafer A, and a final touch-polish step is necessary. This polishing step reduces the surface roughness to less than 0.15 nm and consumes a few hundred angströms of the SOI film. A great advantage is that the seed wafer can be recycled, resulting in just *N+1* silicon wafers processed to produce *N* SOI substrates.

The basic mechanisms involved in the Smart-Cut® process are described in the next sections.

2.9.1.1 Hydrogen / rare gas implantation

Implantation of rare gas ions into materials is long known to lead to the formation of blisters at the material surface. The implantation of alpha particles (He^{++} ions) produced by nuclear reactions into the vacuum walls of fusion reactors causes the wall surface to flake and be covered with blisters.[160] Implantation of a variety of gases in different materials has been shown to cause blister formation, *e.g.*: Ar$^+$ in Ge [161], Ar$^+$ in Si [162], H$^+$ in GaP [163], He$^+$ in Mo and Nb [164], He$^+$ in Ni [165], He$^+$ or Ne$^+$ in Al [166], Ar$^+$ and Xe$^+$ in Si [167] and H$^+$ in Si.[168] In early blistering experiments high-fluence implantation 10^{17}-10^{18} cm^{-2} was used to create gas-filled cavities at a depth near the projected range of the implanted species. If the dose is high enough the gas pressure in the cavities leads to mechanical deformation of the material above the cavity and forms a blister. If the pressure in the cavity is sufficiently large the lids of the blister can break and flaking is observed at the material surface.

It is also possible to form blisters by implanting a moderate dose ($< 10^{17}$ cm^{-2}) of gas ions and then performing a thermal annealing step to promote the coalescence of small gas-containing defects into larger ones. For the fabrication of SOI material it is suitable to use hydrogen rather than helium or any other rare gas because less energy is deposited in the material above the projected range of the ions. As a result fewer defects are created in the silicon surface layer (the future SOI film) and most defects are situated at a depth near the projected range.[169] In the particular case of hydrogen (proton) implantation into silicon the defects created by implantation, or hydrogen-related cavities (HCRs) consist of a mix of hydrogen-decorated vacancies, vacancy clusters, and platelets.[170] Platelets are flat, disk-shaped microcavities containing hydrogen. Their thickness is approximately 2 lattice parameters (1 nm), their diameter is approximately 10 nm, and they are mainly oriented along (100) planes along the (100) surface of the wafer. (Figure 2-28).

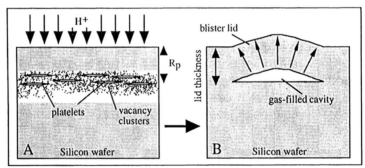

Figure 2-28: A: Hydrogen implantation ad formation of defects (vacancies, vacancy clusters and platelets; B: Formation of a blister upon annealing. Rp is the projected range of the implanted ions.

Typical implantation dose ranges between 1×10^{16} and 7×10^{16} H$^+$/cm^2. Upon annealing, typically above 500°C, the hydrogen atoms in the HRCs sever their bonds with silicon atoms and diffuse in the silicon. The hydrogen atoms aggregate in the larger defects, forcing them to grow in size, through a mechanism called "Ostwald ripening"; the hydrogen atoms lost by the small cavities are captured by larger ones, such that larger cavities grow at the expense of the smaller ones.[171,172] Molecular hydrogen migrating to the larger cavities causes a buildup of pressure and the formation of blisters.[173,174] The coalescence of the small defects into blisters in silicon implanted with hydrogen has been shown to occur at temperatures as low as 250°C (H_2^+, 160 keV, 5×10^{16} cm^{-2}).[175]

2.9.1.2 Bonding to a stiffener

The Smart-Cut® process is based on harnessing the destructive forces produced during blister formation to produce a thin silicon film. This is accomplished by attaching a stiffener to the implanted silicon wafer. Usually the stiffener is an oxidized silicon wafer called the "handle wafer" (Figure 2-29A), but glass, quartz and other materials can also be used. The implanted wafer is usually called the "seed wafer" because it is the wafer from which the thin SOI film will come from.

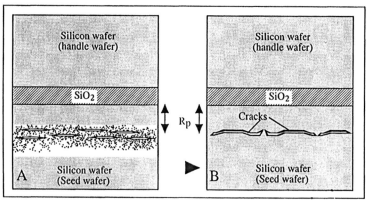

Figure 2-29: Formation of cracks near the projected range of a silicon wafer implanted with hydrogen. A: Bonding of the handle wafer (stiffener) to the seed wafer; B: Formation of a crack network near the projected range upon annealing.

An annealing step is then performed to both strengthen the bond between the two wafers and to make it possible for the smaller HCR to coalesce into larger, hydrogen-filled structures called "cracks" or "microcracks". These cracks correspond to the gas-filled cavities forming blisters in the absence of a stiffener. The role of the stiffener is to prevent mechanical deformation of the thin silicon layer between the defect region and the SiO_2 layer. In addition, the presence of the stiffener provides a restoring force that opposes the vertical lift that leads to blistering and the vertical force is transformed into lateral crack propagation. As a result a network of horizontal cracks filled with pressurized hydrogen is created, which creates a "perforated line"-like separation (Figure 2-30).

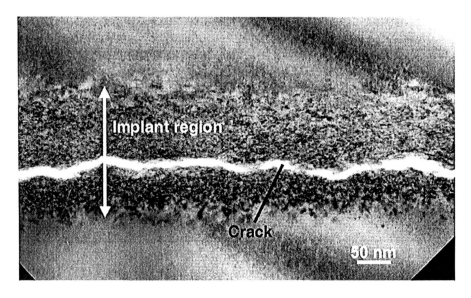

Figure 2-30. TEM micrograph of a crack. (Courtesy C. Maleville)

2.9.1.3 Annealing

The annealing step is responsible for removing hydrogen atoms from the vacancy complexes. The hydrogen diffuses to the growing microcracks. Many different experimental techniques have been used to investigate the coalescence of small HCRs in larger bubbles, blisters or microcracks. Time-to-blister and time-to-splitting experiments have been carried out in order to extract the activation energies involved in the Smart-Cut process. The time-to-splitting of the seed wafer can accurately be measured. It depends of course on the annealing temperature, but also on implantation conditions such as dose, energy, doping species, and bonding parameters. Two types of activation energies can be seen from the Arrhenius-type plot of 1/splitting time versus $1/kT$ (Figure 2-31). At high temperatures the activation energy E_a is 0.5 eV, which is very close to the activation energy for the diffusion of hydrogen in silicon (0.48 eV). At lower temperature the activation energy is approximately 2.2 eV. This corresponds to the sum of activation energies for the diffusion of hydrogen (0.48 eV) and for breaking the H-Si bonds in the HCRs (1.8 ± 0.2 eV). Thus, at high temperature the coalescence of the bubbles is diffusion-limited, while at lower temperatures (<500°C) the reaction is limited by both the extraction of the hydrogen from the hydrogen-related cavities and hydrogen diffusion.

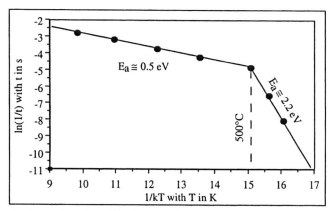

Figure 2-31: Arrhenius plot of the time to split *vs.* annealing temperature. The implant conditions are H⁺, 69 keV, 6x10¹⁶ cm⁻². [176]

The Ostwald ripening mechanism by which hydrogen migrates from the smaller defects into the larger ones has been confirmed by combining infrared spectroscopy and forward recoil scattering (FRS) on wafers annealed without a stiffener. Infrared spectroscopy detects Si-H bonds in the HCRs, while FRS measures the total hydrogen concentration (hydrogen in the bonds and molecular hydrogen in the forming blisters). FRS shows that there is no outdiffusion of hydrogen from the silicon as long as the annealing temperature is below 500°C, and thus that the total amount of hydrogen remains constant (Figure 2-32).

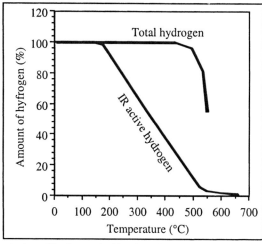

Figure 2-32: Evolution of hydrogen in Si-H bonds and total amount of hydrogen vs. annealing temperature. Annealing time is 30 min.[177]

Infrared spectroscopy indicates that the amount of hydrogen in the Si-H bonds decreases with annealing temperature between 200 and 500°C. The hydrogen in the Si-H bonds disappears almost totally for annealing temperatures above 500°C. This clearly demonstrates the conversion of the hydrogen trapped in the HCRs to an unbound form, i.e. molecular hydrogen. This hydrogen increases both the size of and the pressure in the defects where it accumulates.

Transmission electron microscopy has been used to measure the size and the density of the microcavities as a function of annealing time (T=450°C). Such a study shows that the size of the cavities increases with annealing time, while their density decreases. More importantly the total volume of the cavities (product of their volume by their density) remains constant with annealing time, once again indicating that the total amount of hydrogen is constant and that the gas moves from the smaller defects into the larger ones (Ostwald ripening).[178]

2.9.1.4 Splitting

Splitting of the seed wafer takes place naturally at the end of the annealing process, when the silicon is sufficiently weakened near the projected range depth by the network of microcracks and the buildup of pressure in them. The duration of the splitting itself is extremely short (probably less than one millisecond) and generates a small audible noise. It is, however, possible to induce splitting of the seed wafer by other means, once the silicon has been sufficiently weakened by hydrogen implantation and some annealing. For instance, splitting can be obtained by dipping the wafer pair in liquid nitrogen [179] or applying a mechanical force to the edge of the seed wafer.[180,181] Splitting can also be obtained by supplying microwave energy to the wafer (2-minute microwave "anneal" at 900W, 2.45 GHz) after bonding the implanted wafer and annealing the bonded pair at 150°C (Nova Cut™ process).[182] After splitting, the SOI wafer receives a final mechanical-chemical polish step ("touch polish") to reduce surface roughness and achieve the desired surface flatness for the fabrication of SOI devices. Exposure to an HCl + H_2 gas mixture in an epitaxial reactor can be used to smooth the split surface as well. This technique has been shown to reduce the SOI film RMS surface roughness to less than 0.8 Å.[183,184] Splitting can also be obtained with reduced thermal budget if the hydrogen is implanted into a strained SiGe layer, such that the application of a simple mechanical force applied at the edge of the wafer can induce cleaving at the implant depth (NanoCleave® process).[185]

2.9.1.5 Further developments

The Smart-Cut® process has been successively used to transfer thin films other than silicon. The literature reports the successful transfer of the following materials: SiC (silicon carbide on insulator: SiCOI)[186], GaAs,

InP, LiNbO$_3$ [187], SiGe (SiGe on insulator: SiGeOI) [188], Ge (germanium on insulator: GeOI) [189], etc. The Smart-Cut® process can also be used to fabricate multilayer (Si/SiO$_2$/Si/SiO$_2$/Si/SiO$_2$/etc.) structures [190] and transfer layers with patterned structures.[191] Debonding of the thin silicon overlayer from the handle wafer and its transfer to a third wafer using a polymer bonding technique has be demonstrated as well.[192] Combining Smart-Cut® and debonding, it thus is possible to transfer thin SOI integrated circuits onto flexible plastic films.[193] It has also been shown that in silicon co-implantation of boron with hydrogen allows the Smart-Cut® process to occur at lower temperatures, while co-implantation of hydrogen and helium can reduce the implanted dose required for splitting. Because of the lower dose, the defect density in the active layer is reduced.[194] This improvement has made it possible to transfer a thin silicon film on quartz (silicon on quartz (SOQ)) after annealing at a temperature of only 250°C.[195] The combined use of Smart-Cut and a Si/SiGe/Si etch stop has made it possible to produce SOI films as thin as 5 nm (Ultra-Cut process).[196]

2.9.2 Eltran®

The Eltran® (epitaxial layer transfer) technique combines the formation of a porous silicon layer, epitaxy, and wafer bonding to produce SOI wafers. The properties of porous silicon are discussed in the next section, followed by the description of the Eltran® fabrication process itself.

2.9.2.1 Porous silicon formation

P-type silicon can readily be converted into porous silicon by electrochemical dissolution of p-type silicon in HF; the silicon wafer is immersed in an HF solution and a potential drop is applied between the sample and a platinum electrode dipped into the electrolyte. The degree of porosity of the layer can be controlled by adjusting the current utilized during the reaction. Porous silicon has been used in the 1980's to fabricate SOI wafers. The material was called "FIPOS", which stands for "full isolation by porous silicon".[197,198] Porous silicon basically looks like a silicon "sponge". It is full of wormhole-like cavities, but the silicon that has not been dissolved away by the electrochemical reaction is still single-crystal. In 1985 it was observed that high-quality silicon layers can be grown epitaxially on porous silicon.[199] Porous silicon has two key properties. Firstly the surface pores can be filled by silicon atoms and sealed by baking the material in hydrogen. This property is used to prepare the surface before the growth of epitaxial silicon on porous silicon. Secondly, porous silicon has an extremely large surface to volume ratio (200 m^2/cm^3).[200] As a result it has a very high chemical reactivity and it is possible to etch it in an HF/H$_2$O$_2$ solution with an extremely high selectivity (100,000:1) to silicon.

2.9.2.2 The original Eltran® process

The original Eltran® process was published in 1994.[201] It comprises the following steps (Figure 2-33): the formation of a blanket porous silicon layer on a silicon wafer (seed wafer) followed an hydrogen bake step to seal the surface pores, and the growth of a single-crystal epitaxial silicon film. After thermal growth of an oxide the seed wafer is bonded to an oxidized wafer called the "handle wafer".

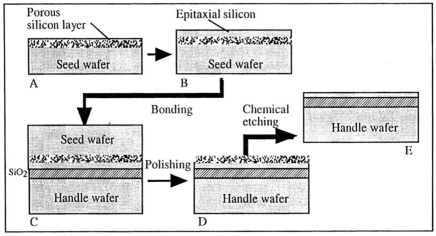

Figure 2-33: The original Eltran® process. A: Formation of a porous silicon layer; B: Growth of epitaxial silicon; C: Bonding to a handle wafer; D: Polishing of the silicon wafer; E: Porous silicon etching.

The bulk of the handle wafer is then removed by grinding and polishing until the porous silicon layer is reached. The porous silicon is removed by etching in an HF/H_2O_2 solution and a final H_2 annealing is applied to smooth the surface. One drawback of the original Eltran® process is that it requires two silicon wafers to produce a single SOI wafer.

2.9.2.3 Second-generation Eltran® process

Similar to the Smart-Cut® process, the second-generation Eltran® process uses $N+1$ silicon wafers to produce N SOI wafers. The process is based on the formation of a double porous silicon layer, or more exactly a layer with two different porosities. The juxtaposition of these two layers generates mechanical stress near the interface between the two types of porous silicon. The application of a high-pressure (20-60 MPa) water jet "unzips" the seed wafer from the handle wafer along the stress region between the two porous silicon layers, thereby producing an SOI substrate and a recyclable seed wafer (Figure 2-34). The important steps of the process are described next.

The porosity of a silicon layer can be modified by changing the current density used during the electrochemical reaction. The porosity increases from 20 to 65% when the current density is increased from 8 to 23 mA/cm². A porous layer with two different porosities can be created by tuning the current density during porous silicon formation. In the Eltran® process a low current density is first used to form a low-porosity layer at the surface of the seed wafer. A low-porosity surface layer is suitable for the subsequent growth of a high-quality epitaxial layer. The current density is then increased to produce a higher-porosity layer bebeath the low-porosity layer. The difference in porosity generates mechanical stress between the two layers. A hydrogen bake step is applied to seal the pores at the surface, and a silicon layer (the future SOI layer) is grown by epitaxy. Hydrophilic bonding is used to attach the seed wafer to the handle wafer. Then a jet of pressurized water with a diameter of 0.1 mm is directed at the edge of the wafer assembly. The water jet acts as a liquid wedge that splits the porous silicon at the region of maximum stress, *i.e.* where the two porosities meet. Because a liquid rather than a solid wedge is used, the splitting easily propagates across the entire wafer assembly, and the porous silicon layer opens like a zipper. Once splitting has been achieved the porous silicon layers can be removed from both the seed and the SOI wafers by etching in an HF/H$_2$O$_2$ solution. A final H$_2$ annealing is applied to smooth the surface.

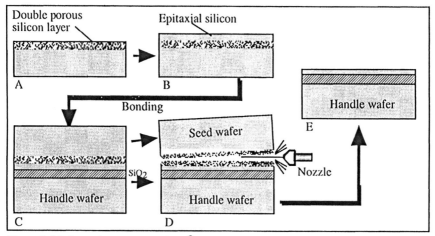

Figure 2-34: Second-generation Eltran® process. A: Formation of a porous silicon layer; B: Growth of epitaxial silicon; C: Bonding to a handle wafer; D: Porous silicon layer splitting using a water jet; E: Etching and H$_2$ annealing.[202,203,204]

2.9.3 Transferred layer material quality

Some of the characteristics of SOI material produced by layer transfer are listed in Table 2-4.

Table 2-4. Transferred layer material properties (in 2003) [1,205,206,207]

Parameter	Unibond®	Eltran®
Wafer diameter	up to 300 mm	up to 300 mm
Silicon film thickness	50 to 300 nm	20 to 145 nm
Silicon film thickness uniformity	down to ± 0.5 nm	±1.6 nm
Buried oxide (BOX) thickness	100 nm to 3 µm	135 or 145 nm
Buried oxide (BOX) thickness uniformity	down to ±0.1 nm	±2 nm
Surface roughness (RMS)	0.15 nm	0.1 nm
Dislocation density	100 cm^{-2}	400 cm^{-2}
HF defect density	< 0.05 cm^{-2}	< 0.05 cm^{-2}
BOX pipe (pinhole) density	none	none
Metallic contamination	5x10^{10} cm^{-2}	5x10^{10} cm^{-2}

2.10 STRAINED SILICON ON INSULATOR (SSOI)

Silicon films with in-plane tensile stress (strain) have a higher electron mobility than relaxed (unstressed) films. Conversely, silicon and SiGe films with in-plane compressive strain have a higher hole mobility.[208] Several techniques have been developed to produced strained silicon-on-insulator (SSOI) material. A first technique employs oxygen implantation and resembles the SIMOX process (Figure 2-35). At first a graded $Si_{1-x}Ge_x$ layer is grown on a silicon wafer. The germanium content, x, is increased from 0 to 0.1 (or higher in [209,210]) during growth. Then a relaxed layer of $Si_{0.9}Ge_{0.1}$ is grown. Next, oxygen ions (180 keV, 4x10^{17} cm^{-2}) are implanted into the relaxed SiGe layer and high-temperature annealing (1350°C for 6 h) is carried out to grow a buried SiO_2 layer inside the top SiGe layer. A strained layer is silicon can directly be grown on the relaxed SiGe layer [211], or a combination of strained SiGe ($Si_{0.82}Ge_{0.18}$) and strained Si films can be grown. The strain in the top SiGe layer is compressive, which increases hole mobility, while the strain in the silicon layer is tensile, which provides enhances electron mobility. SOI MOSFETs have been fabricated using this process, and electron and hole mobility 60% and 30 % higher than in regular SOI devices has been measured, respectively.[212]

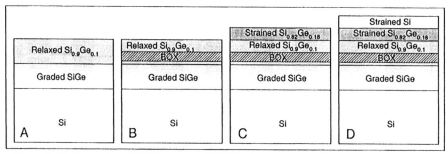

Figure 2-35: Formation of SSOI using a SIMOX-like process. A: Growth of a relaxed SiGe layer; B: Oxygen implantation and annealing to form a buried oxide (BOX) layer; C: Growth of a SiGe layer richer in Ge than the first layer; D: Growth of the strained silicon layer.

Strained silicon-on-insulator can also be produced using the Smart-Cut® process. [213] Figure 2-36 illustrates such a fabrication process. A relaxed SiGe layer is first grown on a silicon wafer, and a strained silicon layer is grown on top of the SiGe. Then hydrogen is implanted through the strained Si (75 keV, 4×10^{16} cm^{-2}) to produce a cleave plane in the SiGe layer. The structure is then bonded to an oxidized handle wafer and the bonded pair is annealed at 570°C to split the SiGe layer. The remaining SiGe on top of the strained silicon layer can then be removed by a combination of low-temperature (<800°C) steam oxidation and dilute HF etching of the resulting (SiGe)O$_2$ film. At temperatures below 800°C the oxidation rate of SiGe is much higher than that of silicon, such that the oxidation front selectively stops at the strained silicon layer.

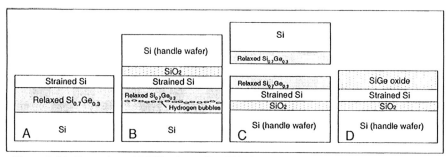

Figure 2-36: Formation of SSOI using a SIMOX-like process. A: Growth of a relaxed SiGe layer and a strained silicon layer; B: Hydrogen implantation, bonding to a handle wafer and annealing; C: Splitting of the SiGe layer near the peak of hydrogen implant; D: Oxidation of the SiGe layer.[214]

2.11 SILICON ON DIAMOND

Diamond can be used as the buried insulator in SOI structures. The main motivation for the use of diamond is its high thermal conductivity which can help evacuating heat from in SOI power devices.

In the temperature range between 100 and 600K undoped diamond is a highly insulating material. A resistivity above 10^{13} Ω.cm and a breakdown electric field above 10^7 V/cm have been reported.[215] At the same time the heat conductivity of diamond is approximately 10 times higher than that of silicon and 1000 times higher than that of SiO_2.

Silicon-on diamond (SOD) can be fabricated by combining chemical vapor deposition (CVD) of diamond and wafer bonding or epitaxial lateral overgrowth of silicon.[216,217] The diamond layer can also be deposited by the hot-filament method using an ambient of 1-2% of CH_4 in H_2 at a temperature of 650-750°C and under a pressure of 30 to 50 Torr.[218] The excellent efficiency of heat evacuation in SOD devices has been demonstrated.[219,220] Compared with bulk silicon, the thermal conductivity of the silicon-on-diamond structure composed of 1 μm of silicon on 300 μm of diamond is increased by 850%.[221] Alternative methods uses aluminum nitride (AlN) which also has a high thermal conductivity.[222,223,224]

2.12 SILICON-ON-NOTHING (SON)

The name "silicon-on-nothing (SON)" does not really apply to a silicon-on-insulator material, but rather to a technique that can be used to fabricate SOI devices.[225] The SON process is shown in Figure 2-37. The SON process consists of the following steps. Firstly a strained layer of SiGe and a layer of silicon are epitaxially grown on a silicon wafer. After the epitaxy, conventional CMOS processing steps are carried out up to the formation of source/drain extensions and the formation of gate spacers. Next, trenches are opened in the source/drain areas using anisotropic plasma etching, which accesses the SiGe layer. The SiGe layer is then removed using a highly selective (100:1) plasma etch process. The removal of the SiGe forms an air tunnel that isolates the device from the substrate. This air tunnel gives the SON name to the process. The device is supported by a bridge structure supported at both ends by the field isolation. The air tunnel can then be filled using LPCVD oxide deposition to produce an SOI structure. The process is completed by epitaxial growth of silicon to form sources and drains.

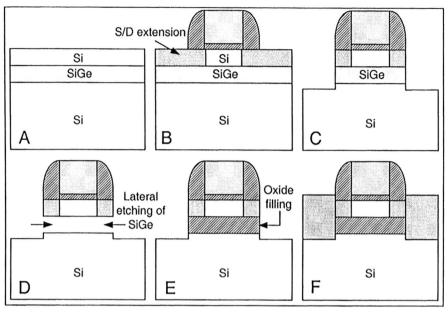

Figure 2-37: SON process; A: Epitaxy of SiGe and Si on a silicon wafer; B: Conventional CMOS process steps until formation of spacers; C: Formation of shallow trenches in the source and drain regions; D: Etching of the SiGe to form a tunnel under the silicon film; E: Filling the tunnel with oxide (optional step); F: Selective epitaxial growth of source and drain regions.[226]

REFERENCES

[1] P.L.F. Hemment, "The SOI odyssey", Electrochemical Society Proceedings, Vol. 2003-05, p. 1, 2003 b.c.

[2] I. Golecki, in "*Comparison of thin-film transistor and SOI technologies*", Ed. by H.W. Lam and M.J. Thompson, Mat. Res. Soc. Symp. Proc., Vol. 33, p. 3, 1984

[3] E.I. Givargizov, A.B. Limanov, Microelectronic Engineering, Vol.8, no.3-4, p. 273, 1988

[4] M.W. Geis, D.A. Antoniadis, D.J. Silversmith, R.W. Mountain, H.I. Smith, Applied Physics Letters, Vol. 37, p.454, 1980

[5] H.M. Manasevit and W.I. Simpson, Journal of Applied Physics Vol. 35, p. 1349, 1964

[6] T. Sato, J. Iwamura, H. Tango, and K. Doi, in "*Comparison of thin-film transistor and SOI technologies*", Ed. by H.W. Lam and M.J. Thompson, Mat. Res. Soc. Symp. Proc., Vol. 33, p. 25, 1984

[7] I. Golecki, in "*Comparison of thin-film transistor and SOI technologies*", Ed. by H.W. Lam and M.J. Thompson, Mat. Res. Soc. Symp. Proc., Vol. 33, p. 3, 1984

[8] S.S. Lau, S. Matteson, J.W. Mayer, P. Revesz, J. Gyulai, J. Roth, T.W. Sigmon, T. Cass, Applied Physics Letters Vol. 34, p. 76, 1979

[9] J. Amano, K.A. Carey, Applied Physics Letters Vol. 39, p. 163, 1981

[10] P.K. Vasudev, D.C. Mayer, Materials Research Society Symposia Proceedings, Vol. 33, p. 35, 1984

[11] M.E. Roulet, P. Schwob, I. Golecki, M.A. Nicolet, Electronics Letters Vol. 15, p. 527, 1979

[12] A.C. Ipri AC (1981), "*The properties of silicon-on-sapphire substrates, devices, and integrated circuits*", Applied Solid-State Sciences, Supplement 2, Silicon Integrated Circuits, Part A, Ed. by. D. Kahng, Academic Press, pp. 253-395, 1981

[13] M. Megahed, M. Burgener, J. Cable, D. Staab, R. Reedy, Topical Meeting on Silicon Monolithic Integrated Circuits in RF Systems, Digest of Papers, p. 94, 1988

[14] R. Reedy, J. Cable, D. Kelly, M. Stuber, F. Wright, G. Wu, Analog Integrated Circuits and Signal Processing Vol. 25, Kluwer academic Publishers, p. 171, 2000

[15] M. Morita, S. Isogai, N. Shimizu, K. Tsubouchi, and N. Mikoshiba, Jpn. J. Appl. Phys, Vol. 20, p. L173, 1981

[16] D.M. Jackson, Jr. and R.W. Howard, Trans. Met. Soc. AIME, Vol. 223, p. 468, 1965

[17] H.M. Manasevit, D.H. Forbes, and I.B. Cardoff, Trans Met. Soc. AIME, Vol. 236, p. 275, 1966

[18] G. Shimaoka and S.C. Chang, J. Vac. Sci. Technol., Vol. 9, p. 235, 1972

[19] I. Golecki, H.M. Manasevit, L.A. Moudy, J.J. Yang, and J.E. Mee, Appl. Phys. Letters, Vol. 42, p. 501, 1983

[20] I. Golecki, in "*Comparison of thin-film transistor and SOI technologies*", Ed. by H.W. Lam and M.J. Thompson, Mat. Res. Soc. Symp. Proc., Vol. 33, p. 3, 1984

[21] I. Golecki, R.L. Maddox, H.L. Glass, A.L. Lin, and H.M. Manasevit, presented at the 26th electronic Materials Conf., Santa Barbara, CA, June 1984

[22] M. Ihara, Y.Arimoto, M. Jifiku, T. Kimura, S. Kodama, H. Yamawaki, and T. Yamaoka, J. Electrochem. Soc., Vol. 129, p. 2569, 1982

[23] Y. Hokari, M. Mikami, K. Egami, H. Tsuya, and M. Kanamori, Technical Digest of IEDM, p. 368, 1983

[24] K. Ikeda, H. Yamawaki, T. Kimura, M. Ihara, and M. Ozeki, Ext. Abstr. 5th International Workshop on Future Electron Devices, Miyagi-Zao, Japan, p. 225, 1988

[25] Asano and H. Ishiwara, Jpn. J. Appl. Phys., Vol. 21, suppl. 21-1, p. 187, 1982

[26] T.R. Harrison, P.M. Mankiewich, and Dayem, Appl. Phys. Lett., Vol. 41, p. 1102, 1982

[27] H. Onoda, M. Sasaki, T. Katoh, and N. Hirashita, IEEE Trans. Electron Dev., Vol. ED-34, p. 2280, 1987

[28] T.I. Kamins, Proceedings of the IEEE, Vol.60, no.7, p.915-16, 1972

[29] K.E. Bean, W.R. Runyan, Electrochemical Society Fall Meeting, Extended abstracts of Battery Division, p.323, 1974

[30] J.F. Krieg, C.J. Neerman, M.W. Savage, J.L. Titus, D. Emily, G.W. Dunham, N. Van Vonno, J. Swonger, IEEE Transactions on Nuclear Science, Vol.47, no.6, pt.3, p.2561, 2000

[31] C.H. Fa and T.T. Jew, IEEE Transactions on Electron Devices, Vol. ED-13, p. 290, 1966

[32] T.I.Kamins, Solid-State Electronics, Vol. 15, p. 789, 1972

[33] S.W. Depp, B.G. Huth, A. Juliana, and R.W. Koepcke, in *"Grain Boundaries in Semiconductors"*,Ed. by H.J. Leamy, G.E. Pike, and C.H. Seager, Mat. Res. Soc. Symp. Proceedings, Vol. 5, p. 297, 1982

[34] T.I. Kamins and P.J. Marcoux, IEEE Electron Device Letters, Vol. EDL-1, p. 159, 1980

[35] H. Shichiro, S.D.S. Malhi, P.K. Chatterjee, A.H. Shah, G.P. Pollack, W.H. Richardson, R.R. Shah, M.A. Douglas, and H.W. Lam, in *"Comparison of thin-film transistor and SOI technologies"*, Ed. by H.W. Lam and M.J. Thompson, Mat. Res. Soc. Symp. Proceedings, North-Holland, Vol. 33, p. 193, 1984

[36] S.D.S. Malhi, in *"Comparison of thin-film transistor and SOI technologies"*, Ed. by H.W. Lam and M.J. Thompson, Mat. Res. Soc. Symp. Proceedings, North-Holland, Vol. 33, p. 147, 1984

[37] G.K Celler, H.J. Leamy, L.E. Trimble, and T.T. Sheng, Appl. Phys Lett., Vol. 39, p. 425, 1981

[38] C. Hill, in *"Laser and Electron-Beam Solid Interactions and Material Processing"*, Ed. by J.F. Gibbons, L.D. Hess, and T.W. Sigmon, Mat. Res. Soc. Symp. Proceedings, Vol. 1, p. 361, 1981

[39] N.M. Johnson, D.K. Biegelsen, H.C. Tuan, M.D. Moyer, and E. Fennel, IEEE Electron Dev. Lett, Vol. EDL-3, p. 369, 1982

[40] E.I. Givargizov, V.A. Loukin, and A.B. Limanov, in *"Physical and Technical Problems of SOI Structures and Devices"*, Kluwer Academc Publishers, NATO ASI Series - High Technology, Vol. 4, p. 27, 1995

[41] G.J. Willems, J.J. Poortmans, and H.E. Maes, J. Appl. Phys., Vol. 62, p. 3408, 1987

[42] T.I. Kamins, M.M. Mandurah, and K.C. Saraswat, J. Electrochem. Soc., Vol. 125, p. 927, 1978

[43] H.W. Lam, R.F. Pinizotto and A.F. Tasch, J. Electrochem. Soc., vol 128, p. 1981, 1981

[44] J.M. Hodé, J.P. Joly, and P. Jeuch, Ext. Abstr. of Electrochem. Soc. Spring Meeting, Vol. 82-1, p. 232, 1982

[45] S. Kawamura, J. Sakurai, M. Nakano, and M. Tagaki, Appl. Phys. Lett., Vol. 40, p. 232, 1982

[46] N.A. Aizaki, Appl. Phys. Lett., Vol. 44, p. 686, 1984

[47] J.P. Colinge, E. Demoulin, D. Bensahel, and G. Auvert, Appl. Phys. Lett., Vol. 41, p. 346, 1982

[48] J.P. Colinge, Ext. Abstracts of 2nd Internat. Workshop on Future Electron Devices, Shuzenji, Japan, p. 13, 1985

49   T. Morishita, T. Miyajima, J. Kudo, M. Koba, and K. Awane, Ext. Abstracts of 2nd Internat. Workshop on Future Electron Devices, Shuzenji, Japan, p. 35, 1985

50   K. Sugahara, S. Kusunoki, Y. Inoue, T. Nishimura, and Y. Akasaka, J. Appl. Phys., Vol. 62, p. 4178, 1987

51   R.C. McMahon, Microelectronic Engineering, Vol. 8, p. 255, 1988

52   T. Hamasaki, T. Inoue, I. Higashinakagawa, T. Yoshii, and H. Tango, J. Appl. Phys., Vol. 59, p. 2971, 1986

53   E.I. Givargizov, *Oriented Crystallization on Amorphous Substrates*, Plenum Press, New York, 1991

54   J.C.C. Fan, M.W. Geis, and B.Y. Tsaur, Appl. Phys. Lett., Vol. 38, p. 365, 1981

55   M.W. Geis, H.I. Smith, B.Y. Tsaur, J.C.C. Fan, D.J. Silversmith, R.W. Mountain, and R.L. Chapman, in *"Laser-Solid Interactions and Transient Thermal Processing of Materials"*, Narayan, Brown and Lemons Eds., (North-Holland), MRS Symposium Proceedings, Vol. 13, p. 477, 1983

56   A. Kamgar and E. Labate, Mat. Letters, Vol. 1, p. 91, 1982

57   T.J Stultz, Appl. Phys. Lett., Vol. 41, p. 824, 1982, and
     T.J. Stultz, J.C. Sturm, and J.F. Gibbons, in *"Laser-Solid Interactions and Transient Thermal Processing of Materials"*, Narayan, Brown and Lemons Eds., (North-Holland), MRS Symposium Proceedings, Vol. 13, p. 463, 1983

58   D.P. Vu, M. Haond, D. Bensahel, and M. Dupuy, J. Appl. Phys., Vol. 54, p. 437, 1983

59   P.W. Mertens, D.J. Wouters, H.E. Maes, A. De Veirman, and J. Van Landuyt, J. Appl. Phys., Vol. 63, p. 2660, 1988

60   E.I. Givargizov, V.A. Loukin, and A.B. Limanov, in *"Physical and Technical Problems of SOI Structures and Devices"*, Kluwer Academc Publishers, NATO ASI Series - High Technology, Vol. 4, p. 27, 1995

61   G.K. Celler and L.E. Trimble, in *"Energy Beam-Solid Interactions and Transient Thermal Processing"*, Fan and Johnson Eds., (North-Holland), MRS Symposium Proceedings, Vol. 23, p. 567, 1984

62   B. Tillack, K. Hoeppner, H.H. Richter, and R. Banisch, Materials Science and Engineering (Elsevier Sequoia), Vol. B4, p. 237, 1989

63   L. Jastrzebski, J.F. Corboy, J.T. McGinn, and R. Pagliaro, Jr., J. Electrochem Soc., Vol. 130, p. 1571, 1983

64   L. Jastrzebski, J.F. Corboy, J.T. McGinn, and R. Pagliaro, Jr., J. Electrochem Soc., Vol. 130, p. 1571, 1983

65   R.P. Zingg, B. Höfflinger, and G.W. Neudeck, IEDM Technical Digest, p. 909, 1989

66   R.P. Zingg, H.G. Graf, W. Appel, P. Vöhringer, and B. Höfflinger, Proc. IEEE SOS/SOI Technology Workshop, p. 52, 1988

67   J.P. Denton and G.W. Neudeck, Proceedings of the IEEE International SOI Conference, p. 135, 1995

68   J.C. Chang, J.P. Denton,and G.W. Neudeck, Proceedings of the IEEE International SOI Conference, p. 88, 1996

69   A. Ogura, Y. Fujimoto, Ext. Abstracts of 8th Internat. Workshop on Future Electron Devices, Kochi, Japan, p. 73, 1990, and
     A. Ogura and Y. Fujimoto, Appl. Phys. Lett., Vol. 55, p. 2205, 1989

70   P.J. Schubert and G.W. Neudeck, IEEE Electron Device Letters, Vol. 11, p. 181, 1990

71   H.-S. Wong, K. Chan, Y. Lee, P. Roper, and Y. Taur, Symposium on VLSI Technology, Digest of Technical Papers, p. 94, 1996

72   S. Venkatesan, C. Subramanian, G.W. Neudeck, and J.P. Denton, Proceedings of the IEEE International SOI Conference, p. 76, 1993

[73] Y. Kunii, M. Tabe, and K. Kajiyama, J. Appl. Phys., Vol. 54, p. 2847, 1983, and
Y. Kunii, M. Tabe, and K. Kajiyama, J. Appl. Phys., Vol. 56, p. 279, 1984

[74] H. Ishiwara, H. Yamamoto, S. Furukawa, M. Tamura, and T. Tokuyama, Appl. Phys. Lett., Vol. 43, p. 1028, 1983

[75] J.A Roth, G.L. Olson, and L.D. Hess, in *"Energy Beam-Solid Interactions and Transient Thermal Processing"*, Fan and Johnson Eds., (North-Holland), MRS Symposium Proceedings, Vol. 23, p. 431, 1984

[76] H. Ishiwara, H. Yamamoto, and S. Furukawa, Ext. Abstracts of 2nd Internat. Workshop on Future Electron Devices, Shuzenji, Japan, p. 63, 1985

[77] Y. Kunii and M. Tabe, Ext. Abstracts of 2nd Internat. Workshop on Future Electron Devices, Shuzenji, Japan, p. 69, 1985

[78] T. Dan, H. Ishiwara, and S. Furukawa, Ext. Abstracts of 5th Internat. Workshop on Future Electron Devices, Miyagi-Zao, Japan, p. 189, 1988

[79] H. Liu, Z. Xiong, and J.O.K. Sin, IEEE Transactions on Electron Devices, Vol. 50, no. 6, p. 1552, 2003

[80] M. Miyao, M. Moniwa, K. Kusukawa, and S. Furukawa, J. Appl. Phys., Vol. 64, p. 3018, 1988

[81] G. Liu, S.J. Fonash, Applied Physics Letters, Vol. 55, p. 660, 1989

[82] G.Radnoczi, A. Robertson, H.T.G. Hentzell, S.F. Gong, M.A. Hasan, J. Applied Physics, Vol. 69, p. 6394, 1991

[83] Seok-Woon Lee, Tae-Hyung Ihn, Seung-Ki Joo, IEEE Electron Device Letters, Vol. 17, no.8, p.407, 1996

[84] Seek-Woon Lee, Seung-Ki Joo, IEEE Electron Device Letters, Vol.17, no.4, p.160, 1996

[85] V.W.C. Chan, P.C.H. Chan, IEEE Electron Device Letters, Vol.22, no.2, p.80, 2001

[86] Yin Chunshan Yin, V.W.C. Chan, P.C.H. Chan, Proceedings of the IEEE International SOI Conference, p.39, 2002

[87] V.W.C. Chan, P.C.H. Chan, Mansun Chan, IEEE Transactions on Electron Devices, Vol.48, no.7, p.1394, 2001

[88] M. Wong, Z. Jin, G.A. Bhat, P.C. Wong, H.S. Kwok, IEEE Transactions on Electron Devices, Vol. 47, p. 1061, 2000

[89] Zhong-He Jin, Yue-Lin Wang, Acta Electronica Sinica, Vol.29, no.8, p.1079, 2001

[90] Z. Jin, K. Moulding, H.S. Kwok, M. Wong, IEEE Transactions on Electron Devices, Vol. 46, p. 78, 1999

[91] A.R. Joshi, T. Krishnamohan, K.C. Saraswat, Journal of Applied Physics, Vol.93, no.1, p.175, 2003

[92] K. Imai, Solid-State Electron., Vol. 24, p.59, 1981

[93] S.S. Tsao, IEEE Circuits and Devices Magazine, Vol. 3, p. 3, 1987

[94] K. Ansai, F. Otoi, M. Ohnishi, and H. Kitabayashi, Proc. IEDM, p. 796, 1984

[95] S. Muramoto, H. Unno, and K. Ehara, Proc. Electrochem. Soc. Meeting, Boston, Vol. 86, p. 124, May 1986

[96] L. Nesbit, Tech. Digest of IEDM, p. 800, 1984

[97] R.P. Holmstrom and J.Y. Chi, Appl. Phys. Lett., Vol. 42, p. 386, 1983

[98] K. Barla, G. Bomchil, R. Herino, and. A. Monroy, IEEE Circuits and Devices Magazine, Vol. 3, p. 11, 1987

[99] E.J. Zorinsky, D. B. Spratt, and R. L. Vinkus, Tech. Digest IEDM, p. 431, 1986

[100] K. Barla, G. Bomchil, R. Herino, J.C. Pfister, and J. Baruchel, J. Cryst. Growth, Vol. 68, p. 721, 1984, and
K. Barla, R. Herino, and G. Bomchil, J. Appl. Phys, Vol. 59, p. 439, 1986

[101] M.I.J. Beale, N.G. Chew, A.G. Cullis, D.B. Garson, R.W. Hardeman, D.J. Robbins, and I.M. Young, J. Vac. Sci. Technol. B, p. 732, 1985

[102] V.P. Bondarenko, A.M. Dorofeev, *Physical and Technical Problems of SOI Structures and Devices*, Kluwer Academic Publishers, NATO ASI Series - High Technology, Vol. 4, p. 15, 1995

[103] K. Izumi, M. Doken, H. Ariyoshi, Electronics Letters, Vol. 14, p. 593, 1978

[104] S.J. Krause and S. Visitserngtrakul, Proceedings of the IEEE SOS/SOI Technology Conference, p. 47, 1990

[105] S. Cristoloveanu, A. Ionescu, T. Wetteroth, H. Shin, D. Munteanu, P. Gentil, S. Hong, and S. Wilson, Journal of the Electrochemical Society, Vol. 144, no. 4, p. 1468, 1977

[106] G.F. Cerofolini, S. Bertoni, L. Meda, C. Spaggiari, Nuclear Instruments & Methods in Physics Research, Section B-84, p. 234, 1993

[107] L. Meda, S. Bertoni, G.F. Cerofolini, C. Spaggiari, Electrochemical Society Proceedings, Vol. 94-11, p. 224, 1994

[108] S. Krause, M. Anc, P. Roitman, MRS Bulletin, Vol. 23, p. 25, 1998

[109] K. Izumi, MRS Bulletin, Vol. 23, p. 20, 1998

[110] S. Nakashima, K. Izumi, Nuclear Instruments & Methods in Physics Research, Vol. B55, p. 847, 1991

[111] B. Aspar, C. Pudda, A.M. Papon, A.J. Auberton-Hervé, J.M. Lamure, Electrochemical Society Proceedings, Vol. 94-11, p. 62, 1994

[112] S. Nakashima, K. Izumi, Electronics Letters, Vol. 26, p. 1647, 1990

[113] A.J. Auberton-Hervé, B. Aspar, J.L. Pelloie, *Physical and Technical Problems of SOI Structures and Devices*, Kluwer Academic Publishers, NATO ASI Series - High Technology, Vol. 4, p. 3, 1995

[114] M.J. Anc, J.G. Blake, T. Nakai, Electrochemical Society Proceedings, Vol. 99-3, p. 51, 1999

[115] M. Chen, X. Wang, Y. Dong, X. Liu, W. Yi, J. Chen, X. Wang, Proceedings of the IEEE International SOI Conference, p. 113, 2002

[116] S. Nakashima, T. Katayama, Y. Miyamura, A. Matsuzaki, M. Imai, K. Izumi, N. Ohwada, Proceedings of the IEEE International SOI Conference, p. 71, 1994

[117] M. Tachimori, S. Masui, T. Nakajima, K. Kawamura, I. Hamaguchi, T. Yano, Y. Nagatake, Electrochemical Society Proceedings, Vol. 96-3 p. 53, 1996

[118] A. Matsamura, K. Kawamura, T. Mizutani, S. Takayama, I. Hamaguchi, Y. Nagatake, Electrochemical Society Proceedings, Vol. 99-3, p. 79, 1999

[119] D.K. Sadana, Electrochemical Society Proceedings, Vol. 2001-2, p. 474, 2001

[120] O.W. Holland, D. Fathy, D.K. Sadana, Applied Physics Letters, Vol. 69, p. 674, 1966

[121] L.P. Allen, M.L. Alles, R.P. Dolan, H.L. Hughes, P. McMarr, Microelectronic Engineering, Vol. 36, p. 383, 1997

[122] J. Blake, K. Dempsey, R. Dolan, Y. Erokhin, P. Powell, S. Richards, Proceedings of the IEEE International SOI Conference, p. 109, 2002

[123] A. Ogura A, Electrochemical Society Proceedings, Vol. 99-3, p. 61, 1999

[124] A. Ogura, Extended Abstracts of the International Conference on Solid-State Devices and Materials, p. 240, 2001

[125] J.P. Colinge, *Low-power HF microelectronics: a unified approach*", IEE circuits and systems series 8, the Institution of Electrical Engineers, p. 139, 1996

[126] G.K. Celler, Electrochemical Society Proceedings, Vol. 90-7, p. 472, 1990

[127] H.J. Hovel, Proceedings of the IEEE International SOI Conference, p. 1, 1996

[128] S. Nakashima, M. Harada, and T. Tsuchiya, Proceedings of the IEEE International SOI Conference, p. 14, 1993

[129] A.G. Revesz and H.L. Hughes, in *"Physical and Technical Problems of SOI Structures and Devices"*, Kluwer Academc Publishers, NATO ASI Series - High Technology, Vol. 4, p. 133, 1995

[130] M. Anc, *Progress in SOI Structures and Devices Operating at Extreme Conditions*, Kluwer NATO Science Series II, Vol. 58, p. 1, 2002

[131] A.G. Revesz, G.A. Brown, and H.L. Hughes, J. Electrochemical Society, Vol. 140, No 11, p. 3222, 1993

[132] B.J. Mrstik, P.J. McMarr, R.K. Lawrence, and H.L. Hughes, IEEE Transactions on Nuclear Science, Vol. 41, no. 6, p. 2277, 1994

[133] R. Stahlbush, H. Hughes, and W. Krull, IEEE Transactions on Nuclear Science, Vol. 40, no. 6, p. 1740, 1993

[134] M.E. Zvannt, C. Benefield, H.L. Hughes, and R.K. Lawrence, IEEE Transactions on Nuclear Science, Vol. 41, no. 6, p. 2284, 1994

[135] G.A. Brown and A.G. Revesz, Proceedings of the IEEE International SOI Conference, p. 174, 1991

[136] http://www.ibis.com

[137] P.L.F. Hemment, in *"Semiconductor-On-Insulator and Thin Film Transistor Technology"*, MRS Symposium Proceedings, Vol. 53, p. 207, 1986

[138] L. Nesbit, S. Stiffler, G. Slusser, and H. Vinton, J. Electrochem. Soc., Vol. 132, p. 2713, 1985

[139] G. Zimmer and H. Vogt, IEEE Transactions on Electron Devices, Vol. 30, p. 1515, 1983

[140] L. Nesbit, G. Slusser, R. Frenette, and R. Halbach, J. Electrochem. Soc., Vol. 133, p. 1186, 1986

[141] C. Serre, A. Pérez-Rodríguez, a. Romano-Rodríguez, J.R. Morante, J. Esteve, M.C. Acero, R. Kögler, W. Skorupa, *Progress in SOI Structures and Devices Operating at Extreme Conditions*, Kluwer NATO Science Series II, Vol. 17, p. 31, 2002

[142] R. Stengl, K.Y. Ahn, and U. Gösele, Japanese Journal of Applied Physics, Vol. 27, p. L2367, 1988

[143] U. Gösele, M. Reich, and Q.Y. Tong, Microelectronic Engineering, Vol. 28, no. 1-4, p. 391, 1995

[144] R. Stengl, T. Tan, U. Gösele, Japanese Journal of Applied Physics, Vol. 28, p. 1735, 1989

[145] W.P. Maszara, G. Goetz, A. Cavigilia, J.B. McKitterick, Journal of Applied Physics, Vol. 64, p. 4943, 1988

[146] C.A. Desmond-Colinge, U. Gösele, MRS Bulletin, Vol. 23, p. 30, 1998

[147] Q.Y. Tong and U. Gösele, *Semiconductor Wafer Bonding Science and Technology*, The Electrochemical Society Series, John Wiley & Sons, 1999

[148] G. Kissinger, W. Kissinger, Sensors & Actuators A-Physical, Vol. A36, no.2, p.149, 1993

[149] P. Amirfeiz, S. Bengtsson, M. Bergh, E. Zanghellini, L. Börjesson, Electrochemical Society Proceedings Vol. 99-35, p. 29, 1999

[150] E. Obermeier, Electrochemical Society Proceedings, Vol. 95-7, p.212, 1995

[151] Q.Y. Tong and U. Gösele, *Semiconductor Wafer Bonding Science and Technology*, The Electrochemical Society Series, John Wiley & Sons, 1999

[152] S. D. Collins, Journal of the Electrochemical Society, Vol. 144, p. 2242, 1997

[153] A. M. Ledger and P.J. Clapis, Proceedings of the IEEE International SOI Conference, p. 64, 1993

[154] P.B. Mumola and G.J. Gardopee, Extended abstracts of the International Conference on Solid-State Devices and Materials, Yokohama, Japan, p. 256, 1994

[155] P.B. Mumola, G.J. Gardopee, T. Feng, A.M. Ledger, P.J. Clapis, E.P. Miller, Electrochemical Society, Proceedings Vol. 93-29, p. 410, 1993

[156] K. Mitani, M. Nakano, T. Abe, Electrochemical Society Proceedings, Vol. 96-3, p. 87, 1996

[157] M. Tricard, P.R. Dumas, D. Golini, J.T. Mooney, Proceedings of the IEEE International SOI Conference, p. 127, 2003

[158] M. Bruel, Electronics Letters, Vol. 31, p. 1201, 1995

[159] M. Bruel, MRS Bulletin, Vol. 23, p. 35, 1998

[160] M. Kaminsky, IEEE Transactions on Nuclear Science, Vol. 18, p. 208, 1971

[161] K. Kamada, Y. Kazumata, K. Kubo, Radiation Effects, Vol. 28, Gordon and Breach Science Publishers Ltd., p. 43, 1976

[162] F.F. Komarov, V.S. Solov'yev, S.Y. Shiryayev, Radiation Effects, Vol. 42, Gordon and Breach Science Publishers Ltd., p. 169, 1979

[163] C. Ascheron, H. Bartsch, A. Setzer, A. Schindler, P. Paufler, Nuclear Instruments and Methods in Physics Research, Vol. B28, p. 350, 1987

[164] J.H. Evans, Journal of Nuclear Materials, Vol. 68, p. 129, 1977

[165] H. Van Swijgenhoven, L.M. Stals, G. Knuyt, Nuclear Instruments and Methods, Vol. 209/210, p. 461, 1983

[166] K. Ono, T. Kino, K. Kamada, H. Osono, Japanese Journal of Applied Physics, Vol. 25, p. 1475, 1986

[167] K. Wittmaack, P. Blank, W. Wach, Radiation Effects, Vol. 39, Gordon and Breach Science Publishers Ltd., p. 81, 1978

[168] E. Ligeon, A. Guivarc'h, Radiation Effects, Vol. 27, Gordon and Breach Science Publishers Ltd., p. 129, 1976

[169] M. Bruel, *"Process for the production of thin semiconductor material films"*, U.S. Patent 5,374,564, 1994

[170] C.F. Cerofolini, L. Meda, R. Balboni, F. Corni, S. Frabboni, G. Ottaviani, R. Tonini, M. Anderle, R. Canteri, Physical Review, Vol. B46, p. 2061, 1992

[171] M. Bruel, MRS Bulletin, Vol. 23, p. 35, 1998

[172] J. Grisolia, G. Ben Assayag, A. Claverie, B. Aspar, C. Lagahe, L. Laanab, Applied Physics Letters, Vol. 76, p. 852, 2000

[173] F.A. Reboredo, M. Ferconi, S.T. Pantelides, Physical Review Letters, Vol. 82, p. 4870, 1999

[174] M.K. Weldon, V.E. Marsico, Y.J. Chabal, A. Agarwal, D.J. Eaglesham, J. Sapjeta, W.L. Brown, D.C. Jacobson, Y. Caudano, S.B. Christman, E.E. Chaban, Journal of Vacuum Science and Technology, Vol. B15, p. 1065, 1997

[175] T.H. Lee, Q.Y. Tong, Y.L. Chao, L.J. Huang, U. Gösele, Electrochemical Society Proceedings, Vol. 97-23, p. 27, 1997

[176] B. Aspar, C. Lagahe, H. Moriceau, E. Jalaguier, A. Mas A, O. Rayssac, A. Soubie, B. Biasse, M. Bruel, Electrochemical Society Proceedings, Vol. 99-35, p. 48, 1999

[177] M.K. Weldon, V.E. Marsico, Y.J. Chabal, A. Agarwal, D.J. Eaglesham, J. Sapjeta, W.L. Brown, D.C. Jacobson, Y. Caudano, S.B. Christman, E.E. Chaban, Journal of Vacuum Science and Technology, Vol. B15, p. 1065, 1997

[178] B. Aspar, H. Moriceau, E. Jalaguier, C. Lagahe, A. Soubie, B. Biasse, A.M. Papon, A. Claverie, J. Grisolia, G. Benassayag, F. Letertre, O. Rayssac, T. Barge, C. Maleville, B. Chyselen, Journal of Electronic Materials, Vol. 30, p. 834, 2001

[179] Q.Y. Tong, R.W. Bower, MRS Bulletin, Vol. 23, p. 40, 1998

[180] K. Hentinnen, I. Suni, S.S. Lau, Applied Physics Letters, Vol. 76, p. 2370, 2000

[181] W.G. En, I.J. Malik, M.A. Bryan, S. Farrens, F.J. Henley, N.W. Cheung, C. Chan, Proceedings of the IEEE International SOI Conference, p. 163, 1998

[182] J.T.S. Lin, J. Peng, T.H. Lee, Proceedings of the IEEE International SOI Conference, p. 189, 2002

[183] A. Thilderkvist, S. Kang, M. Fuerfanger, I. Malik, Proceedings of the IEEE International SOI Conference, p. 12, 2000

[184] M.I. Current, I.J. Malik, M. Fuerfanger, A. Flat, J. Sullivan, S. Kang, H.R. Kirk, M. Norcott, D. Teoh, P. Ong, F.J. Henley, Proceedings of the IEEE International SOI Conference, p. 111, 2002

[185] M.I. Current, S.N. Farrens, M. Fuerfanger, Sien Kang, H.R. Kirk, I.J. Malik, L. Feng, F.J. Henley, Proceedings of the IEEE International SOI Conference, p.11, 2001

[186] J.P. Joly, B. Aspar, M. Bruel, L. Di Coccio, F. Letertre, *Progress in SOI Structures and Devices Operating at Extreme Conditions*, Kluwer NATO Science Series II, Vol. 58, p. 31, 2002

[187] B. Aspar, C. Lagahe, H. Moriceau, E. Jalaguier, A. Mas, O. Rayssac, A. Soubie, B. Biasse, M. Bruel, Electrochemical Society Proceedings, Vol. 99-35, p. 48, 1999

[188] L.J. Huang, J.O. Chu, D.F. Canaperi, C.P. D'Emic, R.M. Anderson, S.J. Koester, H.S.P.Wong, Applied Physics Letters, Vol. 78, p. 1267, 2001

[189] C. Mazuré, Electrochemical Society Proceedings, Vol. 2003-05, p.13, 2003

[190] C. Maleville, T. Barge, B. Ghyselen, A.J. Auberton,. Proceedings of the IEEE International SOI Conference, p. 134, 2000

[191] B. Aspar, M. Bruel, M. Zussy, A.M. Cartier, Electronics Letters, Vol. 32, p. 1985, 1996

[192] C. Colinge, B. Roberds, B. Doyle, Journal of Electronic Materials, Vol. 30, p. 841, 2001

[193] C. Mazuré, Electrochemical Society Proceedings, Vol. 2003-05, p.13, 2003

[194] P. Nguyen, I. Cayrefourcq, B. Blondeau, N. Sousbie, C. Lagahe-Blanchard, S. Sartori, A.M. Cartier, Proceedings of the IEEE International SOI Conference, p. 132, 2003

[195] T.H. Lee, Q.Y. Tong, Y.L. Chao, L.J. Huang, U. Gösele, Electrochemical Society Proceedings, Vol. 97-23, p. 27, 1997

[196] K.D. Hobart, F.J. Hub, M.E. Twigg, M. Fatemi, *Progress in SOI Structures and Devices Operating at Extreme Conditions*, Kluwer NATO Science Series II, Vol. 58, p. 299, 2002

[197] K. Imai K, Solid-State Electronics, Vol. 24, p. 159, 1981

[198] S.S. Tsao, IEEE Circuits and Devices Magazine, Vol. 3, p. 3, 1987

[199] M.I.J. Beale, N.G. Chew, A.G. Cullis, D.B. Gasson, R.W. Hardeman, D.J. Robbins, I.M. Young, Journal of Vacuum Science and Technology, Vol. B3, p. 732, 1985

[200] T. Yonehara, K. Sakaguchi, N. Sato, Electrochemical Society Proceeding, Vol. 99-3, p. 111, 1999

[201] T. Yonehara, K. Sakaguchi, N. Sato, Applied Physics Letters, Vol. 64, p. 2108, 1994

[202] K. Sakaguchi, T. Yonehara, Solid-State Technology, Vol. 43-6, p. 88, 2000

[203] K. Sakaguchi, K. Yanagita, H. Kurisu, H. Suzuki, K. Ohmi, T. Yonehara, Electrochemical Society Proceedings, Vol. 99-3, p. 117, 1999

[204] T. Yonehara, K. Sakaguchi, *Progress in SOI Structures and Devices Operating at Extreme Conditions*, Kluwer NATO Science Series II, Vol. 58, p. 39, 2002

[205] C. Maleville, Electrochemical Society Proceedings, Vol. 2003-05, p. 33, 2003

[206] A.J. Auberton-Hervé, C. Maleville, Proceedings of the IEEE International SOI Conference, p. 1, 2002

[207] N. Sato, Y. Kakizaki, T. Atoji, K. Notsu, H. Miyabayashi, M. Ito, T. Yonehara, Proceedings of the IEEE International SOI Conference, p. 209, 2002

[208] T. Sato, J. Iwamura, H. Tango, and K. Doi, *Comparison of thin-film transistor and SOI technologies*, Materials Research Society Proceedings, Vol. 33, p. 25, 1984

[209] Y. Ishikawa, T. Saito, N. Shibata, Proceedings of the IEEE International SOI Conference,
       p. 16, 1997
[210] S. Fukatsu, Y. Ishikawa, T. Saito, N. Shibata, Applied Physics Letters, Vol. 72, no. 26, p.
       3485, 1998
[211] T. Mizuno, S. Takagi, N. Suiyama, H. Satake, A. Kurobe, A. Toriumi, IEEE Electron
       Device Letters, Vol. 21, no. 5, p. 230, 2000
[212] T. Mizuno, N. Sugiyama, H. Satake, S. Takagi, Symposium on VLSI Technology, p. 210,
       2000
[213] C. Mazuré, Electrochemical Society Proceedings, Vol. 2003-05, p.13, 2003
[214] T.A. Langdo, A. Lochtefeld, M.T. Currie, R. Hammond, V.K. Yang, J.A. Carlin, C.J.
       Vineis, G. Braithwaite, H. Badawi, M.T. Bulsara, E.A. Fitzgerald, Proceedings of the
       IEEE International SOI Conference, p. 211, 2002
[215] A.T. Collins, *Properties and growth of diamond*, Ed. by G. Davies, INSPEC, the
       Institution of Electrical Engineers, London, p. 288, 1994
[216] N.K. Annamalai, J. Sawyer, P. Karulkar, W. Maszara, and M. Landstrass, IEEE
       Transactions on Nuclear Science, Vol. 40, p. 1780, 1993
[217] N.K. Annamalai, P. Fechner, and J. Sawyer, Proceedings of the IEEE International SOI
       Conference, p 64, 1992
[218] M. Matsumoto, Y. Sato, M. Kamo, and N. Setaka, Japanese Journal of Applied Physics,
       Vol. 21, p. L182, 1982
[219] B. Edholm, A. Söderbärg, J. Olsson, and E. Johansson, Japanese Journal of Applied
       Physics, Vol. 34, Part 1, no. 9A, p. 4706, 1995
[220] B. Edholm, A. Söderbärg, and S. Bengtsson, J. Electrochem. Soc., Vol. 143, p. 1326,
       1996
[221] C.Z. Gu, Z.S. Jin, X.Y. Lu, G.T. Zou, J.X. Lu, D. Yao, J.F. Zhang, and R.C. Fang,
       Chinese Physics Letters, Vol. 13, no. 8, p. 610, 1966
[222] S. Bengtsson, M. Bergh, M. Choumas, C. Olesen, and K.O. Jeppson, Japanese Journal of
       Applied Physics, Vol. 38, no. 8, p. 4175, 1996
[223] S. Bengtsson, M. Choumas, W.P. Maszara, M. Bergh, C. Olesen, U. Södervall, and A.
       Litwin, Proceedings of the IEEE International SOI Conference, p. 35, 1994
[224] C. Lin, M. Zhu, C. Men, Z. An, M. Zhang, Electrochemical Society Proceedings, Vol.
       2003-05, p. 51, 2003
[225] M. Jurczak, T. Skotnicki, M. Paoli, B. Tormen, J.L. Regolini, C. Morin, A. Schiltz, J.
       Martins, R. Pantel, J. Galvier, Symposium on VLSI Technology, Digest of Technical
       Papers, p. 29, 1999
[226] M. Jurczak, T. Skotnicki, M. Paoli, B. Tormen, J. Martins, J.L. Regolini, D. Dutartre, P.
       Ribot, D. Lenoble, R. Pantel, S. Monfray, IEEE Transactions on Electron Devices, Vol.
       47, no. 11, pp.217, 2000

Chapter 3

SOI MATERIALS CHARACTERIZATION

3.1 INTRODUCTION

Once Silicon-On-Insulator material has been produced, it is important to characterize it and assess its quality. Accurate measurement of parameters such as defect density, thickness of the top silicon layer and the buried insulator, carrier lifetime, and quality of the silicon-insulator interface is of vital importance.

Some of the techniques used for characterization of SOI materials are very powerful, but destructive (*e.g.*: transmission electron microscopy - TEM). These techniques are well adapted to the thorough examination of the materials, but they can hardly be employed on a routine basis in a production environment. Some other techniques may be less sensitive, but are non-destructive and provide information in a matter of seconds. These may be used to assess the quality of large quantities of SOI wafers. Physical techniques, such as optical film thickness measurement techniques, can be employed on virgin SOI wafers, while some others require the fabrication of devices in the SOI material, and are, therefore, destructive by definition. We will now describe in more detail some of the characterization techniques used to assess the quality of SOI materials (Table 3-1). An extensive review of the characterization techniques developed for SOI materials and devices can be found in the book of S. Cristoloveanu and S.S. Li.[1]

Table 3-1. SOI materials characterization techniques

Parameter characterized	Technique
Film thickness	Spectroscopic reflectometry
	Spectroscopic ellipsometry
	RBS
	dV_{TH}/dV_{G2} technique
Si crystal quality	TEM
	RBS
	UV reflectance
Defect distribution	Chemical decoration
	Laser light scattering
Surface roughness, haze	Visible reflectance
	UV reflectance
	AFM
Stress in the silicon layer	Raman spectroscopy
Bond energy	Crack propagation technique
	Micro indentation
Void distribution	Infrared imaging
Impurity content	SIMS
	TXRF
	Photoluminescence
Carrier lifetime	Surface photovoltage
	Ψ-MOSFET
	Measurement on MOS devices
Si-SiO$_2$ interface	Capacitance measurements
	Ψ-MOSFET
	Charge pumping

3.2   FILM THICKNESS MEASUREMENT

Accurate determination of the thickness of both the silicon overlayer and the buried insulator layer is essential for device processing as well as for the evaluation of parameters such as the threshold voltage. General-purpose film thickness measurement techniques can, of course, be utilized to evaluate SOI structures. These include the etching of steps in the material and the use of a stylus profilemeter to measure the height of the steps. Rutherford backscattering (RBS) can also be used to evaluate the thickness of a silicon film on SiO$_2$, but is not practical for the measurement of the buried oxide thickness. Cross-sectional transmission electron microscopy (XTEM) is probably the most powerful film thickness measurement technique, and atom lattice imaging can provide a built-in ruler for accurate distance measurements. All the above techniques are destructive and time consuming. Routine inspection and thickness mapping of SOI wafers necessitates contactless (non-destructive) methods. Both spectroscopic reflectometry and spectroscopic ellipsometry satisfy this requirement. Electrical film thickness measurement can also be performed on manufactured fully depleted devices.

This measurement is based on the body effect (variation of front threshold voltage with back-gate bias).

3.2.1 Spectroscopic reflectometry

The thickness of both the SOI layer and the buried oxide can be evaluated in a non-destructive way by measuring the reflectivity spectrum of an SOI wafer, usually in the 400 to 800 nm wavelength range. The spectrum can readily be compared with the theoretical reflectivity spectrum calculated from the theory of thin-film optics. For SOI samples with well-defined Si-SiO$_2$ interfaces, a three-layer model can be used (silicon on oxide on silicon), while more complicated multilayer structures have to be employed if the interfaces are not well defined (in the case of as-implanted SIMOX material, for example [2]).

Consider an SOI structure with sharp Si-SiO$_2$ interfaces (Figure 3-1) with normal incidence of the light beam. The following notation will be used: the semi-infinite silicon substrate, the buried oxide, the silicon overlayer and the air ambient are numbered 0,1,2, and 3, respectively. The light-induced electric fields at the different interfaces between the n and the $(n+1)$ layer are expressed by the following equation [3,4,5,6]:

$$\begin{pmatrix} E_R(n+1) \\ E_L(n+1) \end{pmatrix} = W_{n+1,n} \begin{pmatrix} E'_R(n) \\ E'_L(n) \end{pmatrix} \tag{3.2.1}$$

where $E_R(n+1)$ and $E'_R(n)$ are the electric fields of the right-going wave at the left and right of the n-th interface, respectively, and $E_L(n+1)$ and $E'_L(n)$ are the fields of the left-going wave at the left and the right of the n-th interface, respectively. The interface matrix, W, is given by:

$$W_{(n+1,n)} = \frac{1}{2} \begin{pmatrix} 1+\dfrac{y_n}{y_{n+1}} & 1-\dfrac{y_n}{y_{n+1}} \\ 1-\dfrac{y_n}{y_{n+1}} & 1+\dfrac{y_n}{y_{n+1}} \end{pmatrix} \tag{3.2.2}$$

where y_n is the refractive index of the n-th layer. y_n is a complex number composed of a real (non-absorbing) and an imaginary (absorbing) part: $y_n = N_n + jK_n$. For the wavelengths under consideration, the refractive indices of dielectric materials such as SiO$_2$ and Si$_3$N$_4$ can be considered as real and constant, while the refractive index of silicon is complex and wavelength dependent.[7]

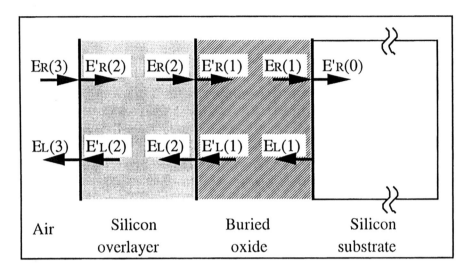

Figure 3-1. Multilayer structure and light-induced electric fields.

The phase change and the absorption occurring within the thickness of a layer (n-th layer) is expressed by the following relationship:

$$\begin{pmatrix} E'_R(n) \\ E'_L(n) \end{pmatrix} = \Phi_n \begin{pmatrix} E_R(n) \\ E_L(n) \end{pmatrix} \tag{3.2.3}$$

where the "phase matrix", Φ, is given by: $\Phi_n = \begin{pmatrix} e^{j\Phi_n} & 0 \\ 0 & e^{-j\Phi_n} \end{pmatrix}$

with $\Phi_n = \dfrac{2\pi}{\lambda} y_n t_n$, where t_n is the thickness of the n-th layer and λ is the wavelength. Considering the full SOI structure, we finally obtain the following expression:

$$\begin{pmatrix} E_R(3) \\ E_L(3) \end{pmatrix} = W_{3,2}\, \Phi_2\, W_{2,1}\, \Phi_1\, W_{1,0} \begin{pmatrix} E'_R(0) \\ E'_L(0) \end{pmatrix} \tag{3.2.4}$$

Since the silicon substrate is semi-infinite there is no returning wave from the substrate and, thus, $E'_L(0) = 0$. The reflectivity of the structure is given by the ratio of reflected power to incident power:

$$R(\lambda) = \frac{[E_L(3)]^2}{[E_R(3)]^2} \qquad (3.2.5)$$

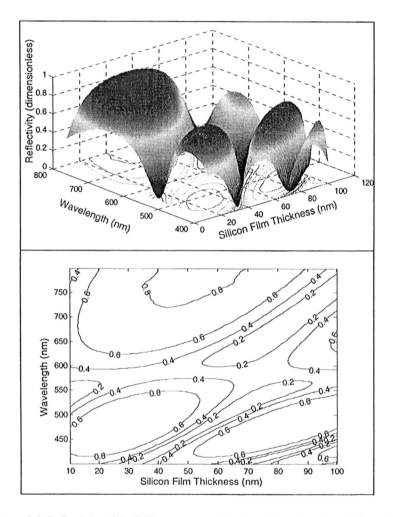

Figure 3-2. Reflectivity of the SOI structure as a function of wavelength and silicon film thickness. The BOX thickness is 400 nm. Top: 3D plot; bottom: contour plot.

The reflectivity is expressed in % of incident power. $E'_R(0)$ can easily be eliminated from equation (3.2.5). $E'_R(0)$ represents the intensity of the incident light and its value can be chosen arbitrarily. Taking $E'_R(0)=1$ equation (3.2.4) yields the values for $E_R(3)$ and $E_L(3)$ from which the reflectivity $R(\lambda)$ can be computed for each wavelength. The above relationships were established for normal incidence of the light on the sample. Angles of incidence up to 15 degrees can, however, be used without introducing any significant error in the measurement. The reflectivity spectrum of an SOI structure is given in Figure 3-2 as a function of the silicon film thickness for a BOX thickness of 400 nm, and for a normal incidence.

The determination of the thickness of both the silicon film and the buried layer is based on a comparison of the measured spectrum and the model.

Figure 3-3 shows an example of thickness mapping made on a SIMOX wafer using spectroscopic reflectometry.[8,9] Commercial tools devoted to this type of measurement are available and are routinely used to monitor film thickness in SOI wafers.

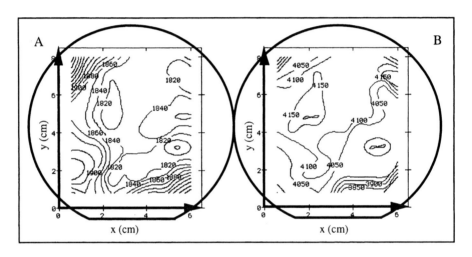

Figure 3-3. Mapping of the thickness of the silicon overlayer (A) and the buried oxide (B) of a 125 mm-diameter SIMOX wafer. Thickness is indicated in angströms (Å). (1Å = 0.1 nm).

3.2.2 Spectroscopic ellipsometry

Ellipsometry is based on the measurement of the change of polarization of a light beam reflected by a sample. In order to maximize the sensitivity of the measurement, ellipsometry is usually carried out at large incidence angles (75°, close to the Brewster angle). The change of polarization can be derived from equations (3.2.1)-(3.2.4) where corrections for non-normal incidence must be introduced. The complete theory of ellipsometry is quite complicated and is outside the scope of this book. The interested reader can, however, refer to [10] for more information. Classical ellipsometry is performed at a single wavelength (usually emitted by an He-Ne laser). After passing through a polarizing filter, the beam is reflected on the sample and is directed into an analyzer, composed of a rotating polarization filter and a photodiode. The polarization direction of the reflected beam is given by the rotation angle of the filter when the amplitude of the light collected by the photodiode reaches a maximum. In spectroscopic ellipsometry, the same analysis is repeated for a large number of wavelengths within the visible spectrum and the near ultraviolet. Typically, about a hundred equidistant measurements are performed at wavelengths ranging between 300 and 850 nm.

The output of the measurement consists of the spectra of *tan Ψ* and *cos δ*. The two angles *Ψ* and *δ* are defined, at each wavelength, by the relationship:

$$tan\,\Psi\ e^{j\delta} = \frac{r_p}{r_s} \qquad (3.2.6)$$

where r_p and r_s are the complex reflection coefficients r for light polarized parallel (*p*), and perpendicular (*s* - from *senkrecht*, in German) to the plane of incidence, respectively. An example of *tan Ψ* and *cos δ* spectra is given in Figure 3-4.

The interpretation of the measured spectra is based on a simulation and regression program that minimizes the difference between the measured data and spectra calculated using thin-film optics theory. Ellipsometry spectra contain much more information than reflectometry spectra. The sensitivity of spectroscopic ellipsometry is such that it can be used to measure complicated multilayer structures such as imperfect SIMOX structures that contain silicon inclusions within the buried oxide and oxide precipitates within the silicon overlayer. It can even measure the thickness of a native oxide layer on top of an SOI wafer.

Spectroscopic ellipsometry is not an analytical technique that stands of its own. Indeed, parameters such as the number of layers to be taken into account as well as their composition have to be fed into the simulator, which will tune each parameter in order to reproduce the measured data. For example, a SIMOX structure with oxide precipitates at the bottom of the silicon overlayer and silicon precipitates at the bottom of the buried oxide will need the following input parameters: estimation of the thickness of the native oxide, silicon overlayer without precipitates, silicon overlayer with precipitates, buried oxide above the silicon precipitates, buried oxide with silicon precipitates, and oxide below the precipitates.

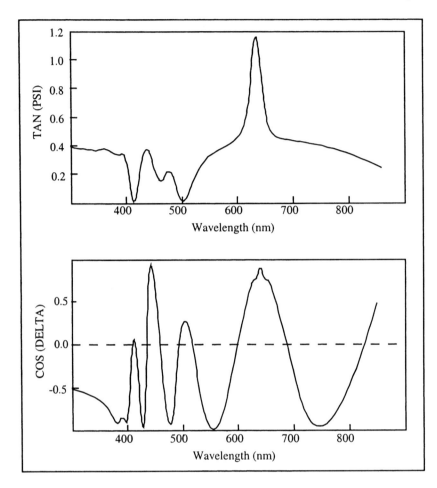

Figure 3-4. Example of the wavelength dependence of the ellipsometry parameters tanΨ and cosδ in a SIMOX structure.

An estimation of the composition of the mixed layers (*e.g.*: silicon precipitates in oxide) must also be given. The refractive indices of mixed layers are calculated using the Bruggeman approximation (or effective medium approximation).[11] This approximation assumes an homogeneous isotropic mixture of spheres of both components (*e.g.*: Si and SiO$_2$) with varying radius in order to obtain complete filling of the layer. In this approximation, the mixed layer is isotropic, and its dielectric constant, ε, is given by:

$$\varepsilon = 0.25 \left(E + \sqrt{E^2 + 8\varepsilon_1\varepsilon_2} \right) \quad \text{with} \quad E = (3c-1)(\varepsilon_2 - \varepsilon_1) + \varepsilon_1 \quad (3.2.7)$$

where c is the fraction of component 2 within the mixed layer, and ε_1 and ε_2 are the dielectric constants of component 1 and component 2, respectively. ε, ε_1 and ε_2 are complex numbers from which the expression of the refractive index $n=\sqrt{\varepsilon}$ can be extracted. The regression analysis during which measured and calculated spectra are compared, endeavors to minimize the

error function $G = \displaystyle\sum_{i=1}^{n} [(D_i^c - D_i^m)^2 + (W_i^c - W_i^m)^2]$ where n is the number

of different wavelengths at which the measurement is performed. The superscripts m and c stand for "measured" and "calculated", respectively. The parameters W and D are given by the following expressions:

$$W_i = \frac{tan^2\,\Psi_i - 1}{tan^2\,\Psi_i + 1} \qquad (3.2.8)$$

and

$$D_i = \frac{2cos\,\delta_i\,tan\,\Psi_i}{tan^2\,\Psi_i + 1} \text{ or } D_i = cos\,\delta_i \text{ if } tan\,\Psi_i \text{ is very small.} \qquad (3.2.9)$$

Spectroscopic ellipsometry is a very sensitive measurement technique. When the layers and interfaces of the SOI structure are well defined, a three-layer model can be used (native oxide/silicon/buried insulator), and convergence of the thickness-finding algorithm can be obtained in a matter of minutes. If the composition of the interfaces is not well defined (*e.g.*: SIMOX with silicon crystallites embedded in the BOX or structure with rough interfaces) mixed layers have to be simulated.[12]

3.2.3 Electrical thickness measurement

The silicon film thickness, t_{si}, is an important parameter in thin-film, fully depleted SOI MOSFETs. It influences all the electrical parameters of thin-film devices (threshold voltage, drain saturation voltage, subthreshold slope). Therefore, it is of interest to measure the thickness of the silicon film after device processing for debugging purposes and to check whether threshold voltage non-uniformities, for example, are due to film thickness variations or not, and whether the final targeted silicon thickness has been reached after device processing.

The dependence of the front threshold voltage, V_{TH1}, of a fully depleted n-channel SOI MOSFET on back-gate voltage, V_{G2}, can be calculated from equation (5.3.19) of Chapter 5 (body effect) if the thickness of the BOX and the gate oxide is known:

$$\frac{dV_{TH1}}{dV_{G2}} = -\frac{C_{si}\,C_{ox2}}{C_{ox1}\,(C_{si} + C_{ox2})} = \frac{-\varepsilon_{si}\,C_{ox2}}{C_{ox1}\,(t_{si}\,C_{ox2} + \varepsilon_{si})} \qquad (3.2.10)$$

from which one can easily derive the following relationship [13]:

$$t_{si} = -\frac{1}{C_{ox2}}\left(\frac{\varepsilon_{si}\,C_{ox2}}{C_{ox1}} - \varepsilon_{si}\right)\left(\frac{dV_{TH1}}{dV_{G2}}\right)^{-1} \qquad (3.2.11)$$

where C_{ox1}, C_{ox2}, and ε_{si} are the gate oxide capacitance, the buried oxide capacitance, and the silicon permittivity, respectively. It is important to note that the above expression is independent of the doping concentration in the silicon film, as long as the device is fully depleted. This measurement technique assumes that C_{ox1} and C_{ox2} (or t_{ox1} and t_{ox2}) are known. The value of t_{ox1} can be obtained from an independent measurement made on bulk silicon monitor wafers undergoing the same gate oxidation process as the SOI wafers. The buried oxide thickness (which is not affected by device processing) can be measured by spectroscopic reflectometry mapping techniques prior to device processing. For this method to be valid, the dV_{TH1}/dV_{g2} measurement must be carried out for back-gate voltages for which the back-gate interface is depleted (*i.e.* neither inverted nor accumulated) (see Figure 5-7).

This technique can be used to measure the thickness of both the silicon film and the buried oxide.[14] Assuming the thickness of the gate oxide is known from an independent measurement, equation (5.3.20) can be used to write:

$$\frac{dV_{TH1}}{dV_{G2}} = -\frac{C_{si}C_{ox2}}{C_{ox1}(C_{si} + C_{ox2})} = \frac{t_{ox1}}{t_{ox2} + \dfrac{t_{si}\,\varepsilon_{ox}}{\varepsilon_{si}}} \qquad (3.2.12)$$

By analogy we can write:

$$\frac{dV_{TH2}}{dV_{G1}} = -\frac{C_{si}C_{ox1}}{C_{ox2}(C_{si} + C_{ox1})} = \frac{t_{ox2}}{t_{ox1} + \dfrac{t_{si}\,\varepsilon_{ox}}{\varepsilon_{si}}} \qquad (3.2.13)$$

Solving these two equations for the two unknowns t_{si} and t_{ox2} we find:

$$t_{ox2} = \frac{\dfrac{dV_{TH2}}{dV_{G1}}\left(1+\dfrac{dV_{TH1}}{dV_{G2}}\right)}{\dfrac{dV_{TH1}}{dV_{G2}}\left(1+\dfrac{dV_{TH2}}{dV_{G1}}\right)} t_{ox1} \qquad (3.2.14)$$

and

$$t_{si} = \frac{\varepsilon_{si}}{\varepsilon_{ox}} \frac{\left(\dfrac{dV_{TH1}}{dV_{G2}}\right)^{-1} - \dfrac{dV_{TH2}}{dV_{G1}}}{1+\dfrac{dV_{TH2}}{dV_{G1}}} t_{ox1} \qquad (3.2.15)$$

3.3 CRYSTAL QUALITY

Although all SOI material producing techniques aim at the fabrication of a perfect single-crystal silicon layer, defects and imperfections are always present in the silicon film. This Section describes some of the various techniques used to assess the quality of the silicon layer (crystal orientation, degree of crystallinity, and crystal defect density).

3.3.1 Crystal orientation

SOI material fabrication techniques are usually designed to produce silicon films with (100) normal orientation. This orientation is automatically obtained when the silicon film is produced by separation of a superficial silicon layer from a (100) silicon substrate by the formation of a buried insulator (SIMOX, SIMNI, FIPOS or wafer bonding) or when the silicon film is epitaxially grown from a single-crystal substrate having normal lattice parameters equal or close to those of (100) silicon (SOS, ELO, LSPE). When the silicon film is recrystallized from the melt over an amorphous insulator (laser and e-beam recrystallization, ZMR), the control of the crystal orientation is more difficult, and substantial deviation from the (100) orientation can be observed. Furthermore, even if the normal orientation is (100), in-plane misorientation can occur. The different single-crystals are then connected by subgrain boundaries.

The crystal orientation can be determined by classical techniques such as X-ray diffraction and electron diffraction in a TEM. More frequently, the electron channeling pattern (ECP) technique (or: pseudo-Kikuchi technique) is employed since it can be carried out in an SEM without any special sample preparation. In this technique, a stationary, defocused electron beam is incident on the sample. The reflection of the electrons depends on the local crystal orientation relative to the incident beam (*i.e.*: it depends on the degree of channeling of the incident electrons in the silicon). The reflected electrons form a pattern indicative of the normal crystal orientation (a cross-shaped pattern is obtained for (100) silicon). The direction of the arms of the cross pattern can be used to determine the in-plane orientation, and a distortion of the pattern indicates a spatial rotation of the crystal axes.[15]

One of the most popular methods used to assess the crystal orientation of SOI films is the etch-pit grid technique.[16] This technique has been widely used to optimize the ZMR process. In this technique, a layer of oxide is grown or deposited on the silicon layer. Using a mask step and HF etch, circular holes are opened in the oxide layers. The holes have a 2-3 μm diameter and are repeated across the entire sample in a grid array configuration, with a pitch of 20 μm, typically. The resist is then stripped, and the holes in the oxide are used as a mask for silicon etching. The silicon is etched in a KOH solution (a mixture of 250g KOH, 800 ml deionized water and 200 ml isopropyl alcohol). The KOH solution etches silicon much more rapidly in the <100> direction than in the <111> direction. As a result, a pattern having the shape of a section of a pyramid is etched in the silicon. When examined inder a microscopw the pits present a square shape if the normal orientation is (100), and the sides of the pits are <111> planes. The pit diagonals indicate [100] directions (Figure 3-5).

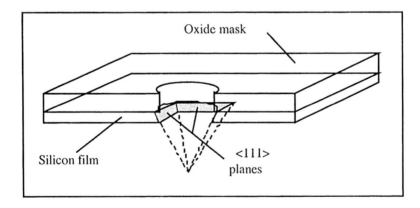

Figure 3-5. The etch-pit grid technique.

If the film orientation is not (100), distorted polyhedral patterns are produced by the intersection of the silicon film plane with the <111> silicon planes. The change of shape between two adjacent patterns indicates the presence of grain boundaries or the rotation of the crystalline axes. A misalignment of the diagonal directions between two patterns without modification of shape is indicative of the presence of subgrain boundaries.

3.3.2 Degree of crystallinity

In cases where the silicon overlayer is damaged during the SOI formation process (e.g.: SIMOX), or when epitaxial growth of silicon is performed, it is sometimes interesting to check the single-crystallinity of the formed layer. Two major techniques are used to determine the degree of crystallinity: Rutherford backscattering (RBS) and UV reflectance.

RBS is a destructive technique based on the impingement of light ions (usually He^+) with mass M_1 on a sample consisting of atoms with mass M_2. The ions are accelerated to an energy E_0 (e.g.: 2 MeV) before reaching the target. These ions loose energy through nuclear and electronic interactions with the target atoms. Most of them will come to rest in the target, but a small fraction of these light ions will be backscattered over an angle Θ with an energy $E_1 = K E_0$. K is the kinetic factor, defined as [17]:

$$K = \frac{\sqrt{M_2^2 - M_1^2 \sin\Theta} + M_1 \cos\Theta}{M_1 + M_2} \qquad (3.3.1)$$

The detection of the energy of the recoiling atoms can thus be used to determine the mass M_2 of the target atoms. The probability for an elastic collision to occur and to result in a scattering event at a certain angle Θ is expressed by the differential scattering cross-section:

$$\frac{d\sigma}{d\Omega} \propto \left(\frac{M_1 M_2 q^2}{4E_0}\right)^2 \frac{1}{\sin^4\Theta} \qquad (3.3.2)$$

where Ω is the detector angle. The average scattering cross-section is then given by: $\sigma = \frac{1}{\Omega} \int_\Omega \frac{d\sigma}{d\Omega} d\Omega$. The stopping cross-section $\varepsilon = \frac{1}{N}\frac{dE}{dx}(E)$ finally accounts for the energy loss of the particle penetrating the target due to electronic collisions or to small-angle collisions with nuclei; x is the depth below the target surface, and N is the volume density of the target atoms. The detected signal is processed by a multichannel analyzer. The output of a measurement session consists of a series of counts, called the backscattering yield, in every channel. To interpret the measured data, one has to convert the channel numbers into an energy scale and, therefore, to determine the energy interval E corresponding to a channel. Actually, the useful information is the depth scale, and E must be correlated with a slab i of thickness τ_i at a depth x_i. It can be shown that the energy difference between a particle backscattered at the surface $(E=KE_0)$ and another one backscattered at a depth x and emerging from the target $(E=E_x)$ is given by:

$$\Delta E = KE_0 - E_x = \left(\frac{K}{\cos\Theta_1} \left. \frac{dE}{dx} \right|_{in} + \frac{K}{\cos\Theta_2} \left. \frac{dE}{dx} \right|_{out} \right) x = [\varepsilon] \, N \, x \qquad (3.3.3)$$

where Θ_1 and Θ_2 are the angles (with respect to normal) of the track of the particle before and after scattering in the target. The latter relationship assumes that dE/dx is constant along each path taken by the particle. This assumption yields a linear relationship between the energy difference, ΔE, and the depth at which scattering occurs. $[\varepsilon]$ is the stopping cross-section factor. In addition to information about cristallinity RBS spectra can give information on the composition of compound materials (such as SiO_2 or, more generally, Si_xO_y).

An RBS spectrum measurement can be performed in two different ways: a crystal direction can be parallel to the incident ion beam, or it can be randomly oriented. In the former case, the ions can penetrate deeper in the crystal by channeling through the lattice, and an "aligned spectrum" is obtained. If the sample is amorphous or randomly oriented, no channeling can occur, and a "non-aligned" spectrum is obtained. Aligned spectra have lower backscattering yield because the ions penetrate deeper in the sample and have a lower probability of escape after a collision. Similarly, the presence of crystalline imperfections (point defects, impurities) increases the backscattering yield of an aligned target. The minimum backscattering yield, χ_{min}, is, therefore, a measure of the lattice disorder. The lower χ_{min}, the better the crystallinity. Single-crystal (100) bulk silicon has a value of χ_{min} equal to 3-4%.

Figure 3-6 shows typical RBS spectra obtained from a SIMOX sample. The most useful information comes from the layers nearest to the surface, where the energy of the backscattered ions is highest. Part (a) of Figure 3-6 corresponds to the silicon overlayer. The non-aligned spectrum shows a high yield and provides information on the thickness of the silicon layer. The aligned spectrum gives information on its crystal quality, through χ_{min}. Part

(b) corresponds to the buried oxide. Both aligned and non-aligned spectra contain information on the thickness of the layer. The non-aligned spectrum provides information on the composition of the layer (the ratio of oxygen to silicon atoms). The yield of the non-aligned spectrum is lower in part (b) than in parts (a) and (c) because the concentration of silicon is lower in the SiO_2 buried layer than in the silicon overlayer or in the substrate. The aligned spectrum, however, has a higher yield for SiO_2 than for Si because SiO_2 is amorphous (no channeling can take place in the buried oxide layer). Part (c) corresponds to the silicon substrate, and the peak that can be observed in part (d) is the signal due to the oxygen present in the buried oxide. The energy of the ions backscattered by the oxygen atoms is relatively low because oxygen is a light element, compared to silicon.

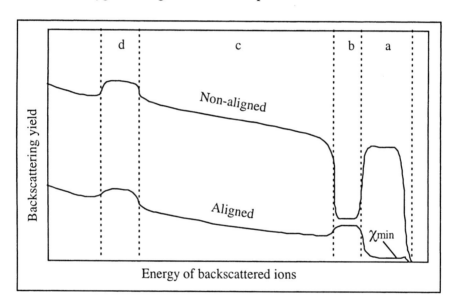

Figure 3-6. Aligned and non-aligned RBS spectra of a SIMOX wafer. From right to left: signal from the surface and the silicon overlayer (a), the buried oxide (b), the substrate (c), and influence of the buried oxide on the ions backscattered from the substrate (d).

UV reflectance is another technique that can be used to assess the crystallinity of SOI samples. Contrary to RBS, it is non-destructive. It has been extensively used to characterize SOS wafers. Both the microtwin density in the film and the fabrication yield of SOS circuits can be correlated to UV reflectance parameters.[18] The quality of the silicon overlayer of SIMOX wafers can also be assessed using UV reflectance.[19] In the case of SOS, measurement are taken at λ=280 nm and, if necessary, λ=400 nm for reference. There are two prominent maxima in the UV reflectance spectrum of single-crystal silicon, at λ=280 nm and λ=365 nm. They are caused by the optical interband transitions at the X point and along the Γ-L axis of the Brillouin zone, respectively. At short wavelengths, in particular at 280 nm, the reflectance is largely determined by the high value of the absorption coefficient $K > 10^6$ cm^{-1} corresponding to a penetration depth of less than 10 nm. Imperfect crystallinity in the near-surface region causes a broadening of the reflectance peak and a reduction of its maximum value. In the case of SIMOX wafers, more wavelengths have to be taken in order to obtain useful information about the quality of the silicon overlayer.

UV measurement of SIMOX wafers has been show to provide information of three morphological features of the material. Firstly, the overall reflectivity reduction (compared to a bulk silicon reference sample) is related to the presence of contamination in the film. This contamination can be due to the presence of carbon or SiO_x. Secondly, Rayleigh scattering caused by surface roughness shows a decrease of the reflectivity as a function of $B\lambda^{-4}$ for the shortest wavelengths (200 nm $< \lambda <$ 250 nm), where B is a constant depending on the rms roughness of the surface. Thirdly, some amorphization of the silicon overlayer reduces the intensity of the reflectance peaks at 280 and 367 nm. The sharpness and intensity of these maxima gives a measure of the degree of crystallinity within the specimen. Hence, semi-quantitative information about contamination, surface roughness and crystallinity can be obtained from UV reflectance measurements.

3.3.3 Defects in the silicon film

3.3.3.1 Most common defects

The most common defects in SOI layers are COPs, dislocations and HF defects.

- **COPs** (crystal-originated particle or crystal-originated pit) are defects found in all silicon wafers, including bulk and SOI wafers. COPs are octahedral voids that form during conventional Czochralski crystal growth through a process of vacancy condensation. In thin SOI films COPs can intersect both the top and the bottom of the silicon film, thereby forming a threading void "pipe" from the bottom to the top of the silicon film.[20] COPS are not found in epitaxial layers, which makes Eltran® essentially COP-free.

- **Dislocations** are the main defect found in many SOI materials. In the case of SIMOX, the dislocations are threading dislocations running vertically from the Si/buried oxide interface up to the surface of the silicon overlayer. The presence of such dislocations may pose yield and reliability hazard problems. Indeed, metallic impurities readily diffuse to dislocations upon annealing, and dislocations decorated with heavy metal impurities can cause weak points in gate oxides, so that low breakdown voltage is observed. Early studies show, nevertheless, that the integrity of gate oxides grown on SIMOX is comparable to that of oxides grown on bulk silicon.[21] In addition, SOI technology has since then brought about steady improvement of both dislocation density and metal contamination level.

- **HF defects** consist of inclusions of metallic silicides or silicates in the silicon film. Unlike pure silicon, these materials react with HF. Small pits, corresponding to these inclusions, are produced in the silicon film when the wafer is immersed in HF. If the etch time is long enough the BOX underneath the SOI film can also be etched, as shown in Figure 3-7.

Transmission electron microscopy (TEM) is one of the most powerful techniques for the analysis of crystal defects. It is, nevertheless, limited by the size of the samples that can be analyzed. In cross-section TEM (XTEM), the dimensions of a sample are limited to a width of 20 μm and a depth of 0.7 μm. This means that the maximum observable area is on the order of 10^{-7} cm^{-2} and, consequently, that the minimum measurable defect density is approximately 10^7 defects/cm^2. Plane-view TEM allows one to observe larger sample areas. Areas of the size of the sample holder (7 mm^2 grid) can be analyzed. In practice, it becomes difficult to observe dislocations with a magnification lower than 10,000 and it is more realistic to consider that an observation session yields 10 micrographs, each with a 10.000X magnification. In that case, the observed area is equal to 10^{-5} cm^2, and the minimum observable defect density is 10^5 defects/cm^{-2}. TEM observation often necessitates a lengthy and delicate sample preparation. Defect decoration techniques, combined with optical microscopy are, therefore, preferred to TEM if the nature of the sample and the defects allow it.

3.3.3.2 Chemical decoration of defects

The most common etch mixtures used for SOI defect decoration are listed in Table 3-2. All these solutions, with the exception of the electrochemical etch, are based on the mixture of HF with an oxidizing agent (CrO_3, $K_2Cr_2O_7$ or HNO_3). Defect decoration is a result of the preferential etch (higher etch rate) of the defects with respect to silicon. Decoration is most effective for high-disorder defects such as grain and subgrain boundaries. The etch rate of silicon is approximately 1 μm/min for most mixtures (Dash,

Secco, Stirl and Wright etch solutions), while the Schimmel etch rate is substantially lower. The decoration of dislocations in thin-film SOI material is almost impossible using classical etch mixtures, since all the silicon is removed before efficient decoration of the defects is achieved. Lower etch rates can be obtained by diluting the etch solutions with water. Another decoration technique, based on the electrochemical etching of silicon in diluted (5%) HF has been developed to reveal crystal defects in thin SOI films without etching of the silicon overlayer itself.[22] This technique necessitates the use of n-type ($N_d \cong 10^{15}$ cm^{-3}) doped silicon overlayers, and an ohmic contact must be provided to both the front side of the sample (*i.e.* to the SOI layer) and to the back. Electrochemical etching is performed for 10-30 minutes in 5 wt.% HF using a three-electrode configuration with the silicon controlled at +3 volts *vs.* a Cu/CuF$_2$ reference electrode. This decoration technique does not etch defect-free silicon. Defects such as dislocations, metal contamination-related defects, and oxidation-induced stacking faults (OSF) produce pits in the silicon film during this electrochemical etch procedure, and optical microscopy is used to observe and count the pits after decoration.[23]

Table 3-2. Decoration solutions, by volume, used to reveal defects in SOI films

Name of etch solution	Composition	Reference
Dash etch	HF:HNO$_3$:CH$_3$COOH 1:3:10	[24]
Schimmel etch	HF:1M CrO$_3$ 2:1	[25]
Secco etch	HF:0.15M K$_2$Cr$_2$O$_7$ 2:1	[26]
Stirl etch	HF:5M CrO$_3$ 1:1	[27]
Wright etch	60ml HF:30 ml HNO$_3$:30 ml 5M CrO$_3$: 2 grams Cu(NO$_3$)$_2$:60 ml H$_2$O	[28]
Electrochemical etch	5% wt HF	[29,30]
Iodine etch	2M KI:0.5M I$_2$:2.5M HF:28.5M CH$_3$OH:66.5M H$_2$O	[31,32]

Note: M = "mole"

The etch rate of a solution such as Secco etch is too high for use with thin SOI films and the entire silicon film can be etched away before defects are revealed with sufficient contrast for optical microscope observation. Diluting the Secco solution helps reducing the etch rate. A combination of Secco and HF (hydrofluoric acid) etch has been proposed to facilitate the observation of defects in thin SOI films. Dilute Secco is first used to decorate defects in the silicon layer (Figure 3-7 B). At this stage the pit created by the Secco etch does not have enough contrast for easy microscope observation. The sample is then dipped in HF. The acid penetrates through the pit in the silicon film and etches the buried oxide. A circular pattern is then etched in the oxide and it can easily be seen through the silicon layer. A more elaborate etch scheme, called a "transferred layer etch", and involving bonding of the silicon film under observation to a bulk silicon wafer and the transfer of the etch pit pattern to the bulk wafer has been proposed.[33]

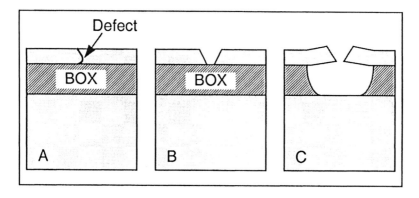

Figure 3-7. Defect in thin SOI layer (A) revealed by Secco etch (B) followed by HF etch (C).[34]

3.3.3.3 Detection of defects by light scattering

Defects can also be detected using laser light scattering at the surface of a wafer. Defect-mapping systems based on that principle are commercially available and are routinely used to monitor wafers in production lines.[35] They are designed to detect scattering centers such as defects in the silicon film or particulates. These centers are be discriminated against the background scattering noise, often called the "haze" of the wafer. These measurement systems, are calibrated for bare silicon wafers, however, SOI wafers have a reflectivity that is different from that of silicon. As discussed in Section 2.1 the reflectivity of an SOI wafer depends on BOX and silicon film thickness (Figure 3-8). The reflectivity of an SOI wafer can be higher or lower than that of bare silicon. High-reflectivity wafers are called "shiny" wafers and low-reflectivity wafers are called "dark" wafers. When calibrated for use with bare silicon wafers, laser scattering defect detection systems overestimate the size of defects on shiny SOI wafers and underestimate their size on dark wafers. Furthermore, the level of background haze appears to be inversely proportional to the reflectivity of the wafer.[36] To avoid this problem the use of UV lasers has been proposed. The absorption depth of UV light in silicon being very small, internal reflections and reflectivity dependence on film thickness should be eliminated by the use of UV lasers.[37]

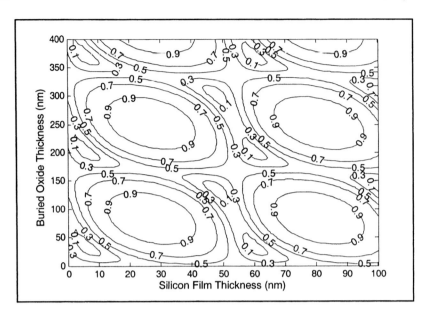

Figure 3-8. Reflectivity of an SOI wafer as a function of silicon film and BOX thickness for a laser wavelength of 488 nm under normal incidence (adapted from [38]).

3.3.3.4 Other defect assessment techniques

Impulsive Stimulated Thermal Scattering (ISTS) can be used to evaluate the defect (dislocation) density in SOI wafers. ISTS is a contactless, non-destructive, opto-acoustic technique that employs a pulse of laser light (λ=532 nm) from a pair of crossed excitation beams, which creates an interference pattern at the surface of the SOI layer. The absorption of laser energy creates a thermal grating in the sample, which in turn, produces an acoustic wave generated by sudden thermal expansion. At the same time the sample is probed by a low-intensity light beam and the diffracted signal reflected from the sample is analyzed. This signal is time-dependent and it is analyzed with a model from acoustic wave physics, which permits determination of material properties. When used on an SOI wafer the diffracted signal decays exponentially after the initial thermal excitation and the amplitude of the signal peak can be correlated to the defect density. The technique has been calibrated against Secco etch pit microscopic observation and was found to be reproducible within 10%.[39]

Photoluminescence (PL) is based on the photo-excitation of electrons and the observation of the wavelength and intensity of photons re-emitted after excitation. PL measurements are usually carried out at temperatures ranging between liquid helium (4.2K) and room temperature (300K). Electrons are injected from the valence band into the conduction band using a laser beam. The depth of the region being probed depends on the laser wavelength. It ranges from 5-10 nm when a UV laser is used to 1-3 μm for visible light. After generation, the electrons recombine from band edge to band edge or through one or several deep levels. Defects can be directly identified if they act as radiative recombination centers. Most of the photo-excited carriers, however, recombine non-radiatively via defect levels by a Shockley-Read-Hall mechanism. This non-radiative process governs the effective recombination lifetime. Since the intensity of the band-to-band recombination photoluminescence signal is proportional to the lifetime, the variation of its intensity reflects the distribution of defects. The deep-level photoemission signal is proportional to the product of the lifetime and the concentration of the deep-level defects. As a result, the variation of intensity of the deep-level emission provides information on the distribution of specific defects.[40,41] Photoluminescence systems can be used to map defects in SOI wafers. Defects such as dislocations, stacking faults and metal precipitates can be observed due to the localized lifetime degradation around the defect.[42]

3.3.3.5 Stress in the silicon film

The stress induced in the silicon film by the SOI fabrication process or by device processing can be measured using the Raman microprobe technique.[43] In this technique an argon laser (λ=457.9 nm) beam is focused on the sample. The measured area can be as small as 0.6 μm.[44] The spectrum of the reflected beam is analyzed and compared to the spectrum provided by a virgin bulk silicon reference. The shift of the spectrum peak and the value of the full width half maximum (FWHM) of the spectrum provide information on the amount of stress in the silicon film.[45]

3.3.4 Defects in the buried oxide

Buried oxides produced by SIMOX typically contain defects called "pipes". These are silicon filaments that run from the SOI layer to the substrate, through the buried oxide. These defects can be decorated using a copper solution. A distribution map of the pipes can be produced, and their diameter can be estimated using the technique described in Figure 3-9.

A Texwipe™ (white piece of cloth used in clean rooms) is soaked in a copper sulfate ($CuSO_4$) solution and sandwiched between the SOI wafer and a copper electrode. The silicon film side of the wafer is in contact with the Texwipe™. Contact is made to the back of the wafer using an aluminum electrode (Figure 3-9). When a voltage is applied to the structure, current flows through the BOX along the pipes.

The electrolytic reaction $CuSO_4 \rightarrow Cu^{++} + SO_4^-$ and $Cu^{++} + 2e^- \rightarrow Cu(s)$ takes place in the Texwipe™ wherever it receives current from a pipe. Stains (dots) of metallic copper are, therefore, created on the Texwipe™, each corresponding to a pipe in the BOX.

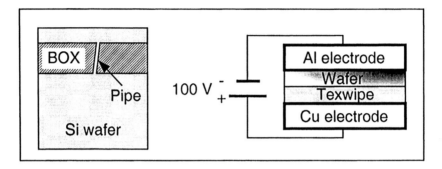

Figure 3-9. BOX pipe (left) and decoration set-up (right).[46]

The diameter of the stain is proportional to the current in each pipe, and the wider the pipe, the larger the stain. Assuming a cylindrical shape for the pipes the diameter of a pipe, d, can be found as a function of the current running trough it using Ohm's law:

$$d = 2\sqrt{\frac{I \rho t_{BOX}}{\pi V}} \qquad (3.4.1)$$

where I, ρ, t_{BOX} and V are the current through the pipe, the silicon resistivity, the BOX thickness, and the applied voltage, respectively. A calculation based on the fact that two electrons are needed to create one copper atom allows one to estimate that the smallest diameter of a pipe that can create an observable copper stain is approximately 5 nm. Similar results can be obtained by passing current through an SOI wafer immersed in a copper nitrite solution mixed with methanol. In that case the copper stains deposit directly on the wafer and can be observed on the SOI layer itself.[47]

3.3.5 Bond quality and bonding energy

Wafer bonding is a key technique used in the preparation of several SOI materials such as BESOI, Smart-Cut® and Eltran®. It is important to assess the quality and the strength of the bond between wafers. Infrared transmission imaging is routinely used to look for the presence of voids between bonded wafers. The experimental set-up for IR imaging is very simple (Figure 3-10) and consists of an incandescent light bulb and an infrared video camera. Figure 3-11 as well as pictures in Figure 2-25 were taken using such a setup.

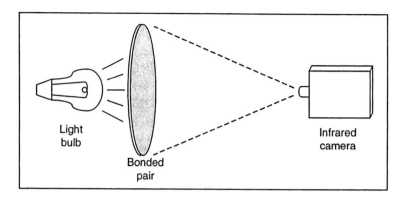

Figure 3-10. Infrared imaging technique.

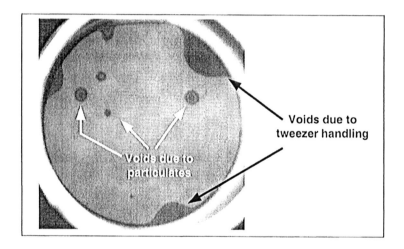

Figure 3-11. Voids between bonded vafers revealed by infrared (IR) imaging (Courtesy J.P. Raskin). [48]

The strength of the bond (or bonding energy) is measured in erg/cm² or J/m² (1 erg/cm² = 10^{-3} J/m²). The most common technique used to measure bond energy is the so-called "crack propagation technique" based on the insertion of a razor blade between the bonded wafers (Figure 3.12). The insertion of a blade of thickness t_b causes the formation of a "crack" or a de-bonded region that extends over a length L past the edge of the blade. The extension of the crack can be measured by infrared imaging. The bond energy, γ, is given by:

$$\gamma \;=\; \frac{3\,E\,t_w^3\,t_b^2}{32\,L^4}\;\;(\mathrm{J/m^2})\tag{3.5.1}$$

where t_w is the thickness of each individual wafer (m), E is the silicon Young's modulus (166 GPa) and L is the crack length (m).[49,50] This measurement is very easy to perform, but any inaccuracy in measuring the crack length induces large variations of the calculated bonding energy, because of the fourth-power dependence of γ on L. A 10% inaccuracy in crack length measurement, for instance, leads to a near 50% error in bonding energy.

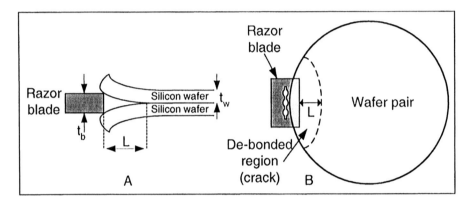

Figure 3-12. Crack propagation measurement; A: lateral view; B: top view.

Another technique based on the classical Vickers indentation technique has been proposed to measure the bonding energy.[51] The Vickers indenter is a diamond-tipped, square-based pyramidal tip, which can be pressed into a solid with a given force to produce a pyramid-shaped indentation in the solid. Such a tool is routinely used to measure the hardness of metals and the toughness of brittle materials such as ceramics and glass. To measure bond strength the indenter is pressed into the bonding interface with one diagonal of the pyramid parallel to the bonding interface and the other perpendicular to it. As pressure is applied cracks develop along the bonding interface and perpendicular to it (Figure 3-13).

The length of the crack along the bond interface is a measure of the bond energy. Theory shows that the toughness of the material, K_{ic}, is proportional to $P\beta\sqrt{c}$, where P is the force applied to the Vickers tool, $2c$ is the length of the crack (tip to tip), and β is a geometrical factor.[52] Once the toughness is known, the fracture energy, γ, can be calculated using the following relationship:[53]

$$\gamma = \frac{K_{ic}^2(1-v^2)}{E} \qquad (3.5.2)$$

where v and E are the material's Poisson ratio and Young's modulus, respectively.

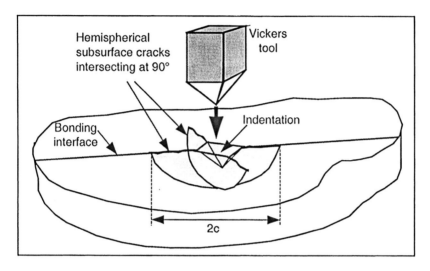

Figure 3-13. Vickers micro-indentation technique.[51]

3.4 CARRIER LIFETIME

The lifetime of minority carriers is a measure of the quality of the silicon films. The lifetime is affected by the presence of both crystal defects and metallic impurities. There exists no non-destructive method as such for measuring the lifetime in the silicon overlayer of an SOI structure. The measurement techniques rely either on the measurement of the lifetime in devices such as MOSFETs realized in the SOI material or on the measurement of the lifetime in the underlying silicon substrate and on the correlation between the lifetime in the substrate and the level of metal contamination within the silicon overlayer.

3.4.1 Surface Photovoltage

The surface photovoltage (SPV) technique relies on the generation of a voltage at the surface of a silicon sample upon illumination. SPV uses a chopped beam of monochromatic light of photon energy hv slightly larger than the band gap, E_G, of silicon. Electron-hole pairs are produced by the absorbed photons. Some of these pairs diffuse to the illuminated surface where they are separated by the electric field of the surface space-charge region whose thickness is *w*, thereby producing a surface photovoltage *ΔV*. A portion of *ΔV* is capacitively coupled to a transparent conducting electrode adjacent to the illuminated surface (Figure 3-14). This signal is then amplified to produce a quasi-dc analog output proportional to *ΔV*. The magnitude of *ΔV* is a function of the excess minority carrier (holes in the case of n-type silicon) density *Δp(0)* at the edge of the surface space-charge region. This density *Δp(0)* is, in turn, dependent on the incident light flux I_O, the optical absorption coefficient *α*, the optical reflectance at the illuminated surface, *ρ*, the recombination velocity at the illuminated surface *S*, and the diffusion length *L*. A steady-state solution of the one-dimensional diffusion equation for the excess carrier density is given by $\Delta p(0) = \dfrac{I_O\,(1\text{-}\rho)}{D/L + s}\ \dfrac{\alpha L}{1+\alpha L}$

[54] if one assumes *αw<<1*, *w<<L* and *Δp<<n_O*, where n_O is the majority carrier density. *D* is the minority carrier diffusion coefficient and *W* is the thickness of the silicon wafer with *αW>>1*.

A series of different wavelength values is selected to yield different values of *α*. At each wavelength, I_O is adjusted to give the same value (constant magnitude) of *ΔV*. As a consequence, the value of *Δp(0)* is constant as well. If *ρ* is essentially constant over the wavelength region of interest, the above equation may be rewritten as: $I_O = C[1 + (\alpha L)^{-1}]$, where *C* is a constant. If I_O is plotted against α^{-1} for each constant-magnitude *ΔV* point, the result is a linear graph whose extrapolated intercept on the negative α^{-1} axis is *L* (Figure 3-14).[55] The carrier lifetime, τ_p, can be deduced from the relationship $L^2 = D\tau$, with $D = \dfrac{kT}{q}\ \mu$, where *μ* is the carrier mobility.

The SPV technique cannot be directly used to measure the carrier recombination lifetime in the silicon overlayer of an SOI sample because the silicon film is too thin for the condition, *αW>>1*, to be met.[56] SPV can, nevertheless, be employed as an indirect measurement of the heavy metal contamination of SOI material. An excellent correlation is observed between the metallic impurity concentration (Fe, Cr, Ni and Cu) in the silicon overlayer of as-implanted SIMOX wafers, measured by SIMS or by spark source mass spectrometry, and the SPV diffusion length measured in the silicon substrate. The actual relationship between the lifetime, *L*, and the heavy metal concentration, *C*, is in the form $L \cong 1/C$. The relationship between the metal concentration and the lifetime in the substrate can be explained as follows. Metal impurities are introduced in the wafer during the oxygen implantation process. In the experiment reported in Ref. [57], the implantation took 6 hours, and was carried out at a wafer temperature of 580°C. In 6 hours, and at this temperature, iron, nickel, chromium and copper can diffuse in silicon at distances of 650, 3000, 200 and 6500 micrometers, respectively. Therefore, one can assume that the metal concentration is constant throughout the depth of the substrate, and that the concentration in the silicon film is correlated to that in the substrate. SPV is used to routinely monitor the heavy metal impurity level in SIMOX wafers.[58] The surface photovoltage technique can be used to monitor the interface charges at the silicon-buried insulator interface as well.[59] Other optical techniques, such as the measurement of the decay of free carriers generated by pulsed optical excitation, can be found in the literature.[60,61]

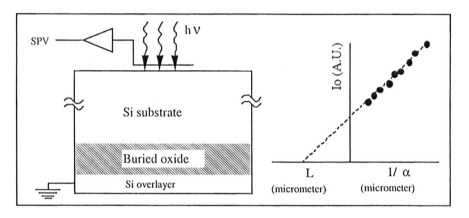

Figure 3-14. Experimental setup for SPV measurement of an as-implanted SIMOX wafer (left) and graphical method for extracting the carrier diffusion length (right).

3.4.2 Photoluminescence

The principles behind the photoluminescence technique is explained in section 3.3.3.4. The recombination lifetime, τ_{eff}, in the silicon layer of an SOI wafer is given by:[40]

$$\frac{1}{\tau_{eff}} = \frac{1}{\tau_{SRH}} + \frac{1}{\tau_{rad}} + \frac{1}{\tau_{Auger}} + \frac{S_{top}}{t_{si}} + \frac{S_{bot}}{t_{si}} \qquad (3.6.1)$$

where τ_{SRH}, τ_{rad} and τ_{Auger} are the lifetimes of the Shockley-Read-Hall, radiative and Auger recombination processes, respectively. S_{top} and S_{bot} are the recombination velocities at the top and bottom interfaces of the SOI film, and t_{si} is the thickness of the SOI film. The quality of the silicon layer is characterized by τ_{SRH}. Determination of τ_{SRH} from a τ_{eff} measurement is, however, quite difficult in *thin* SOI films because of the large influence of surface recombination on τ_{eff}. Free carriers quickly reach the top and bottom interface, such that the photoluminescence carrier decay process is predominatly governed by S_{top} and S_{bot}, which makes it very difficult to extract information of defects in the silicon layer.

3.4.3 Measurements on MOS transistors

The minority carrier lifetime can be extracted from junction leakage current measurements [62,63,64] and measured using Zerbst-like techniques.[65,66] More appropriate for the evaluation of the minority carrier generation lifetime in thin silicon films are techniques based on MOS transistor measurements.

3.4.3.1 Accumulation-mode transistor

A first technique is based on the measurement of the time required to create an inversion layer in an accumulation-mode MOSFET. The minority carriers needed to create the inversion layer can only come from generation in the bulk of the silicon film or at the Si-SiO$_2$ interfaces. The dc characteristics of such a device are described in Section 5.12. We will consider the case of an n-channel (N+-N--N+) device.[67,66] Using equation 5.12.7 and taking into account both the presence of interfaces traps and the formation of an inversion layer (for large negative gate biases), one finds:

$$V_G\text{-}V_{FB} = \frac{qN_d x_{depl}^2}{2\varepsilon_{Si}} + \frac{qN_d x_{depl}}{C_{ox}} + \frac{Q_i + Q_{it}}{C_{ox}} \qquad (3.6.2)$$

where Q_i is the inversion charge, Q_{it} is the surface-state charge, and N_d is the n-type doping density in the channel region. One can also write:

$$x_{depl}(t) = t_{si} - \frac{L}{\sigma W V_{DS}} I_D(t) = t_{si} - B\, I_D(t) \qquad (3.6.3)$$

where $B = \dfrac{L}{\sigma W V_{DS}}$ if the device is operating in the linear regime (V_{DS} is

small), and σ is the conductivity of the N-type silicon in the channel region. L is the gate length and W is the device width. Equation (3.6.2) can be rewritten as:

$$Q_i(t) + Q_{it}(t) = C_{ox}(V_G\text{-}V_{FB}) - \frac{qN_dC_{ox}}{2\varepsilon_{Si}}(x_{depl}(t))^2 - qN_dx_{depl}(t)$$

or

$$Q_i(t) + Q_{it}(t) = C_{ox}(V_G\text{-}V_{FB}) - \frac{q\,N_d\,C_{ox}}{2\,\varepsilon_{si}}\left[\left(x_{depl}(t)+\frac{\varepsilon_{si}}{C_{ox}}\right)^2 - \left(\frac{\varepsilon_{si}}{C_{ox}}\right)^2\right]$$

$$(3.6.3)$$

Using the classical model for generation of carriers in a semiconductor and the definition of surface generation velocity, we can write [68]:

$$\frac{d(Q_i(t) + Q_{it}(t))}{dt} = \frac{q\,n_i}{\tau_{gen}}\left(x_{depl}(t) - x_{depl}(t=\infty)\right) + q\,n_i\,S_o \qquad (3.6.4)$$

where τ_{gen} is the generation lifetime in the silicon film, and S_o is the surface generation velocity. Taking the derivative of (3.6.3) and combining with (3.6.2) and (3.6.4), one obtains:

$$F(t) = \frac{B}{\tau_{gen}}\left(I_D(t=\infty) - I_D(t)\right) + S_o \qquad (3.6.5)$$

where

$$F(t) = -\frac{N_d\,C_{ox}}{2\,n_i\,\varepsilon_{Si}}\frac{d}{dt}\left(\left(t_{Si}+\frac{\varepsilon_{Si}}{C_{ox}}\right) - B\,I_D(t)\right)^2 \qquad (3.6.6)$$

Plotting $F(t)$ as a function of $\left(I_D(t=\infty) - I_D(t)\right)$ yields the values of both τ_{gen} (slope of the dotted line) and S_o (y-axis intercept) (Figure 3-15).

This model assumes that the surface generation velocity is constant, which is true only when the silicon interface underneath the gate is in inversion, which yields low values of S_o. A better value of the surface generation is obtained by taking S_o equal to the difference between the measured data and the extrapolated line of Figure 3-15.B. A graph of $S_o(t)$ can then be produced (Figure 3-16.A).

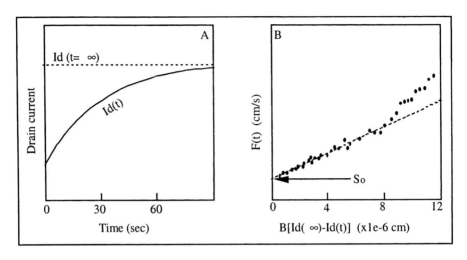

Figure 3-15. Drain current as a function of time in an accumulation-mode SOI MOSFET pulsed into inversion (A) and extraction of τ_{gen} and S_o from the slope and the intercept of equation (3.6.5) (B).

The surface generation velocity is highest right after the application of the gate bias (no inversion layer is present to act as a screen between the interface and the depletion layer). Its value then decreases with time, as the inversion layer forms (Figure 3-16.A). Similarly, the effective generation lifetime can be plotted as a function of the depth in the silicon film. Indeed, the depth of the depletion zone varies with the magnitude of the applied gate bias. Small gate biases allow one to measure the effective lifetime near the top surface, where the value of τ_{gen} is typically low because of the presence of an interface - Figure 3-16.B); larger biases measure τ_{gen} deeper in the silicon film, *and, as a result,* τ_{gen} increases. At even larger biases the back interface is probed and low values of τ_{gen} are measured.

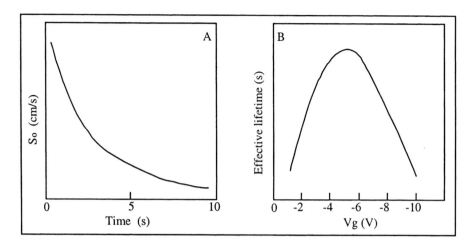

Figure 3-16. Surface generation as a function of time after gate biasing (A) and effective generation lifetime as a function of gate bias (B).

The above relationships were developed for edgeless devices. If a device with edges is measured, generation at the edges has to be taken into account, and equation (3.6.5) must be written as [69,70]:

$$F(t) = \frac{B}{\tau_{gen}}\left(I_D(t=\infty)-I_D(t)\right) + \frac{2 L S_{edge}}{L W} B \left(I_D(t=\infty)-I_D(t)\right) + S_O$$

or

$$F(t) = \frac{B}{\tau_{eff}}\left(I_D(t=\infty)-I_D(t)\right) + S_O \qquad (3.6.7)$$

such that one obtains:

$$\frac{1}{\tau_{eff}} = \frac{1}{\tau_{gen}} + \frac{2 S_{edge}}{W} \qquad (3.6.8)$$

where W is the width of the device and S_{edge} is the surface generation velocity at the edges of the device.

3.4.3.2 Inversion-mode transistor

It is also possible to measure the generation lifetime from the transient response of a partially depleted, inversion-mode transistor.[71] If the gate voltage is suddenly stepped from an above-threshold value (*e.g.*: 1.8 V in Ref. [71]) to a below-threshold value (*e.g.*: 0.4 V in Ref. [71]), with 0.1 V at the drain, a drain current transient is observed. The current is suppressed immediately after the negative voltage step and it gradually increases with time to a steady-state value. Because of capacitive coupling effects the potential of the floating body becomes negative when the gate voltage step is applied. This forces the threshold voltage to increase, resulting in a reduction of drain current. With time the drain current reaches the steady-state value as the body potential relaxes back to zero volts as carriers are generated. During the transient carriers are generated to replenish the depletion region between the steady-state width, $x_{d\infty}$, and the maximum depletion depth, xd_{max}. The variation of depletion depth with time is given by:

$$qN_a \frac{dx_d(t)}{dt} = q\frac{n_i}{\tau}\left(x_d(t) - x_{d\infty}\right) + q\, n_i\, S \tag{3.6.9}$$

where N_a is the doping concentration, n_i is the intrinsic carrier concentration, τ is the effective generation lifetime, and S is the surface generation velocity. Assuming that surface generation is negligible ($S=0$) the solution to (3.6.9) is:

$$x_d(t) = \left(x_{d\,max} - x_{d\infty}\right)\exp\left(-\frac{n_i}{N_a}\frac{t}{\tau}\right) + x_{d\infty} \tag{3.6.10}$$

The subthreshold current is proportional to $\exp\left(\dfrac{q[V_G - V_{TH}(t)]}{n\, kT}\right)$ where n is the body effect coefficient (or: body factor). In a partially depleted device, n is equal to $1 + C_D/C_{ox}$ where C_D is the depletion capacitance and C_{ox} is the gate oxide capacitance. In the transient case, C_D is time dependent and the body factor is equal to $n(t) = 1 + C_D(t)/C_{ox}$. The transient current is, therefore, given by:

$$I_D(t) = I_D(t = \infty)\, \frac{\exp\left(\dfrac{q[V_G - V_{TH}(t)]}{n(t)\, kT}\right)}{\exp\left(\dfrac{q[V_G - V_{TH}(t = \infty)]}{n(t = \infty)\, kT}\right)} \tag{3.6.11}$$

To simplify the calculation one can assume that the body factor does not vary with time.[71] Under this assumption equation (3.6.11) becomes:

$$I_D(t) = I_D(t = \infty)\exp\left[\frac{q\gamma\left(\sqrt{2\Phi_F} - \sqrt{2\Phi_F - V_B(t)}\right)}{n\quad kT}\right] \tag{3.6.12}$$

where $\gamma = \dfrac{\sqrt{2q\varepsilon_{si}N_a}}{C_{ox}}$ and $V_B(t) = -\dfrac{qN_a}{2\varepsilon_{si}}\left([x_d(t)]^2 - x_{d\infty}^2\right)$. If we define T_o

as the time needed to reach 90% of the steady-state drain current (Figure 3-17) the generation lifetime, τ, can be calculated using equations (3.6.10) and (3.6.12):

$$\tau = T_o \dfrac{n_i \Big/ N_a}{\sqrt{x_{d\infty}^2 + \dfrac{2\varepsilon_{si}}{qN_a}\left(\left[\sqrt{2\Phi_F} + 0.1\dfrac{kT}{q}\dfrac{n}{\gamma}\right]^2 - 2\Phi_F\right)} - x_{d\infty}} \qquad (3.6.13)$$

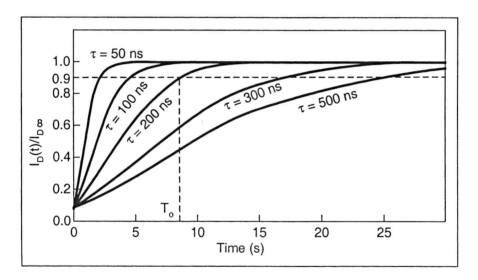

Figure 3-17. Transient drain current for various lifetime values. T_o is shown for $\tau = 200$ ns.

3.4.3.3 Bipolar effect

The recombination lifetime, or more exactly the effective recombination lifetime (which includes the influence of interface recombination), is a critical parameter for all characteristics involving parasitic bipolar effects. Lifetime can be estimated through the measurement of the common-emitter current gain, β_F, of lateral bipolar transistors with different base widths. From bipolar junction transistor theory, we know tha $\beta_F \cong 2\,(L_n/L_B)^2 - 1$ (equation 5.10.3), where L_B is the base width, which can be assumed, in first approximation, to be equal to the effective channel length, L_{eff}, and L_n is the electron diffusion length (we consider here the case of a NPN (n-channel) device). From the relationship $L_n^2 = D_n \tau_n$, where D_n and τ_n are the diffusion coefficient and the effective recombination lifetime of the minority carriers (electrons) in the base, respectively.

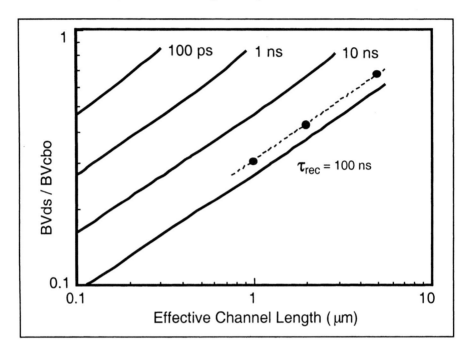

Figure 3-18. Determination of the effective recombination lifetime by means of drain breakdown voltage measurements. The circles represent experimental data.[72]

If lateral bipolar transistor structures (MOSFETs with body contacts) are not available, an estimation of the effective recombination lifetime can be obtained from a graph representing the drain breakdown voltage of n-channel SOI MOSFETs as a function of the effective gate length. Experimental values of BV_{DS}/BV_{CBO}, can be plotted as a function of channel length. BV_{DS} is the measured drain breakdown voltage, and BV_{CBO} is the intrinsic breakdown voltage of the drain junction, which is approximately equal to the BV_{DS} of a very long channel device ($L_{eff} \cong L = 50...100$ μm). Figure 3-18 illustrates the method and gives the example of measurements carried out on devices with an effective recombination lifetime τ_{rec} of approximately 70 nsec.

3.5 SILICON/INSULATOR INTERFACES

3.5.1 Capacitance measurements

Classical C-V techniques can be employed to measure the charges in the oxide layers and at the Si-SiO_2 interfaces of SOI devices, but the interpretation of the data is rather difficult. Indeed, the direct measurement of the capacitance of the whole SOI structure yields a complicated C-V curve (Figure 3-19) in which one can find contributions from the accumulation, depletion and inversion states at the front and back interfaces of the silicon film as well as at the top of the silicon substrate. Although analytical models of the SOI capacitor are available [73], it is usually easier to simulate the metal-insulator-semiconductor-insulator-semiconductor (MISIS) structure numerically and to compare the simulation results with the measurements in order to obtain an estimation of the charges in the oxide and at the Si/SiO_2 interfaces.[74]

It is possible to separately measure the different capacitance components from MOS capacitors fabricated using a conventional SOI-CMOS process. Interdigitated capacitors are often used to minimize the parasitic resistance of the lightly doped channel region. Figure 3-20 presents the schematic configuration of such a capacitor. One assumes that the silicon film underneath the gate is not fully depleted. C_1 is the gate capacitance, C_2 is the capacitance across the buried oxide underneath the gate region, C_3 is the capacitance across the buried oxide underneath the P$^+$ diffusion, and C_4 is the capacitance between the metal patterns (line + pad) and the substrate (Figure 3-20). The capacitance C_4 is actually composed of two capacitances, C_{4F} and C_{4G} corresponding to the metal lines needed to contact the P$^+$ diffusion (film contact) and the gate, respectively. The equivalent circuit representing the structure can be reduced to three capacitors, C_A, C_B and C_C, where

$$C_A = C_1$$
$$C_B = C_2 + C_3 + C_{4F}$$
$$C_C = C_{4G} \qquad (3.5.1)$$

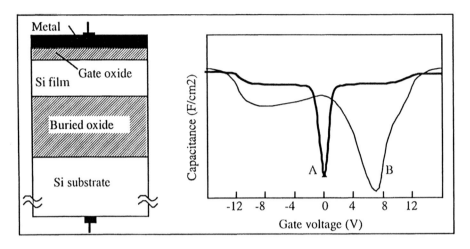

Figure 3-19. The SOI structure and an example of quasistatic C-V curves measured between the top metal electrode and the substrate. A: curve with one minimum; B: curve with two minima.

Access to the capacitors is obtained through three terminals: the gate pad, the film contact pad, and the silicon substrate (Figure 3-19).[75]

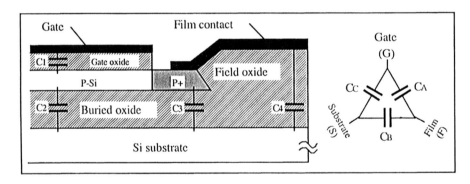

Figure 3-20. Schematic configuration of an SOI capacitor fabricated using a standard CMOS process and equivalent capacitor network.

If capacitance measurement is carried out between two of these terminals, leaving the third one floating, (*e.g.* measuring C_1 between the gate and film electrodes), altered C-V curves are obtained due to the presence of the two other capacitors. Better results are obtained by measuring C-V curves between one contact and the two others connected to each other. For instance, $C_{G/FS}$ is measured between the gate terminal and the substrate and film connected together.

From the equivalent circuit of Figure 3-20, we have three possible measurement configurations:

$$C_{G/FS} = C_C(V_{GS}) + C_A(V_{GF})$$
$$C_{F/SG} = C_A(-V_{GF}) - C_B(V_{FS})$$
$$C_{S/GF} = C_B(-V_{FS}) + C_C(-V_{GS}) \qquad (3.5.2)$$

Combining these equations, one obtains:

$$C_A(V_{GF}) + C_A(-V_{GF}) = C_{G/FS}(V_{G/FS}) + C_{F/SG}(V_{F/SG}) - C_{S/GF}(-V_{S/GF}) \quad (3.5.3)$$

as well as two similar expressions for C_B and C_C, obtained by circular permutations. Unfortunately, C_A, C_B and C_C cannot be derived independently from the above relationships because they are asymmetrical with respect to the applied bias voltage (*i.e.*: $C_A(V_{GF}) \neq C_A(-V_{GF})$). Nevertheless, $C_C = C_{4G}$ can be measured independently using a metal field capacitor, such as an unconnected metal pad, of capacitance C_T. We then obtain: $C_C(V_{GS}) = C_{4G} = C_T(V_{GS}) \dfrac{A_G}{A_T}$ and $C_{4F} = C_T \dfrac{A_F}{A_T}$, where A_G, A_F, and A_T are the areas of the capacitors C_{4G}, C_{4F}, and C_T, respectively. Now, C_A and C_B can easily be found using (3.5.2) where $C_1 = C_A$, and $C_2 + C_3 = C_B - C_{4F}$ can be obtained from (3.5.1). Separation of $C2$ and $C3$ can be achieved by measuring two test structures with different area ratios for the capacitors C_2 and C_3. Finally, C-V analysis can be performed on the C_1 and C_2 curves.

3.5.2 Charge pumping

The charge pumping technique [76,77] is very efficient for characterizing Si/SiO_2 interfaces. It can be used with small area devices and can yield the distribution of interface states in the band gap. In the case of SOI devices, the front and back interfaces can be characterized independently from one another. The principle of the charge pumping technique in a bulk device is the following. Source and drain are connected together and slightly reverse biased, with respect to the substrate. A periodic triangular or trapezoidal signal, ΔV_G, with frequency f is applied to the gate. ΔV_G is sufficiently large to switch the silicon surface underneath the gate from accumulation to strong inversion (Figure 3-21). When the device is in inversion, minority carriers are provided by the source and the drain to form the inversion channel. Some of these carriers are trapped by interface states. When the gate voltage is switched to produce an accumulation layer in the device, the inversion carriers (electrons in an n-channel device) move towards source and drain, and the minority carriers (electrons) trapped in the surface states now recombine with majority carriers from the substrate. This hole current constitutes the charge-pumping current I$_{cp}$.

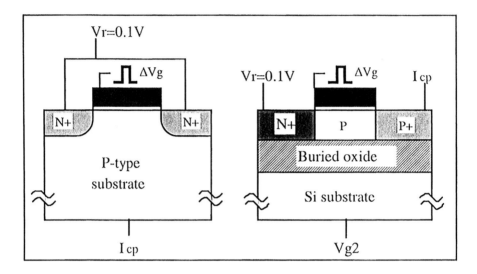

Figure 3-21. Experimental set-up for charge-pumping measurement in a bulk MOSFET (left) and an SOI gated PIN diode (right). I$_{cp}$ is the charge-pumping current.

If the gate is pulsed at a frequency *f*, the charge-pumping current is given by:

$$I_{cp} = q^2 \; \overline{N_{it}} \; A \; \Delta\Phi_s f \tag{3.5.4}$$

where $\overline{N_{it}}$ is the average interface trap density, $A = WL$ is the channel area,

and $q\Delta\Phi_s$ is the energy range scanned within the bandgap. This basic expression can be extended to more sophisticated pumping techniques using gate offset bias and trapezoidal gate voltage pulses with different rise and fall times, in which case the energy distribution of the surface states across the bandgap can be obtained.

In the case of an SOI MOSFET, the charge-pumping current can be measured through a substrate contact. Better results are, however, obtained by measuring a P+P-N+ gated diode, also known as a Lubistor (Figure 3-21).[78] In inversion, minority carriers are provided by the N+ cathode, and the charge-pumping current is measured at the P+ anode. Measurement of the front-interface trap density is obtained by pulsing the front gate, while back-interface trap density can be measured by pulsing the silicon substrate (back gate).

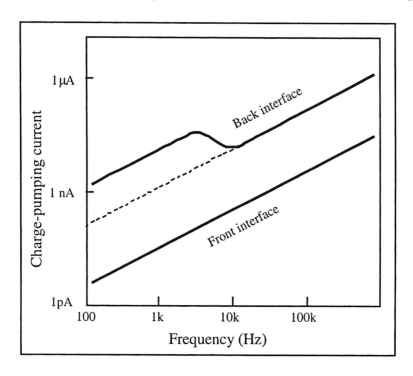

Figure 3-22. Charge-pumping current measured on a SIMOX PIN diode, as a
function of frequency.

Figure 3-22 presents the charge-pumping current measured on a SIMOX
PIN gated diode as a function of the pulse frequency. Front and back
interfaces were measured separately, by pulsing either the front gate or the
back gate. The charge pumping current is larger when the back interface is
probed, indicating a larger density of traps, N_{it} . One can also observe that
the back interface states encompass two components: a high density of slow
states having a cutoff frequency of $\cong$1-10 kHz and a lower density of fast
states, indicated by the dotted line in Figure 3-22. Using devices with
different channel widths, the charge-pumping technique can also be used to
measure the trap density at the interface between the silicon islands and the
LOCOS isolation.[79]

It is worth mentioning that the surface photovoltage technique (see
Section 3.4.1) can be used to measure the interface trap density at the silicon
film / buried oxide interface. Being non-desructive, it can be employed to
map the interface trap density on SOI wafers before device processing.[80]

3.5.3 Ψ-MOSFET

The acronym Ψ-MOSFET (Psi-MOSFET) stands for "pseudo-MOSFET" or "point-contact MOSFET". It allows one to perform transistor-like measurements on SOI wafers without the need for actual device processing.[81,82] This method can, therefore, be used as a quick-turnaround tool for SOI wafer characterization. It is based on the fact that it is possible to obtain well-behaved transistor-like characteristics from an SOI wafer using a 2-point probe tool as source and drain contacts, and using the substrate as a gate (Figure 3-23).

Depending on the polarity of the gate voltage, either accumulation or inversion can be induced at the bottom of the silicon film, such that n-channel and p-channel-like characteristics can be obtained from the same wafer. Increasing the probe pressure reduces the series resistance of the source and drain contacts. For example, an increase in probe pressure from 10 to 30 grams improves the measured transconductance by 35%. The transconductance then saturates for higher probe pressures.

The drain current of a Ψ-MOSFET can flow in an inversion or accumulation channel at the bottom of the SOI layer as well as through the neutral region of the film. The simplest case is that of a *fully depleted* film, where the current flowing through the silicon film itself is negligible. The measured current follows standard MOSFET theory:

$$I_D = f_g \, C_{ox2} \frac{\mu_0}{1 + \theta(V_{G2} - V_{T2/FB2})} (V_{G2} - V_{T2/FB2}) \, V_D \qquad (3.5.5)$$

where f_g is a geometrical factor replacing the classical W/L term and which accounts for the non-parallel distribution of the current lines.

In most practical configurations, the value of f_g is approximately equal to $0.75 \times V_{T2/FB2}$, which is either the threshold voltage in the case of inversion channel, or the flat-band voltage in the case of an accumulation channel.

θ is a surface mobility reduction factor which depends on the source and drain resistance according to the relationship $\theta = \theta_o + f_g \mu_o C_{ox2} R_{SD}$. In the backside channel of a regular SOI MOSFET the value of θ can be very small (0.01 V^{-1}). In the Ψ-MOSFET the resistance of the metal-semiconductor contacts are significant, and θ reaches values higher than 0.05 V^{-1}. C_{ox} is the buried oxide capacitance.

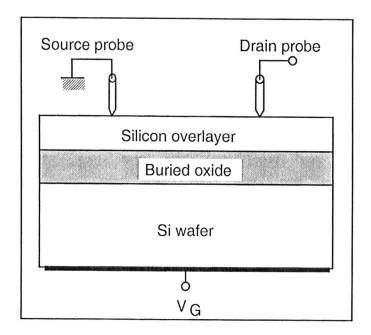

Figure 3-23. Principle of the Ψ-MOSFET.

In the case of a *partially depleted* film (P-type, for instance) one can distinguish three regions of operation:

- For a sufficiently positive back gate bias the back interface is inverted ($V_{G2}>V_{T2}$) and the back depletion depth is maximum. The measured drain current is the sum of the current given by Equation (3.5.5) and the current in the quasi-neutral ("bulk") region of the silicon film.

- For $V_{FB2}<V_{G2}<V_{T2}$ current flows only in the quasi-neutral ("bulk") region of the silicon film, the thickness of which is equal to $t_{si}-x_d$. According to equation 5.12.9 the extension of the depletion zone is given by:

$$x_d = \frac{-\varepsilon_{si}}{C_{ox2}} + \sqrt{\varepsilon_{si}^2/C_{ox2}^2 + \frac{2\varepsilon_{si}}{qN_a}(V_{G2} - V_{FB2})}$$

It is, however, convenient to assume that the bulk current decreases linearly with the gate voltage and to write:

$$I_D = q f_g \mu_0 N_a V_D (t_{si}-x_d) = f_g \mu_0 C_{ox2} V_D (V_0-V_{G2})$$

where $V_0 = V_{FB2} + \dfrac{qN_a t_{si}}{C_{ox2}}$. V_0 is the gate voltage that *would* induce full depletion (Figure 3-24).

- For a negative gate bias ($V_{G2} < V_{FB2}$) the current flows both in the accumulation channel (Equation 3.5.5) and through the neutral "bulk" of the silicon film ($x_d = 0$).

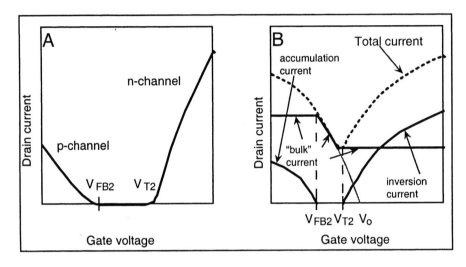

Figure 3-24. I-V_G characteristics of a Ψ-MOSFET. A: fully depleted film; B: partially depleted film.[82]

Different parameters of the SOI wafer can be extracted from Ψ-MOSFET measurements. From $I_D(V_{G2})$ and $\dfrac{I_D}{\sqrt{g_m}}(V_{G2})$ characteristics the mobility, the back threshold voltage and the back flat-band voltage can be extracted.[83, 84, 85] Variations of fixed oxide charges (from wafer to wafer, from batch to batch, or after exposure to ionizing radiations) can be monitored through the variation of V_{FB}. The back interface trap density can be determined from the subthreshold swing:

$$S = \frac{kT}{q}\, ln(10)\left(1 + \frac{C_d}{C_{ox2}} + \frac{qN_{it}}{C_{ox2}} \right) \qquad (3.5.6)$$

In addition, the Ψ-MOSFET can be used in the pulsed mode to measure carrier generation lifetime in the SOI layer, using a Zerbst-like technique similar to that described by Equations (3.6.2) to (3.6.6).[86,87,88]

The geometrical factor, f_g, in equation (3.5.5) is approximately equal to 0.75. This value was established both empirically from experimental results and theoretically by numerical simulations.[89] If the circular, concentric electrodes of a mercury probe are used as source and drain electrodes, instead of in-line needles of a four-point-probe, another geometrical factor must be used (Figure 3-25). In such a geometry the total drain current $I_d = 2\rho J_\rho$ is a constant at any distance, ρ, from the center of the drain electrode. There the current density, J_ρ, can be expressed as a function of radial electric field, E_ρ, and, thus, as a function of the potential $V(\rho)$ using the relationship $J_\rho = \sigma E_\rho = \sigma dV/d\rho$. The resulting differential equation for the potential is:

$$dV = (I_d/2\pi\sigma)(d\rho/\rho) \qquad (3.5.7)$$

This relationship includes the dependence of the Si-film conductivity, σ, on gate voltage. In the linear region of operation, where $\sigma = \mu C_{ox}(V_g - V_{TH2/FB2})$, the integration of equation 3.5.7 from R_1 to R_2 yields the potential drop between the source and drain contacts

$$V_D = (I_d/2\pi\sigma)ln(R_2/R_1) \qquad (3.5.8)$$

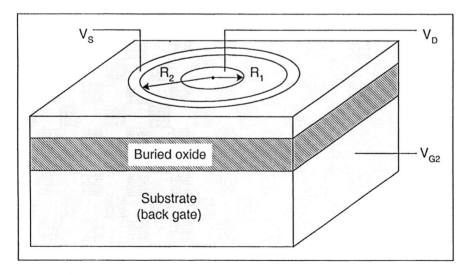

Figure 3-25. Ψ-MOSFET with circular, concentric source and drain electrodes.[90]

The drain current can now be expressed as a function of the gate and drain voltage:

$$I_D = \frac{2\pi}{\ln(R_2/R_1)} \mu C_{ox} \left(V_{G2} - V_{TH2/FB2}\right) V_D \qquad (3.5.9)$$

Comparison between equations (3.5.5) and (3.5.10) yields the geometrical factor for the circular, concentric source and drain electrode configuration:

$$f_g = W/L = 2\pi/\ln(R_2/R_1) \qquad (3.5.10)$$

Note that the pseudo-MOSFET technique can also be used with "horseshoe" mercury probes, provided correction is made for the geometrical factor.[91]

REFERENCES

[1] S. Cristoloveanu and S.S. Li, *Electrical Characterization of Silicon-On-Insulator Materials and Devices*, Kluwer Academic Publishers, 1995

[2] S.N. Bunker, P. Sioshansi, M.M. Sanfacon, S.P. Tobin, Appl. Phys. Lett. Vol. 50, p. 1900, 1987

[3] Z. Knittl, *Optics of Thin Films*, Wiley, New York, p. 37, 1976

[4] F. Van de Wiele, *Solid-State Imaging*, NATO Advanced Study Institutes Series, Noordhoff, Leyden, p. 29, 1976

[5] J.P. Colinge and F. Van de Wiele, Journal of Applied Physics, Vol. 52, p. 4769, 1981

[6] T.I. Kamins, J.P. Colinge, Electronics Letters, Vol. 22, no. 23, p. 1236, 1986

[7] D.E. Aspenes, *Properties of Silicon*, Published by INSPEC (IEE), p. 59, 1988

[8] J. Vanhellemont, J.P. Colinge, A. De Veirman, J. Van Landuyt, W. Skorupa, M. Voelskow, and H. Bartsch, Electrochemical Society Proceedings, Vol. 90-6, p. 187, 1990

[9] T.Wetteroth, S.R. Wilson, H. Shin, S. Hong, N.D. Theodore, W.M. Huang, M. Racanelli, Proceedings of the IEEE International SOI Conference, p. 50, 1996

[10] R.M.A. Azzam and N.M. Bashara, *Ellipsometry and Polarized Light*, Elsevier Science Publishers, North-Holland Personal Edition, chapter 1, 1987

[11] D.A.G. Bruggeman, Annalen der Physik, Vol. 5, p. 636, 1935

[12] D.E. Aspenes, Thin Solid Films, Vol.89, p. 249, 1982

[13] J. Whitfield and S. Thomas, IEEE Electron Device Letters, Vol. 7, p. 347, 1986

[14] M. Haond, M. Tack, IEEE Transactions on Electron Devices, Vol.38, no.3, p. 674, 1991

[15] D.C. Joy, D.E. Newbury, and D.L. Davidson, J. Appl. Phys., Vol. 53, p. R81, 1982

[16] K.A. Bezjian, H.I. Smith, J.M. Carter, M.W. Geis, J. Electrochem. Soc, Vol. 129, p. 1848, 1982

[17] W.K. Chu, J.W. Mayer, M.A. Nicolet, *Backscattering Spectrometry*, Academic Press, N.Y., 1978

[18] M.T. Duffy, J.F. Corboy, G.W. Cullen, R.T. Smith, R.A. Soltis, G. Harbeke, J.R. Sandercock, and M. Blumenfeld, J. Crystal Growth, Vol. 58, p. 10, 1982

[19] G. Harbeke and L. Jastrzebski, J. Electrochem. Society, Vol. 137, p. 696, 1990

[20] G.K. Celler, S. Cristoloveanu, Journal of Applied Physics, Vol. 93, no. 9, p. 4955, 2003

[21] T.I. Kamins, Electronics Letters, Vol. 23, p. 175, 1987

[22] T.R. Guilinger, M.J. Kelly, J.W. Medernach, S.S. Tsao, J.O. Steveson, and H.D.T. Jones, Proceedings of the IEEE SOS/SOI Technology Conference, p. 93, 1989

[23] M.J. Kelly, T.R. Guilinger, J.W. Medernach, S.S. Tsao, H.D.T. Jones, and J.O. Steveson, Electrochemical Society Proceedings, Vol. 90-6, p. 120, 1990

[24] W.C. Dash, J. Appl. Phys, Vol. 27, p. 1993, 1956

[25] D.G. Schimmel, J. Electrochem. Soc, Vol. 126, p. 479, 1979

[26] F. Secco d'Aragona, J. Electrochem. Soc., Vol. 119, p. 948, 1972

[27] E. Sirtl and A. Adler, Zeitung für Metallkunde, Vol. 52, p. 529, 1961

[28] M. Wright Jenkins, J. Electrochem. Soc., Vol. 124, p. 757, 1977

[29] T.R. Guilinger, M.J. Kelly, J.W. Medernach, S.S. Tsao, J.O. Steveson, and H.D.T. Jones, Proceedings of the IEEE SOS/SOI Technology Conference, p. 93, 1989

[30] M.J. Kelly, T.R. Guilinger, J.W. Medernach, S.S. Tsao, H.D.T. Jones, and J.O. Steveson, Electrochemical Society Proceedings, Vol. 90-6, p. 120, 1990

[31] K. Imamura, K. Daido, K. Mimegishi, H. Nakanishi, Japanese Journal of Applied Physics, Vol.16, suppl.1, p. 547, 1977

[32] Y. Moriyasu, T. Morishita, M. Matsui, A. Yasujima, M. Ishida, Electrochemical Society Proceedings, Vol. 99-3, p. 137, 1999

[33]   M. Sudou, M. Kainuma, T. Nakai, Tomizawa, Electrochemical Society Proceedings, Vol. 99-3, p. 119, 1999

[34]   H. Moriceau, B. Aspar, M. Bruel, A.M. Cartier, C. Morales, A. Soubie, T. Barge, S. Bressot, C. Malevillee , A.J. Auberton, Electrochemical Society Proceedings, Vol. 99-3, p. 173, 1999

[35]   M. Alles, J. Dunne, C. Treadwell, B. Fiordalice, R. Nguyen, Proceedings of the IEEE International SOI Conference, p. 17, 2001

[36]   C. Maleville, E. Neyret, L. Ecarnot, T. Barge, A.J. Auberton, Proceedings of the IEEE International SOI Conference, p. 19, 2001

[37]   C. Maleville, Electrochemical Society Proceedings Vol. 2003-05, p. 33, 2003

[38]   C. Maleville, C. Moulin, E. Neyret, Proceedings of the IEEE International SOI Conference, p. 194, 2002

[39]   M.J. Anc, M. Gostein, M. Banet, L.P. Allen, Electrochemical Society Proceedings, Vol. 2001-3, p. 51, 2001

[40]   M. Tajima, Electrochemical Society Proceedings, Vol. 2003-03, p. 413, 2003

[41]   Z.Q. Zi, M. Tajima, M. Warashina, S. Sumie, H. Hashizume, A. Ogura, Proceedings of the IEEE International SOI Conference, p. 135, 2003

[42]   V. Higgs, M.J. Anc, Electrochemical Society Proceedings, Vol. 2001-3, p. 63, 2001

[43]   I. De Wolf, J. Vanhellemont, H.E. Maes, A. Romano-Rodriguez, and H. Norström, Electrochemical Society Proceedings, Vol. 92-13, p. 307, 1992

[44]   E. Martin, A. Pérez-Rodríguez, J. Jimenez, and J.R. Morante, Electrochemical Society Proceedings Vol. 94-11, p. 185, 1994

[45]   J. Macía, T. Jawhari, A. Pérez-Rodríguez, and J.R. Morante, Electrochemical Society Proceedings Vol. 94-11, p. 148, 1994

[46]   L.P. Allen, M.J. Anc, M. Duffy, J.H. Parechanian, J.H. Yap, Electrochemical Society Proceedings Vol. 96-3, p. 18, 1996

[47]   K. Notsu, N. Homma, T. Yonehara, Electrochemical Society Proceedings Vol. 2001-3, p. 57, 2001

[48]   X.X. Zhang, J.P. Raskin, Electrochemical Society Proceedings, Vol. 2003-19, p. 233, 2003

[49]   W.P. Maszara, G. Goetz, A. Caviglia, J.B. McKitterick, Journal of Applied Physics, Vol. 64, p. 4943, 1988

[50]   Q.Y. Tong and U. Gösele, *Semiconductor Wafer Bonding Science and Technology*, The Electrochemical Society Series, John Wiley & Sons, p. 25, 1999

[51]   C.A. Colinge, M.C. Shaw, R.H. Esser, K.D. Hobart, Electrochemical Society Proceedings Vol. 2001-27, p. 85, 2001

[52]   G.R. Anstis, P. Chantikul, B.R. Lawn, D.B. Marshall, Journal of the American Creamic Society, Vol. 64, no. 9, p. 533, 1981, and
       P. Chantikul, G.R. Anstis, B.R. Lawn, D.B. Marshall, Journal of the American Ceramic Society, Vol. 64, no. 9, p.539, 1981

[53]   F. Ebrahimi, Materials Science and Engineering A, Vol. A268, p. 116, 1999

[54]   T.S. Moss, *Optical Properties of Semiconductors*, Butterworths, London, Chapter 4, 1959

[55]   A.M. Goodman, J. Appl. Phys, Vol. 53, p. 7561, 1982

[56]   L. Lukasiak, E. Kamieniecki, A. Jakubowski, J. Ruzyllo, Electrochemical Society Proceedings, Vol. 2003-05, 2003

[57]   L. Jastrzebski, G. Cullen, and R. Soydan, J. Electrochem. Society, Vol. 137, p. 303, 1990

[58]   M.A. Guerra, Electrochemical Society Proceedings, Vol. 90-6, p. 21, 1990

[59]   K. Nauka, M. Cao, and F. Assaderaghi, Proceedings of the IEEE International SOI Conference, p. 52, 1995

60 J.L. Freeouf and S.T. Liu, Proceedings of the IEEE International SOI Conference, p. 74, 1995

61 J.L. Freeouf, N. Braslau, and M. Wittmer, Applied Physics Letters, Vol. 63, p. 189, 1993

62 H.S. Chen, F.T. Brady, S.S. Li, and W.A. Krull, IEEE Electron Device Letters, Vol. 10, p. 496, 1989

63 H.S. Chen and S.S. Li, Proceedings of the 4th International Symposium on Silicon-on-Insulator Technology and Devices, Ed. by D. Schmidt, the Electrochemical Society, Vol. 90-6, p. 328, 1990

64 H.S. Chen and S.S. Li, Electrochemical Society Proceedings, Vol. 90-6, p. 328, 1990

65 D.P. Vu and J.C. Pfister, Appl. Phys. Letters, Vol. 47, p. 950, 1985

66 T. Elewa, H. Haddara, and S. Cristoloveanu, in *"Solid-State Devices"*, Ed. By. G. Soncini and P.U. Calzolari, Elsevier Science Publishers (North-Holland), p. 599, 1988

67 D.P. Vu and J.C. Pfister, Appl. Phys. Letters, Vol. 47, p. 950, 1985

68 M. Zerbst, Z. Angew. Phys., Vol. 22, p. 30, 1966

69 P.K. McLarty, T. Elewa, B. Mazhari, M. Mukherjee, T. Ouisse, S. Cristoloveanu, D.E. Ioannou, and D.P. Vu, Proceedings of the IEEE SOS/SOI Technology Conference, p. 54, 1989

70 T. Elewa, Ph.D. Thesis, ENSERG-LPCS, Grenoble (France), p. 90, July 1990

71 H. Shin, M. Racanelli, W.M. Huang, J. Foerstner, S. Choi, D.K. Schroder, IEEE Transactions on Electron Devices, Vol. 45, no. 11, p. 2378, 1998

72 M. Haond and J.P. Colinge, Electronics Letters, Vol. 25, p. 1640, 1989

73 D. Flandre and F. Van De Wiele, IEEE Electron Device Letters, Vol. 9, p. 296, 1988

74 M. Gaitan and P. Roitman, Proceedings of the IEEE SOS/SOI Technology Conference, p. 48, 1989

75 J.H. Lee and S. Cristoloveanu, IEEE Electron Device Letters, Vol. 7, p. 537, 1986

76 J.S. Brugler and P.G.A. Jespers, IEEE Transactions on Electron Devices, Vol. 16, p. 297, 1969

77 G. Groeseneken, H.E. Maes, N. Beltran, and R.F. Dekeersmaecker, IEEE Transactions on Electron Devices, Vol. 31, p. 42, 1984

78 T. Elewa, H. Haddara, S. Cristoloveanu and M. Bruel, J. de Physique, Vol. 49, No 9-C4, p. C4-137, 1988

79 Y. Li and T.P. Ma, International Symposium on VLSI Technology, Systems, and Applications, Proceedings of Technical Papers, p. 123, 1997

80 K. Nauka, Microelectronic Engineering, Vol. 36, no. 1-4, p. 351, 1997

81 S. Cristoloveanu and S. Williams, IEEE Electron Device Letters, Vol. 31, p. 102, 1992

82 S. Cristoloveanu and S.S. Li, *Electrical Characterization of Silicon-On-Insulator Materials and Devices*, Kluwer Academic Publishers, p. 104, 1995

83 T. Ouisse, P. Morfouli, O. Faynot, H. Seghir, J. Margail, and S. Cristoloveanu, Proceedings of the IEEE International SOI Conference, p. 30, 1992

84 S. Cristoloveanu, A. Ionescu, C. Maleville, D. Munteanu, M. Gri, B. Aspar, M. Bruel, and A.J. Auberton-Hervé, in "Silicon-On-Insulator Technology and Devices VII", Ed. by. P.L.F. Hemment, S. Cristoloveanu, K. Izumi, T. Houston, and S. Wilson, Electrochemical Society Proceedings Vol. 96-3, p. 142, 1996

85 S. Wiliams, S. Cristoloveanu, and G. Campisi, Materials Science Engineering, Vol. B12, p. 191, 1992

86 A.M. Ionescu, S. Cristoloveanu, S.R. Wilson, A. Rusu, A. Chovet, and H. Seghir, Nuclear Instr. and Methods in Phys. Res., Vol. 112, p. 228, 1996

87 A.M. Ionescu, S. Cristoloveanu, D. Munteanu, T. Elewa, and M. Gri, Solid-State Electronics, Vol. 39, no. 12, p. 1753, 1996

[88]   S.G. Kang, D.K. Schroder, Transactions on Electron Devices, Vol. 49, no. 10, p. 1742, 2002

[89]   S. Cristoloveanu, D. Munteanu, M.S.T. Liu, IEEE Transactions on Electron Devices, Vol. 47, no. 5, p. 1018, 2000

[90]   D. Munteanu, S. Cristoloveanu, and H. Hovel, Electrochemical and Solid-State Letters, Vol. 2, p. 242, 1999.

[91]   H. Kirk, S. Bedell, and M. Current, Electrochemical Society Proceedings Vol. 2001-3, p. 103, 2003

Chapter 4

SOI CMOS TECHNOLOGY

Silicon-on-Insulator substrates are being used for many applications such as micro electro-mechanical systems (MEMS),[1] integrated optics [2,3,4] and sensors [5,6] but the most important consumer of SOI wafers is the CMOS integrated circuit industry. This chapter will compare CMOS processing on bulk silicon and on SOI wafers. Processing of fully and partially depleted devices will be discussed. SOI wafers contain only silicon and silicon dioxide, and their appearance is very similar to that of bulk silicon wafers. As a consequence, SOI circuit processing can be carried out in standard bulk silicon processing lines.

4.1 SOI CMOS PROCESSING

Processing techniques for the fabrication of CMOS circuits in bulk silicon and in SOI are very similar. Figure 4-1 presents cross-sections of CMOS inverters made with simple bulk (N-well technology), partially depleted SOI, and fully depleted SOI processes.

Figure 4-1 is, of course, quite schematic. For instance, bulk CMOS can be much more complicated by making use of an epitaxial substrate, twin wells, or retrograde wells. From the cross sections, it is obvious that SOI processing, and more specifically fully depleted SOI processing, is simpler than bulk processing. For instance, there is no need to create diffused wells in SOI. Anti-punchthrough implant such as HALO are commonly used in bulk CMOS and can be used in both fully depleted and partially depleted SOI as well.[7,8,9,10] In thin-film, fully-depleted SOI devices, well formation is unnecessary, and the entire impurity profile in the channel area is determined by a single shallow implant. Table 4-1 compares simplified CMOS process flows for bulk, partially depleted (PD) SOI, and fully depleted (FD) SOI. The similarity between bulk and SOI process flows is striking, yet one can notice that SOI processing requires fewer steps.

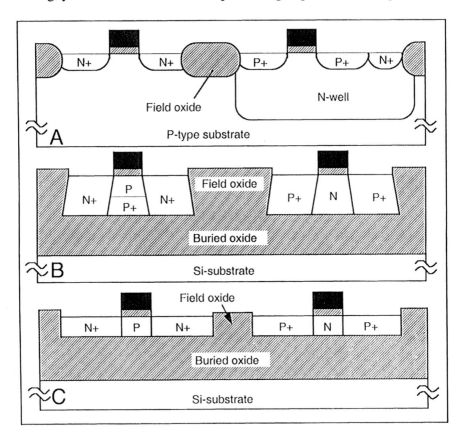

Figure 4-1. Cross-section of A: bulk, B: partially depleted SOI, and C: fully depleted SOI CMOS inverters.

There might be a concern regarding the quality of the gate oxide of SOI devices. Early reports on the properties of gate oxides indicated higher leakage and charge trapping characteristics in oxides grown on SIMOX than on bulk silicon.[11] This was due to the poor quality of the material, including carbon and metal contamination. In 1990, the material quality had improved, such that it was found that the breakdown field of thin gate oxides (12.5 nm) grown on SIMOX was 10-15% lower than that of similar oxides grown on bulk silicon (10 MV/cm *vs.* 11.5 MV/cm). Additionally the spread of the distribution around the mean value was quite similar to what was measured in bulk.[12] This problem seems to have disappeared in modern SIMOX material. For example, an oxide breakdown field of 11.5 MV/cm in

250Å SIMOX gate oxides and cumulative oxide breakdown plots in SIMOX have been found to be similar to those measured on bulk silicon.[13,14]

Table 4-1. Comparison between bulk, partially depleted (PD) SOI, and fully depleted (FD) SOI basic process flows.

Bulk CMOS	PD SOI CMOS	FDSOI CMOS
P-type (100) bulk wafer	P-type (100) SOI wafer	P-type (100) SOI wafer
-	SOI film thinning (if needed)	SOI film thinning (if needed)
Pre-well oxidation	-	-
Well lithography	-	-
Well doping and drive-in	-	-
Oxide strip	-	-
Nitride deposition	Nitride deposition	Nitride deposition
Active area lithography	Active area lithography	Active area lithography
Nitride etch	Nitride etch	Nitride etch
Field implant	Field implant	Field implant
Field oxide growth	Field oxide growth	Field oxide growth
Nitride strip	Nitride strip	Nitride strip
P-channel lithography	P-channel lithography	- ▫
P-channel N-type doping	P-channel N-type doping	- ▫
Gate oxide growth	Gate oxide growth	Gate oxide growth
P-ch. blanket V_{TH} implant	P-ch. blanket V_{TH} implant	P-ch. blanket V_{TH} implant
N-channel V_{TH} lithography	N-channel V_{TH} lithography	N-channel V_{TH} lithography
N-ch. antipunchthru implant	N-ch. back-channel implant	-
N-channel V_{TH} implant	N-channel V_{TH} implant	N-channel V_{TH} implant
Poly deposition and doping*	Poly deposition and doping*	Poly deposition and doping*
Gate litho and etch	Gate litho and etch	Gate litho and etch
P-channel S&D lithography	P-channel S&D lithography	P-channel S&D lithography
P-channel S&D implant	P-channel S&D implant	P-channel S&D implant
N-channel S&D lithography	N-channel S&D lithography	N-channel S&D lithography
N-channel S&D implant	N-channel S&D implant	N-channel S&D implant
Source and drain anneal	Source and drain anneal	Source and drain anneal
Oxide deposition	Oxide deposition	Oxide deposition
Contact lithography	Contact lithography	Contact lithography
Contact hole etch	Contact hole etch	Contact hole etch
Metal deposition	Metal deposition	Metal deposition
Metal lithography	Metal lithography	Metal lithography
Metal etch	Metal etch	Metal etch
Sintering anneal	Sintering anneal	Sintering anneal

▫ P-channel devices are assumed to be accumulation-mode

* N⁺ polysilicon gate is used for all devices.

4.1.1 Fabrication yield and fabrication cost

In a bulk silicon device a defect can produce a yield hazard if it is located in the active area, *i.e,* if it is located in the channel region, the source or the drain. If λ is the minimum design rule for all levels, the area where the presence of a defect can affect a device is given by: *A = device length x device width = [2(field isolation to contact hole spacing + contact hole length + contact hole to gate spacing) + gate length)] × [field isolation to contact hole spacing + contact hole width + field isolation to contact hole spacing] = [2(3λ)+λ][3λ]=21λ²*. In the case of an SOI device with reach-through junctions, the area where the presence of a defect can be felt is given by the area of the channel region only, because of the full dielectric isolation of the source and drain junctions. This area is equal to *3λ²* in our example, and is, therefore, significantly smaller than in a bulk device. According to such considerations, the use of SOI wafers allows higher fabrication yield over bulk silicon if the defect densities in both materials are comparable.

The cost of SOI wafers is higher than that of bulk silicon wafers, but the increased demand stimulates competition between suppliers and the increased throughput of fabrication techniques will force the wafer price to drop significantly. For example, the price of a UNIBOND® wafer is predicted to be equal to that of a bulk epi wafer. The cost of front-end CMOS processing, on the other hand, increases steadily as smaller and smaller devices are fabricated. This cost increase is significantly larger in bulk CMOS processing than in thin-film SOI processing. Indeed, deep-submicron CMOS processing necessitates heavy investments in terms of R&D efforts in order to produce adequate device isolation, shallow junctions, anti-latchup and anti-punchthrough techniques. SOI devices, on the other hand, are scalable in their current form to deep-submicron dimensions without the need for any major new technique development. As a result the need for costly R&D investments is reduced. Front-end thin-film SOI processing is already currently cheaper than front-end bulk processing, and it will become even more so in the future. In addition, higher packing density (and, therefore, higher yield) as well as increased performance can be obtained when SOI is used. When the considerations of wafer and processing cost, potential higher yield and higher performance are added one can conclude that higher benefits can be made by switching from bulk to SOI.[15,16,17,18]

We will now describe in more detail some particularities of SOI processing such as isolation techniques, doping profiles, and source and drain resistance issues.

4.2 FIELD ISOLATION

There are many different ways of isolating the active silicon islands from one another. All these are simpler than the isolation schemes used in bulk silicon technology, owing to the presence of the buried insulator, which provides an intrinsic vertical isolation. In a sense, one could say that a part of the complexity of the device isolation process used in bulk technology has been transferred to the wafer manufacturing stage (the fabrication of silicon-on-insulator material). The main isolation techniques are described next.

4.2.1 LOCOS

The LOCOS (local isolation of silicon) technique is a very popular isolation scheme used in bulk and SOI CMOS. It is a well-known, reliable and well-controlled process presenting little yield hazards. It consists of the following steps: a thin "pad" oxide is grown, and a silicon nitride layer is deposited. Using a mask step, the nitride is patterned to define the active silicon areas. Boron is then implanted around those islands that will contain n-channel devices (channel-stop implant), and the field oxide is thermally grown. No oxide is grown where the silicon is protected by nitride. Finally the nitride layer and the pad oxide are stripped. The exact same process can be used in SOI for lateral device isolation. During the field oxide growth step, the silicon that is not covered with nitride is completely consumed, and the thermally grown field oxide reaches through to the buried oxide (Figure 4-2).

The lateral encroachment of the field oxide, called "bird's beak", is proportional to the thickness of the grown oxide. This encroachment has typical values of a fraction of a micrometer, and hampers the use of the LOCOS technique for submicron bulk processing, where 0.5...0.8 mm-thick field oxides are grown. The local stress in the silicon is also related to the thickness of the field oxide. In an SOI process the thickness of the oxide grown to fully isolate the silicon islands is 2.5 - 3 thicker than the silicon film. If a 100 nm-thick SOI film is used, a field oxide of only 250-300 nm must be grown. This implies less stress in the silicon, and, most importantly, a significantly smaller bird's beak. As a consequence, the LOCOS isolation can be used to realize devices with smaller geometries in SOI than in bulk.

In bulk technology the P$^+$ implant (channel-stop implant) that is carried out prior to growing the field oxide prevents surface inversion of the silicon underneath the field oxide. Such inversion would lead to leakage between n-channel devices. In SOI devices, the same implant is used for preventing source-to-drain leakage from taking place along the edges of the silicon island. Indeed, an inversion layer can form at the tip of the bottom corners of the islands (Figure 4-2). This inversion layer degrades the subthreshold slope of the device and is the source of leakage current when the device is turned off. The edge leakage current is caused by the presence of a parasitic edge transistor in parallel with the main transistor (Figure 4-3). It has the same gate length as the main device, but has a much smaller width.

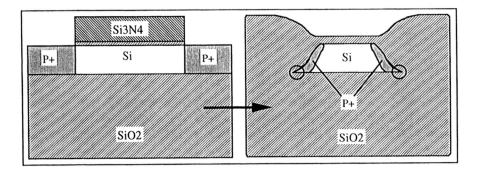

Figure 4-2. LOCOS isolation. The circles at the edges of the silicon island (left) indicate where source-to-drain edge leakage may occur.

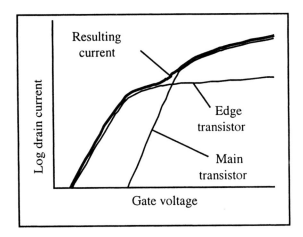

Figure 4-3. $I_D(V_G)$ characteristics of a transistor with edge leakage.

If the P$^+$ impurity concentration at the edges of the silicon island is high enough, the threshold voltage of the edge device is higher than that of the main device, and the edge current is overshadowed by the main device current for all values of gate voltage, such that the presence of the edge transistor has no detrimental effect on the overall device characteristics. If, on the other hand, the P$^+$ impurity concentration at the edges of the silicon island is too low, the threshold voltage of the edge device is lower than that of the main device, and the edge transistor current dominates the current characteristics of the overall device at low gate voltage values (Figure 4-3). This results in a degradation of the subthreshold characteristics and an increase of the off-state leakage current. One problem of LOCOS isolation is that most of the boron implanted in the silicon segregates into the oxide. As a result, the doping at the tip of the bottom corners of the islands is quite low and edge leakage appears. One solution is to use a recessed LOCOS technique (half the silicon thickness is etched in the field regions before the oxide is grown). This procedure allows one to reduce the oxidation time, and, therefore, the boron segregation, and to eliminate the tip of the bottom corners of the islands. Sacrificial oxidation can be used as an alternative to silicon etching prior to field oxide growth.[19]

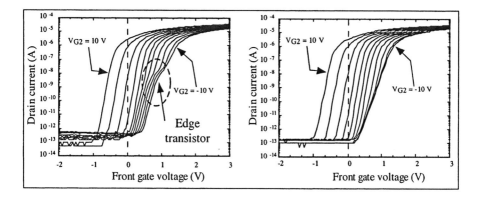

Figure 4-4. $I_D(V_G)$ characteristics of n-channel transistors with LOCOS isolation (left) and recessed LOCOS isolation (right). V_D=50 mV. The back-gate voltage ranges from -10 to 10 volts in 2-volt steps. W/L=20µm/5µm.[20]

Figure 4-4 presents the $I_D(V_G)$ characteristics of fully depleted n-channel transistors made using a regular LOCOS and a recessed LOCOS isolation. Although the effect of edge conduction can hardly be seen when the back-gate voltage is equal to 0 volt, edge leakage rapidly occurs when a negative back-gate bias is applied, which gives rise to a kink in the subthreshold $I_D(V_G)$ characteristics. This kink increases the values of both the subthreshold slope and the OFF leakage current. When the recessed LOCOS process is used no kink can be seen in the subthreshold $I_D(V_G)$ characteristics, even at large negative back-gate bias, and near-ideal subthreshold characteristics are observed until back accumulation is induced, at the highest negative gate biases. Another way of minimizing the segregation of boron into the oxides consists of performing the edge implant after LOCOS growth (Local Implantation post Field oxidation, LIF)[21] or even after gate patterning. [22,23] Poly-encapsulated LOCOS (PELOX) with high dose boron implantation dose can also be used to eliminate the edge transistor current.[24]

4.2.2 Mesa isolation

The mesa isolation technique is another way of isolating silicon islands from one another. This technique is attractive because of its simplicity. It simply consists of patterning the silicon into islands -or "mesas"- using a mask step and a silicon etch step. Passivation of the island edges is performed at the gate oxidation step, where the gate dielectric is grown not only on top of the silicon islands, but on their edges as well (Figure 4-5). This technique has been extensively used in SOS processing, where a KOH solution was used to etch the silicon and produce sloped island edges.

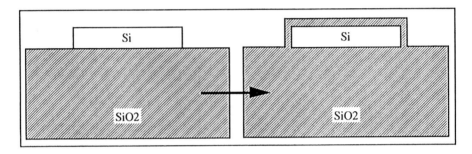

Figure 4-5. Mesa isolation.

Several problems are associated with the mesa isolation. It is well known that the oxidation of silicon corners produces SiO_2 layers with non-uniform thickness. Indeed, the thickness of the oxide grown on the corners of an island can be 30 to 50% thinner than that grown on its top surface.[25] This thinning of the oxide depends on the oxidation temperature (the effect is more pronounced if oxidation is carried out below or close to the SiO_2 viscous flow temperature (965°C) [26]). Oxidation is also known to sharpen silicon corners [27], such that corners sharper than an angle of 45 degrees are produced. This effect is enhanced if more than a single oxidation step is performed (*i.e.*: if a sacrificial oxide is grown and stripped prior to gate oxidation). The thinning of the oxide and the sharpening of the corners both contribute to a reduction of the gate oxide breakdown voltage observed when mesa isolation is used. Sidewall leakage may also be observed, as in the case where LOCOS isolation is used. In a mesa process, the gate oxide and the gate material covers both the top of the silicon island and its edges. Therefore, there exist lateral (edge) transistors in parallel to the main (top) device. Furthermore, due to charge sharing between the main and the edge devices, the threshold voltage is reduced at the corner of the island [28]. This can produce a kink in the subthreshold characteristics as well as leakage currents similar to those described in Figure 4-3. Both the oxide breakdown and the leakage currents can be improved by using P+ sidewall doping and mesa edge rounding techniques [29,30].

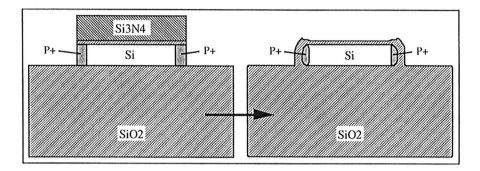

Figure 4-6. Oxidized mesa isolation.

The oxidized mesa technique results from the combination of mesa and LOCOS processes. As in the LOCOS formation, a nitride layer is patterned to define the active silicon areas. Boron is implanted around the islands that will contain n-channel devices. Some of the boron is then driven laterally in the silicon located underneath the nitride, after which the silicon is etched away to form mesas. The sidewalls of the silicon islands are then oxidized (Figure 4-6), and the nitride is stripped. The oxidized mesa process has several advantages. The corners of the silicon islands are rounded during the lateral oxidation step. This increases the breakdown voltage of gate oxide by over 30%, compared to the mesa isolation process.[31,32] In addition, there are no regions where the silicon is extremely thin, as it is the case when LOCOS isolation is used. This contributes to improved subthreshold characteristics and reduce leakage currents.

4.2.3 Shallow trench isolation

Shallow trench isolation (STI) has been adapted from bulk processing technology to SOI in 1997.[33] The STI process bears similarities with the Mesa process, and is described in Figure 4-7. A thin pad oxide is first grown on the silicon film, and a layer of silicon nitride (Si_3N_4) is deposited. Photolithography and plasma etching are then used to pattern the nitride, pad oxide, and SOI layer Figure 4-7A. Thermal oxidation is then used to grow oxide liners on the sidewalls of the silicon islands (Figure 4-7B). CVD oxide is then deposited to fill the trenches between the SOI islands, and chemical-mechanical polishing (CMP) is used to planarize the structure (the nitride is used as an etch stop for the CMP step – Figure 4-7C). The nitride and pad oxide are then removed by chemical etch (Figure 4-7D).

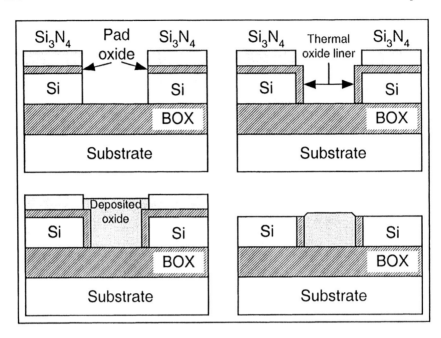

Figure 4-7. Shallow trench isolation; A: lithography and nitride/pad oxide/silicon etch; B: growth of sidewall thermal oxide, C:CVD oxide deposition and CMP; D: nitride and pad oxide strip.

As in the LOCOS and Mesa isolations parasitic sidewall conductance can happen when STI is used, and subthreshold characteristics similar to that shown in Figure 4-3 can occur. This problem can be solved by doping the sidewalls of the n-channel islands using tilted implantation. Parasitic conduction can also take place along the bottom corners of the silicon island. That particular type of conduction can be eliminated using a deep retrograde implant.[34] One disadvantage of the STI process is that the growth of the thin sidewall liner oxide causes compressive stress in the silicon island.[35] This stress slightly pushes the sides of the islands upward (Figure 4-8).[36] It causes the generation of defects in the silicon and lowers electron mobility (hole mobility, on the other hand, is enhanced by the compressive stress).[37] These problems can be taken care of by modifying the STI process and either growing the sidewall liners after depositing the trench filling oxide [38], forming the liners by medium temperature oxide deposition [39], or by limiting the trench to a depth less than the thickness of the silicon film (partial trench isolation – PTI).[40]

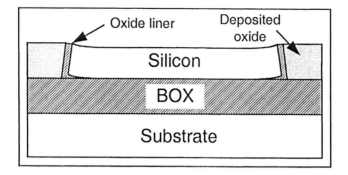

Figure 4-8. Silicon island bending caused by oxide liner growth.

4.2.4 Narrow-channel effects

The importance of the lateral interface between silicon islands and field isolation becomes increasingly important as device width is scaled down. We have seen that the sidewalls and corners present in LOCOS, Mesa and STI isolation can be the source of leakage current (parasitic edge transistor). Charge sharing between the top of the silicon island top and the sidewalls, or between the bottom interface and the sidewalls explains why the the threshold voltage of parasitic edge/corner transistors is lower than that on the main transistor.[41] Furthermore, boron tends to segregate out of the silicon into the oxide, giving rise to boron depletion near the Si/SiO_2 interfaces. As a result the overall threshold voltage of the device decreases as the channel width, W, is decreased. This effect is called the "narrow-channel effect".[42] In addition, surface recombination at the sidewall interfaces reduces carrier mobility [43] and carrier lifetime [44] in narrow devices, which, in turn, reduces floating-body effects. In very narrow devices quantum effects due to carrier confinement gives rise to an increase of threshold voltage similar to that observed when the silicon film thickness is reduced to very small dimensions (Figure 5-23).[45]

4.3 CHANNEL DOPING PROFILE

The optimization of the channel doping profile in SOI MOSFETs serves two main purposes: the adjustment of the front threshold voltage and the elimination of back-channel leakage. Different gate materials can be used. The doping profile in a device realized with a P^+ polysilicon gate will, of course, be totally different from that of a device having an N^+ polysilicon as gate material.[46,47,48] We will only consider here the case of N^+ poly gate, which is, by far, the most common.

In partially depleted devices, the silicon film is thick enough ($t_{si} \geq$ ...80... nm) such that two distinct implants can be performed for the front and back interfaces. The result is that the front and back threshold voltages can be adjusted separately. The doping profile of the (enhancement-mode) n-channel transistors presents a double hump (Figure 4-9A). Indeed, a superficial boron implant is used to adjust the threshold voltage to the desired value, and a higher energy implant is carried out to give the back interface a high enough threshold voltage (*e.g.* $\geq$10V) to avoid back-channel leakage problems.

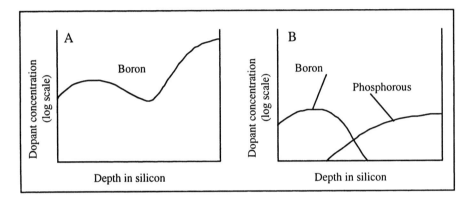

Figure 4-9. Doping profiles in partially depleted SOI MOSFETs. A: n-channel device. B: p-channel device.

The doping profile of a partially depleted p-channel transistor is shown in Figure 4-9B. It is similar to the profile found in a buried-channel bulk device. A deep n-type (usually phosphorous) implant is used to avoid leakage between source and drain and to control the drain punchthrough voltage, while a shallow p-type implant is used to adjust the front threshold voltage.

In the case of thin, fully depleted ($t_{si} \leq$ ...80... nm) MOS devices, the thickness is such that an almost flat doping profile is created by any implant. This "flat" profile is only the center portion of the Gaussian-like profile produced by the implantation, and front and back threshold voltages cannot be adjusted independently. The doping profiles of thin-film n-channel and p-channel devices are presented in Figure 4-10. In this example the p-channel transistor is an accumulation-mode device, the channel region of which is fully depleted of holes (and, therefore, non-conducting) when the device is turned off. An accumulation channel forms at the top Si-SiO$_2$ interface when a negative gate voltage is applied. The n-channel transistor is a standard enhancement-mode device.

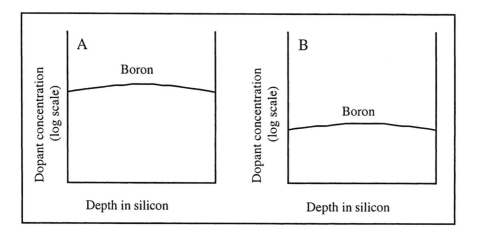

Figure 4-10. Doping profiles in thin, fully depleted SOI MOSFETs. A: n-channel device. B: p-channel device.

The front threshold voltage of fully depleted SOI MOSFETs is a function of the silicon film thickness. Indeed, the total depletion charge, Q_{depl}, is equal to $-qN_a t_{si}$. Using Equations 5.3.10 and 5.3.12 the dependence of the threshold voltage on the film thickness can be established. Typically the dV_{th}/dt_{si} variation is on the order of 10 mV per nanometer. The front threshold voltage varies also with the amount of charge in the BOX and the interface state density at the $Si-SiO_2$ back interface. For this reason, partially depleted MOSFETS have often been preferred over fully depleted devices. The variation of threshold voltage with film thickness in fully depleted devices, can, however, be minimized:

- Using the concept of "constant dose" doping instead of that of "constant concentration" doping, it is possible to significantly reduce the dV_{th}/dt_{si} variation. This approach optimizes the channel implantation energy in such a way that the dependence of the implanted dose (cm^{-2}) on the film thickness is minimized.[49,50]

- In a fully depleted device the variation of drain current with film thickness (dI_D/dt_{si}) is significantly lower than the dV_{th}/dt_{si} variation. Indeed, when the film thickness is increased the body factor decreases, which partially compensates for the current decrease caused by the increase of threshold voltage.[51]

- The channel doping concentration can be reduced to virtually zero if midgap material is used for the gate electrode. In that case the threshold voltage is controlled by the work function of the gate material and virtually independent on silicon film thickness.[52]

A statistical analysis of threshold voltage and drain current variations with processing parameters such as film thickness and channel doping concentration can be found in the literature.[53] It is important to point out that the silicon film thickness variation found in modern SOI wafers is low enough to support fabrication of fully depleted VLSI circuits [54] and that threshold voltage variations (σ_{VTH}) lower than 10 mV are routinely obtained in fully depleted MOSFETs across 300-mm wafers.[55,56]

The doping concentration in the channel of a fully depleted transistor can also be varied from source to drain. Decreasing the doping concentration from source to drain brings about an improvement (decrease) of the output conductance and reduces parasitic bipolar effects (graded-channel MOSFET).[57,58]

4.4 SOURCE AND DRAIN ENGINEERING

Lightly doped drain (LDD) and HALO structures are commonly used in bulk CMOS and can readily be adapted to SOI MOSFETs.[59,60,61,62] The real problem facing SOI MOSFETs is the high resistance values of the source and drain. In a MOSFET the source and drain resistance is inversely proportional to the film thickness. This resistance can reach high values in thin-film devices, which jeopardize circuit speed performance. The source and drain resistance can be lowered by several techniques: silicide formation, elevated source and drain, metal clad deposition, and the use of Schottky junctions.

4.4.1 Silicide source and drain

Formation of a silicide layer on source and drain is a routine operation in bulk silicon processes and it is widely used in SOI processing as well. The most popular silicides are titanium silicide ($TiSi_2$), cobalt silicide ($CoSi_2$) and nickel silicide ($NiSi$).[63,64,65,66] The metal layers deposited on thin SOI devices to form silicides must be thinner than those used in bulk devices, however.

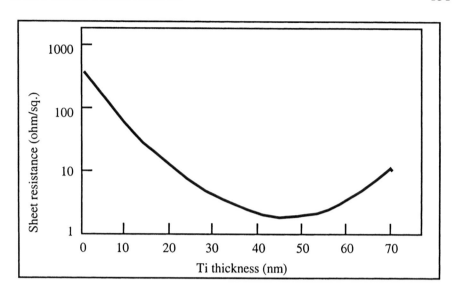

Figure 4-11. Source and drain sheet resistance as a function of the thickness of the deposited titanium thickness (100 nm-thick silicon film).[67,68]

Figure 4-11 presents the source and drain sheet resistance in silicided (TiSi$_2$) SOI MOSFETs as a function of the deposited titanium thickness. The silicon film thickness is 100 nm. The source and drain sheet resistance can be reduced from 300 Ω/square to 2 Ω/square by sputtering titanium and forming the silicide in a two-step annealing process. The lowest resistance is obtained when a 45nm-thick metal layer is deposited, and the use of thicker titanium layers leads to increased resistance, as well as shorts between the gate, and the source, and drain.

More recent results have confirmed that there exists a process window for the thickness of the deposited titanium. For 100 nm-thick devices, the optimal titanium thickness is between 35 to 45 nm. Thinner metal yields a non-continuous silicide layer with high resistivity, and thicker metal layers tend to consume all the silicon in the source and drain regions. Since silicon is the diffusing species in titanium silicide, silicon from the channel region diffuses into the silicide once the entire silicon thickness has been consumed, which leads to the formation of Kirkendall voids underneath the gate edges, resulting in non-functional devices.[69]

In addition to the formation of voids in the silicon film, there is another reason to keep the silicide from reaching the buried oxide. The series resistance of a silicided SOI junction abruptly increases when the thickness of the silicide approaches that of the silicon film. This effect is due to a reduction of the contact area once the silicide consumes the silicon layer because the horizontal portion directly underneath the silicide is no longer available for contact.[70] The effective area through which current can flow is then drastically reduced and the resistance increases (Figure 4-11). The optimum silicide thickness appears to be approximately 80% of the total silicon film thickness.

The contact resistance R_c between the silicide and the source/drain junction can be modeled by two resistors in parallel (Figure 4-12):

$$\frac{1}{R_c} = \frac{1}{R_L} + \frac{1}{R_B} \qquad (4.4.1)$$

where the first resistor R_L representing the lateral contact resistance between the silicide and the source/drain junction:

$$R_L = \frac{\rho_c}{W\,t_M} \qquad (4.4.2)$$

and where R_B represents the distributed resistance below the silicide, which can be calculated using transmission line theory:

$$R_B = R_\square \frac{L_T}{W} \coth\left(\frac{L_c}{L_T}\right) \quad \text{with } L_T = (\rho_c/R_\square)^{1/2} \qquad (4.4.3)$$

where W is the device width, L_C is the length of the silicide region, ρ_c is the specific (vertical) contact resistance between the silicide and the junction and $R_\square = \rho_{si}/(t_{si}-t_M)$ is the sheet resistance of the silicon under the silicide.[71] When the thickness of the silicide, t_M, is increased the overall contact resistance decreases because R_L decreases. However, when the silicon thickness underneath the silicide becomes too small, R_B increases and the overall contact resistance increases. Table 4-2 lists the values of the sheet resistance of silicides (R_{sheet}), ρ_c and $R_\square$ for CoSi$_2$ and TiSi$_2$.[72]

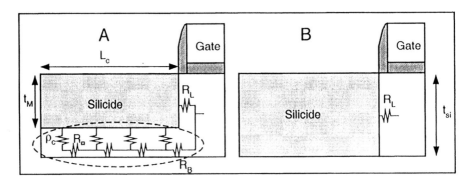

Figure 4-12. Current flow path in a silicide junction. A: the silicide is thinner than the silicon film; B: the silicide reaches the BOX.

Table 4-2. Resistive elements in silicide contacts

	Junction type	$CoSi_2$	$TiSi_2$	No silicide
R_{sheet} ($\Omega/\square$)	N^+	14	8	118*
	P^+	14	11	174*
$R_\square$ (mΩ.cm)	N^+	0.70	0.80	0.59
	P^+	1.11	1.28	0.87
ρ_c ($\mu\Omega$.cm^2)	N^+	0.31	0.38	0.74**
	P^+	0.49	0.42	3.40**

* sheet resistance of the N^+ or P^+ silicon

** contact resistance between metal 1 and the N^+ or P^+ silicon

When the film is completely converted to silicide (Figure 4-12B) the overall contact resistance is simply equal to R_L. Initial publications ruled out full silicidation of source and drain junctions because of the formation of Kirkendall voids. Recent improvements on silicide formation, however, have demonstrated the achievability of full silicidation of thin (12-27 nm) source and drain junctions. It has been shown that the formation of cobalt silicide using a two-step annealing process can be used to create thin, fully silicide junctions with no Kirkendall voids and low leakage current.[73,74] Contact resistances as low as 4×10^{-8} Ω.cm^{-2} and 10^{-7} Ω.cm^{-2} have been obtained on N^+ and P^+ junctions, respectively. To obtain the contact resistance in ohms one should multiply those numbers by the area of the junction $A = W\,t_{si}$.[75]

4.4.2 Elevated source and drain

Another way of fabricating thin SOI devices with good source and drain (S&D) resistance consists of using different silicon film thicknesses for the channel and the S&D regions: the use of a thinner silicon film in the channel region provides the desired fully depleted SOI MOSFET features, and the use of thicker silicon for the sources and drains decreases the S&D resistance. This can be achieved in two ways. In a first approach, a thin silicon film is used, and selective epitaxial growth is used to increase the thickness of the S&D regions (elevated source and drain technique) (Figure 4-13).[76,77,78,79,80,81,82] It is important to note that while it reduces the device resistance the presence of elevated source and drain structures increases the gate-to-source and gate-to-drain capacitances, such that increased CV/I (intrinsic time delay) characteristics can seriously deteriorate when thick raised extensions and thin liners, separating the gate from the source and drain, are used (Figure 4-13).[83]

A second technique to fabricate thin-body devices with thicker source and drain starts from a silicon film having the thickness desired for source and drain. The channel area is then thinned using a LOCOS-like technique (recessed channel technique, presented in Figure 4-14).[84,85,86,87,88] It is even possible to self-align the recessed region to the polysilicon gate.[89]

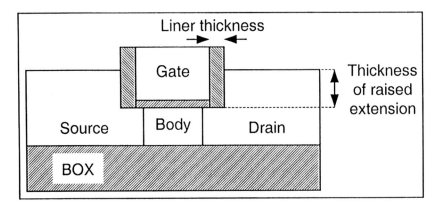

Figure 4-13. Elevated source and drain.

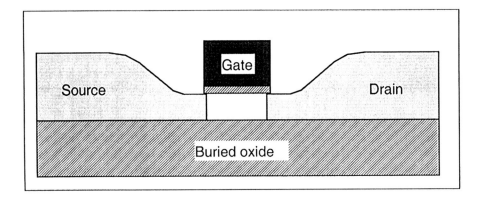

Figure 4-14. Thin-film SOI MOSFET with recessed channel.

4.4.3 Tungsten clad

The source and drain resistance can be reduced as well by using selective tungsten deposition on the source and drain (Figure 4-15).[90] Tungsten can be selectively deposited on silicon by chemical vapor deposition based on the reduction of WF_6 by SiH_4 at 250°C followed by a 30-second rapid thermal annealing step at 550°C. When carried out on a 20 nm-thick N^+ junction the operation forms a 48 nm-thick tungsten clad layer on 6 nm of remaining silicon. On a P^+ junction it forms a 38 nm-thick tungsten layer on 9 nm of remaining silicon. The sheet resistance of such a tungsten layer is 10

$\Omega/\square$, compared to the 15 and 30 $\Omega/\square$ measured in $TiSi_2$ and $CoSi_2$ films of comparable thickness, respectively.[91]

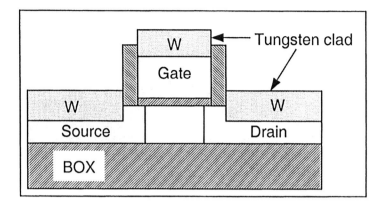

Figure 4-15. Transistor with tungsten clad.

4.4.4 Schottky source and drain

Schottky sources and drains in SOI CMOS provide an alternative to standard S&D engineering.[92] Schottky S&D have a smaller junction RC constant than classical junctions, and, therefore can outperform standard p-n junction S/D MOSFETs. The Schottky barrier formed between the metal or metal silicide should be as low as possible to minimize the contact resistance.[93] Platinum silicide (PtSi) has been used by several groups to form Schottky junctions in p-channel SOI MOSFETs [94,95], while erbium silicide ($ErSi_{1.7}$) has been used in n-channel devices.[96] Theoretical studies show that the use of Schottky junctions in very short-channel MOSFETs could essentially eliminate most of the S&D resistance, provided that negative or very small positive Schottky barriers are formed.[97] The lowest published Schottky formed on p-type silicon are 0.2-0.22, 0.09 and 0.05 eV for platinum silicide, iridium silicide and PtGeSi, respectively.[98] The lowest barriers formed on n-type silicon are 0.27 and 0.06 eV for erbium silicide and platinum/erbium silicide.[99] It is worth noting that Schottky junctions have been used to form the collector of SOI bipolar transistors.[100]

4.5 GATE STACK

Traditional MOSFET design makes use of N^+ and P^+ polysilicon as gate material and SiO_2 as gate insulator. Other gate materials are, however, being explored to eliminate polysilicon depletion effects and adjust threshold voltage. In a parallel research effort, high-k dielectric materials are being used to increase current drive while limiting the gate current due to tunneling.

4.5.1 Gate material

The work function of the gate material influences the threshold voltage since Φ_{MS1} in expressions 5.3.10 to 5.3.12 is equal to $\Phi_M - \Phi_{si}$, where Φ_{si} is the workfunction of the silicon in the channel. Classical CMOS use N^+ polysilicon as gate material for n-channel devices and P^+ polysilicon for p-channel transistors. Under those conditions relatively high doping concentrations are used in the channel region to obtain adequate threshold voltages. The threshold voltage of a fully depleted inversion-mode SOI MOSFET is given by 5.3.10 and 5.3.12:

$$V_{TH1,depl2} = \Phi_{MS1} - \frac{Q_{ox1}}{C_{ox1}} + \left(1 + \frac{C_{si}}{C_{ox1}}\right) 2\Phi_F - \frac{Q_{depl}}{2C_{ox1}} - \frac{C_{si}\,C_{ox2}}{C_{ox1}\,(C_{si} + C_{ox2})} \ (V_{G2} - V_{G2,acc}) \quad (4.5.1)$$

with $Q_{depl} = q\,N_a\,t_{si}$. The workfunction of N^+ polysilicon is 4.1 eV. Equation 4.5.1 shows that one can lower the doping concentration N_a while maintaining a constant threshold voltage by using a gate material with a higher Φ_M. Such a material is P^+ polysilicon with a workfunction of 5.2 eV, a value that is too high for practical n-channel devices. In N^+ polysilicon the Fermi level coincides with the conduction band minimum while in P^+ polysilicon it coincides with the maximum of the valence band. Materials whose Fermi level is near the intrinsic energy level of silicon are called "midgap materials" and have workfunction values near 4.61 eV that are suitable for use as gate material in both n-channel and p-channel devices. Among such material are titanium nitride (TiN) [101], tantalum (Ta) [102], TaSiN [103], P^+SiGe [104,105], NiSi [106] and molybdenum (Mo). Molybdnemum and TiN are especially attractive since their workfunction can be tuned by nitrogen implantation and annealing.[107,108] Midgap gate materials are very attractive for fully depleted single- and multiple-gate SOI MOSFETs where short-channel effects are controlled by the use of ultrathin silicon films rather than by high channel doping concentrations. They are not useful in partially depleted devices where strong short-channel effects appear unless high channel doping concentrations are used.[109]

4.5.2 Gate dielectric

The integrity of gate oxides grown on SOI is similar to that of oxides grown on bulk silicon.[110] As devices dimensions are scaled down, gate oxides are approaching a thickness of 1.5 nm or below [111] to increase current drive (through an increase of C_{ox} in equation 5.4.25). Electrons can tunnel through such oxides, giving rise to a gate current. If the gate oxide is very thin the gate current may no longer be negligible.[112] One can increase C_{ox} and avoid gate current by using gate dielectrics that have a higher dielectric constant, κ, than SiO_2 (κ_{SiO2} =3.9). Such materials are referred to as "high-k dielectrics". Gate dielectrics such as hafnium oxide (HfO_2)[113] and tantalum oxide (Ta_2O_5)[114] have been used in SOI MOSFETs. One drawback due to the lack of maturity of high-k dielectric technology, is the poor quality of the silicon/dielectric interface, which causes mobility degradation.

4.5.3 Gate etch

The formation of the gate electrode involves a plasma etch, resist ashing and ion implantation steps. During the etch, plasma unbalance can create an excess of ion current and a charge build-up at the wafer surface, which in turn can damage the gate oxide.[115] This phenomenon is called the "antenna effect".

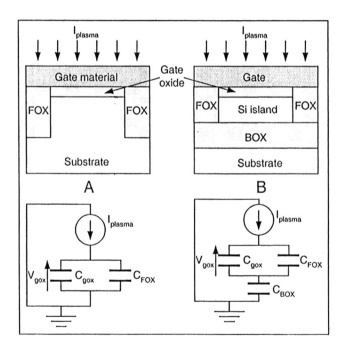

Figure 4-16. Antenna effect during plasma etch step in bulk (A) and SOI (B).[116]

In bulk wafers the charge at the surface of the wafer gives rise to a voltage V_{gox} across the gate oxide. The value of that voltage is determined by the intensity of the plasma current (I_{plasma}), the value of the gate and field oxide capacitances, C_{gox} and C_{FOX}, and the amplitude of the Fowler-Norheim current flowing through the gate oxide (Figure 4-16A). In SOI an additional capacitor corresponding to the buried oxide, C_{BOX}, must be added to the equivalent circuit (Figure 4-16B). One can readily see that the presence of C_{BOX} reduces the value of the plasma-induced voltage drop across the gate oxide, V_{gox}. As a result SOI wafers are less prone to gate oxide damage caused by the antenna effect than bulk wafers, unless the buried oxide capacitor, C_{BOX}, is short-circuited by electrically conductive defects such as silicon pipes, running through the BOX.

4.6   SOI MOSFET LAYOUT

Several layouts can be adopted for SOI MOSFETs, depending on the application. The densest and most common layout is presented in Figure 4-17A. It consists of a rectangular active area, a gate, and contact holes. When the application in which a circuit is used may lead to edge leakage problems (such as in devices submitted to ionizing radiations where generation of oxide charges in the field oxide at the edges of the silicon islands can give rise to large leakage currents), "edgeless" device designs can be utilized (Figure 4-17B).[117] In such a device, the silicon island (active area) presents no edge underneath the gate between the source and the drain. It is, however, worth noting that edgeless devices occupy much more silicon real estate than conventional devices, and are not used where integration density is a prime concern.

4.6.1 Body contact

Some applications require devices with body contacts because grounding the silicon underneath the gate effectively suppresses the kink effect as well as parasitic lateral bipolar effects. Several schemes exist to provide the transistor body with a contact.

The conventional approach is presented in Figure 4-18I. It consists of a P+ diffusion contact with the P-type silicon underneath the gate. Such a device can also be used as a lateral bipolar transistor, the P+ diffusion being the base contact, and the source and drain being used as emitter and collector. The T-gate design (Figure 4-18T) is a variation on the same body contact structure. In transistors with large gate width, the presence of a single body contact at one end of the channel region may not be sufficient to suppress kink or bipolar effects. These effects can indeed take place underneath the gate, "far" from the body contact, the efficiency of which is reduced by the high resistance of the weakly doped channel region. The H-gate MOSFET design helps solve this problem, since body contacts are present at both ends of the channel (Figure 4-18H). Furthermore, the H-gate device offers no direct edge leakage path between source and drain (the edges run only from N⁺ to P⁺ diffusions). [118]

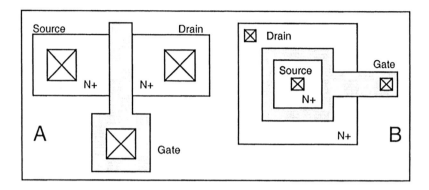

Figure 4-17. Layout of SOI MOSFETs. A: "regular" device. B: edgeless device.

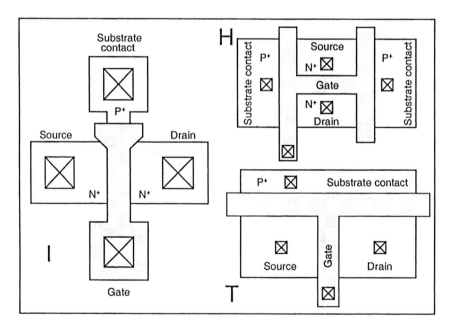

Figure 4-18. Transistors with body contacts. I: Basic body contact (I-gate) ; H: H-gate device; T: T-gate device.[119]

Another type of body contact, more compact than the previous ones, is shown in Figure 4-19. The P⁺ body ties are created on the side of the N⁺ source diffusion. As in the case of the H-gate device, there is no direct edge leakage path between source and drain (the edges of the active area under the gate run only from N⁺ to P⁺ diffusions). If the device is very wide, additional

P+ regions can be formed in the source, such that a P+-N+-P+...N+-P+ structure is produced. This device has the drawback of being asymmetrical (source and drain cannot be swapped), and the effective channel width, W_{eff}, is smaller than the width of the active area.[120]

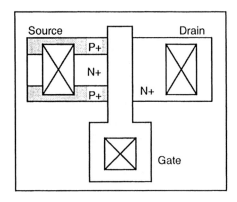

Figure 4-19. N-channel transistor with source body ties.

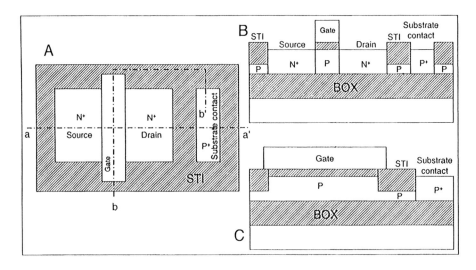

Figure 4-20. Body tie using partial trench isolation: A: top view (layout); B: a-a' cross section; C: b-b' cross section.

Body contacts can be achieved by adapting the fabrication process instead of modifying the individual transistor layout. Figure 4.20 shows an n-channel SOI transistor where a body contact is formed by reducing the depth of the field isolation (STI) in such a way that there remains silicon between the STI and the BOX. This reduced-depth STI process is called the "partial trench isolation" technique (PTI). The silicon between the field isolation and the BOX can be used as a link connecting the body of the transistors to a "substrate contact". This technique allows for the use of straight bulk CMOS layout for SOI circuits.[121,122]

4.7 SOI-BULK CMOS DESIGN COMPARISON

Generally speaking, SOI CMOS technology offers a higher integration density than bulk CMOS. This becomes evident from comparison between the layout of a bulk CMOS inverter and that of an SOI CMOS inverter (Figure 4-21). This higher density results mainly from the absence of wells in SOI. A second cause of density increase is that a direct contact between P^+ and N^+ junctions (such as the drains of the n-channel and the p-channel devices of Figure 4.6.2) is possible. The number of contact holes per gate is also lower in SOI than in bulk. This reduces a source of fabrication yield hazard, compared to bulk.

One of the major differences between SOI and bulk design is the difference of body effect and of body/back gate bias conditions. The body effect induced by the back gate (dV_{TH1}/dV_{G2}) is negligible in partially depleted devices. The expressions for the body effect in fully depleted devices (the dependence of threshold voltage on back-gate bias) can be derived from section 5.3.1 and 5.12.1 for inversion-mode and accumulation-mode devices, respectively. Furthermore, the back-gate bias configuration of SOI MOSFETs is different from the substrate bias used in bulk. Let us take the example of a simple CMOS inverter (Figure 4-22). In bulk CMOS, the body of the n-channel device is connected to ground (V_{SS}), while the body of the p-channel transistor is connected to V_{DD}. Hence, the potential of the body is the same as that of the source in both types of devices. In the SOI inverter, the back gate (the underlying silicon wafer) is common to both n- and p-type devices. It is usually grounded. Hence, the back-gate voltage is 0V for the n-channel device, but it is equal to $-V_{DD}$ for the p-channel transistor, where the source voltage being is as a reference. As a consequence, SOI p-channel transistors must be designed for operation with a back-gate bias, V_{G2}, which is equal to $-V_{DD}$.

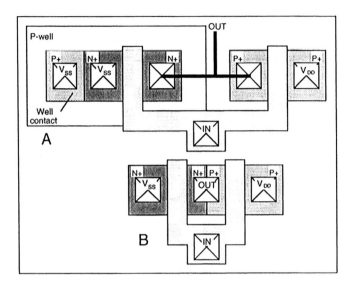

Figure 4-21. Layout of A: a bulk CMOS inverter, and B: an SOI CMOS inverter.

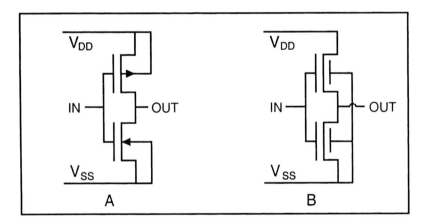

Figure 4-22. Back-gate (body) bias configuration in bulk (A) and SOI (B) CMOS inverters.

4.8   ESD PROTECTION

The input and output pads of MOS circuits have to be protected against electrostatic discharge (ESD). Specifications are that the circuit should, for instance, survive a 2,000 to 4,000 volt pulse delivered by a RC network corresponding to the capacitance and the resistance of a human body (*human body model,* HBM). Under these conditions significant power is dissipated in the input/output stages of the circuit, and device destruction (or: "zapping") can occur. Since SOI devices have a reduced junction area and are thermally insulated from the substrate, there is a concern about the efficiency and the reliability of SOI ESD protection structures. The snapback characteristics of the $I_D(V_D)$ curve of an SOI MOSFET can be used to realize ESD protection structures (Figure 4-23).[123,124] If the gate of an n-channel SOI MOSFET is grounded and the drain voltage is increased up to a certain value, called snapback voltage, V_{sb}, drain current suddenly flows in the device and the drain voltage drops to a holding value, V_h. If one attempts to further increase the drain voltage, the current further increases. Snapback is due to classical avalanche multiplication and impact ionization mechanisms and to the presence of a parasitic bipolar transistor in the device (see section 5.10).[125] Modelling of the drain breakdown characteristics of SOI MOSFETs can be found in the literature.[126,127,128]

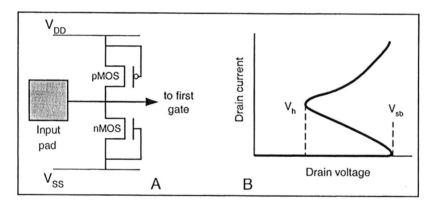

Figure 4-23. A: Input protection based on snapback transistors, and B: drain snapback characteristics of an n-channel SOI MOSFET.[129]

These snapback characteristics can be used to clamp ESD pulses and can, therefore, be used as protection devices (Figure 4-23A). The level of protection reached by SOI snapback n-channel MOSFETs is expressed in ESD pulse volts per micron of protection device width. The protection level obtained from snapback transistors typically ranges from 5 to 10 volts per micrometer of device width, which is similar to the level of protection achieved in bulk silicon. Higher protection levels can be reached by using more elaborate structures combining snapback MOSFETs and Zener diodes.[130] Another alternative consists of opening a window in the buried oxide and fabricating the ESD protection devices in the bulk silicon substrate.[131] 4000-V ESD protection (HBM) of fully depleted SOI CMOS circuits ($L = 0.35$ μm) has been demonstrated.[132] The wide transistors used in ESD input protection structures are usually wrapped around the I/O pads of the chip.

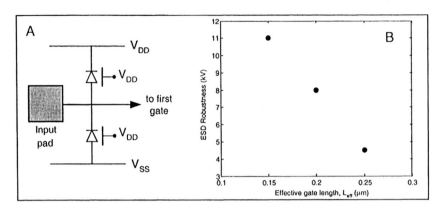

Figure 4-24. A: First-stage of gated-diode (Lubistor)-based input protection and B: Robustness of Lubistor ESD protection structure as a function of technology generation.[133]

ESD protection can also be achieved using SOI lateral gated diodes (or Lubistors - see Section 6.5 and Figure 4-24). [134, 135, 136] The robustness Lubistor ESD protections is typically on the order of 5 V/μm, such that a protection level of 4 kV is achieved using a 800 μm-wide device with a gate length of 1.2 μm.[137] Figure 4-24 presents the improvement of the robustness of Lubistor ESD protection in IBM SOI circuits as a function of technology generation. From this graph it is clear that the technology migration choices and scaling of the technology have led to improved ESD robustness in the SOI Lubistor structure. These include the transition from $TiSi_2$ to $CoSi_2$, Al to Cu interconnects and dimensional scaling. Figure 4-24 shows the ESD robustness of the SOI Lubistor network as the structure migrated from the 0.25 to 0.12 μm technology generation. In the 0.25 μm technology generation, ESD robustness of 5 V/μm was demonstrated. ESD protection of 4 kV was achieved in structures with 800 μm structures. Since designs have been scaled to 0.12 μm L_{eff} SOI technology, ESD results continue to show an improvement. In the 0.12 μm L_{eff} SOI technology, ESD results over 8 kV have been demonstrated in high-pin-count microprocessor chips with structures smaller in design perimeter and in physical area.

REFERENCES

1   A.Y. Usenko, W.N. Carr, Electrochemical Society Proceedings, Vol. 99-3, p. 347, 1999
2   T.W. Ang, P.D. Hewitt, A. Vonsovici, G.T. Reed, A.G.R. Evans, P.R. Routley, T. Blackburn, M.R. Josey, Electrochemical Society Proceedings, Vol. 99-3, p.353, 1999
3   A. Scherer, M. Loncar, T. Yoshie, K. Okamoto, B. Maune, J. Witzens, Proceedings of the IEEE International SOI Conference, p. 9, 2003
4   Y. Iida, Y. Omua, M. Tsuji, Proceedings of the IEEE International SOI Conference, p. 117, 2003
5   A. Lehto, Electrochemical Society Proceedings, Vol. 99-3, p. 11, 1999
6   B.R. Takulapalli, T.J. Thornton, D. Gust, B. Ashcroft, S.M. Lindsay, H.Q. Zhang, N.J. Tao, Proceedings of the IEEE International SOI Conference, p. 114, 2003
7   H. van Meer, K. De Meyer, Proceedings of the IEEE International SOI Conference, p. 45, 2001
8   L. Vancaillie, V. Kilchitska, D. Levacq, S. Adriaensen, H. van Meer, K. De Meyer, G. Torrese, J.P. Raskin, D. Flandre, Proceedings of the IEEE International SOI Conference, p. 161, 2002
9   M.A. Pavanello, J.A. Martino, E. Simoen, A. Mercha, C. Claeys, H. van Meer, K. De Meyer, Electrochemical Society Proceedings. Vol. 2003-05, p. 389, 2003
10 V. Kilchytska, A. Nève, L. Vancaillie, D. Levacq, S. Adriaensen, H. van Meer, K. De Meyer, S. Raynaud, M. Dehan, J.P. Raskin, D. Flandre, IEEE Transactions on Electron Devices, Vol. 50, no. 3, p. 577, 2003
11 C.T. Lee and A. Burns, IEEE Electron Device Letters, Vol. 9,p. 235, 1988
12 P.H. Woerlee, C. Juffermans, H. Lifka, W. Manders, F. M. Oude Lansink, G.M. Paulzen, P. Sheridan, and A. Walker, Technical Digest of IEDM, p. 583, 1990
13 P.K. Karulkar, IEEE Electron Device Letters, Vol. 14, p. 80, 1993
14 I.K. Kim, W.T. Kang, J.H. Lee, S. Yu, S.C. Lee, K. Yeom, Y.G. Kim, D.H. Lee, G. Cha, B.H. Lee, S.I. Lee, K.C. Park, T.E. Shim, and C.G. Hwang, Technical Digest of IEDM, p. 605, 1996
15 T. Stanley, Proceedings of the IEEE International SOI Conference, p. 166, 1992
16 T. Stanley, Electrochemical Society Proceedings Vol. 94-11, p. 441, 1994
17 T.D. Stanley, Electrochemical Society Proceedings Vol. 93-29, p. 303, 1993
18 Y. Kado, H. Inokawa, Y. Okazaki, T. Tsuchiya, Y. Kawai, M. Sato, Y. Sakakibara, S. Nakayama, H. Yamada, M. Kitamura, S. Nakashima, K. Nishimura, S. Date, M. Ino, K. Takeya, and T. Sakai, Technical Digest of IEDM, p. 635, 1995
19 T. Ohno, Y. Kado, M. Harada, and T. Tschuiya, IEEE Transactions on Electron Devices, Vol. 42, no. 8, p. 1481, 1995
20 J.P. Colinge, A. Crahay, D. De Ceuster, V. Dessart, and B. Gentinne, Electronics Letters, Vol. 32, no. 19, p. 1834, 1996
21 H.S. Kim, S.B. Lee, D.U. Choi, J.H. Shim, K.C. Lee, K.P. Lee, K.N. Kim, and J.W. Park, Technical Digest of the Symposium on VLSI Technology, p. 143, 1995
22 M. Racanelli, W.M. Huang, H.C. Shin, J. Foerstner, J. Ford, H. Park, S. Cheng, T. Wetteroth, S. Hong, H. Shin, and S.R. Wilson, in "Silicon-on-Insulator Technology and Devices VII", Ed. by P.L.F. Hemment, S. Cristoloveanu, K. Izumi, T. Houston, and S. Wilson, Proceedings of the Electrochemical Society, Vol. 96-3, p. 422, 1996
23 M. Racanelli, W.M. Huang, H.C. Shin, J. Foerstner, B.Y Hwang, S. Cheng, P.L. Fejes, H. Park, T. Wetteroth, S. Hong, H. Shin, and S.R. Wilson, Technical Digest of the IEDM, p. 885, 1995

[24] H. van Meer, K. De Meyer, Electrochemical Society Proceedings, Vol. 2001-3, p. 301, 2001

[25] S.S. Tsao, D.M. Fletwood, V. Kaushik, A.K. Datye, L. Pfeiffer, and G.K. Celler, Proceedings of the IEEE SOS/SOI Technology Workshop, p. 33, 1987

[26] R.K. Smeltzer and J.T. McGinn, Proceedings of the IEEE SOS/SOI Technology Workshop, p. 32, 1987

[27] R.B. Marcus and T.T. Sheng, J. Electrochem. Soc., Vol. 129, p. 1278, 1982

[28] M. Matloubian, R. Sundaresan, and H. Lu, Proceedings of the IEEE SOS/SOI Technology Workshop, p. 80, 1988

[29] M. Haond and O. Le Néel, Proceedings of the IEEE SOS/SOI Technology Conference, p. 132, 1990

[30] O. Le Néel, M.D. Bruni, J. Galvier, and M. Haond, in "ESSDERC 90", Adam Hilger Publisher, Ed. by. W. Eccleston and P.J. Rosser, p. 13, 1990

[31] M. Haond, O. Le Néel, G. Mascarin, and J.P. Gonchond, Proceedings of the IEEE SOS/SOI Technology Conference, p. 68, 1989

[32] M. Haond, O. Le Neel, G. Mascarin, and J.P. Gonchond, in ESSDERC'89, European Solid-State Device Research Conference, Berlin, Ed. by. A. Heuberger, H. Ryssel and P. Lang, Springer-Verlag, p. 893, 1989

[33] K. Mistry, G. Grula, J. Sleight, L. Bair, R. Stephany, R. Flatley, P. Skerry, Technical Digest of the International Electron Device Meeting, p. 583, 1997

[34] C. Fenouillet-Béranger, O. Faynot, C. Tabone, T. Colladant, V. Ferlet, C. Jahan, J. du Port de Pontcharra, G. Lecarval, J.L. Pelloie, Proceedings of the IEEE International SOI Conference, p. 87, 2001

[35] M. Mendicino, L. Yang, N. Cave, S. Veeraraghavan, P. Gilbert, Procedings of the IEEE International SOI Conference, p, 132, 1997

[36] K.W. Su, K.H. Chen, T.X. Chung, H.W. Chen, C.C. Huang, H.Y. Chen, C.Y. Chang, D.H. Lee, C.K. Wen, Y.M. Sheu, S.J. Yang, C.S. Chiang, C.C. Huang, F.L. Yang, Y.L. Chiang, Proceedings of the IEEE International SOI Conference, p. 80, 2003

[37] Y. Jeon, G.C.F. Yeap, P. Grudowski, T. Van Gompel, J. Schmidt, M. Hall, B. Melnick, M. Mendicino, S. Venkatesan, Proceedings of the IEEE International SOI Conference, p. 164, 2003

[38] W. G. En, D.Y. Ju, D. Chan, S. Chan, O. Karlsson, Proceedings of the IEEE International SOI Conference, p. 132, 1997

[39] T.J. Lee, Y.H. Roh, B.S. Kim, D.H. Ahn, E.H. Kim, C.H. Jeon, Y.W. Kim, S.C. Lee, C.S. Choi, K.P. Suh, Proceedings of the IEEE International SOI Conference, p. 83, 2001

[40] T. Iwamatsu, T. Ipposhi, T. Uchida, S. Maegawa, M. Inuishi, Proceedings of the IEEE international SOI Conference, p. 80, 2000

[41] J.B. Kuo, K.W. Su, Proceedings of the IEEE International SOI Conference, p. 84, 1997

[42] D.E. Ioannou, Electrochemical Society Proceedings, Vol. 2001-3, p. 369, 2001

[43] H. Lee, J.H. Lee, H. Shin, Y.J. Park, H.S. Min, IEEE Transactions on Electron Devices, Vol. 49, no. 4, p. 605, 2002

[44] J. Pretet, D. Ioannou, N. Subba, S. Cristoloveanu, W. Maszara, C. Raynaud, Solid-State Electronics, Vol. 46, no. 11, p. 1699, 2002

[45] H. Majima, H. Ishikuro, T. Hiramoto, IEEE Transactions on Electron Devices, Vol. 21, no. 8, p. 396, 2000

[46] T. Aoki, M. Tomizawa, and A. Yoshii, IEEE Transactions on Electron Devices, Vol. 36, p. 1725, 1989

[47] T. Nishimura, Y. Yamaguchi, H. Miyatake, and Y. Akasaka, Proceedings of the IEEE SOS/SOI Technology Conference, p. 132, 1989

48 M. Haond, in ESSDERC'89, European Solid-State Device Research Conference, Berlin, Ed. by. A. Heuberger, H. Ryssel and P. Lang, Springer-Verlag, p. 881, 1989

49 M.J. Sherony, L.T. Su, J.E. Chung, and D.A. Antoniadis, IEEE Electron Device Letters, Vol. 16, no. 3, p. 100, 1995

50 D.A. Antoniadis, Proceedings of the IEEE International SOI Conference, p. 1, 1995

51 P. Smeys, U. Magnusson, J.P. Colinge, Proceedings of ESSDERC'92, Microelectronic Engineering, Vol 19, p. 823, 1992

52 V.P. Trivedi, J.G. Fossum, IEEE Transactions on Electron Devices, Vol. 50, no. 10, p. 2095, 2003

53 N. Miura, H. Hayashi, K. Fukuda, K. Nishi, Proceedings of the International Conference on Simulation of Semiconductor Processes and Devices. (SISPAD'99), p. 87, 1999

54 Y. Fukuda, S. Ito, M. Ito, OKI Technical Review, Vol. 68, no. 4, p. 54, 2001

55 M. Harada, T. Douseki, T. Tsuchiya, Symposium on VLSI Technology Digest of Technical Papers, p. 96, 1996

56 C. Raynaud, O. Faynot, J.L. Pelloie, S. Deleonibus, D. Vanhoenacker, R. Gillon, J. Sevenhans, E. Compagne, G. Fletcher, E. Mackowiak, Proceedings of the IEEE International SOI Conference, p. 67, 1998

57 M.A. Pavanello, J.A. Martino, V. Dessard, D. Flandre, Solid-State Electronics, Vol. 44, no. 7, p. 1219, 2000

58 M.A. Pavanello, J.A. Martino, V. Dessard, D. Flandre, Electrochemical Society Proceedings, Vol. 99-3, p. 293, 1999

59 H. van Meer, K. De Meyer, Proceedings of the IEEE International SOI Conference, p. 45, 2001

60 L. Vancaillie, V. Kilchitska, D. Levacq, S. Adriaensen, H. van Meer, K. De Meyer, G. Torrese, J.P. Raskin, D. Flandre, Proceedings of the IEEE International SOI Conference, p. 161, 2002

61 M.A. Pavanello, J.A. Martino, E. Simoen, A. Mercha, C. Claeys, H. van Meer, K. De Meyer, electrochemical Society Proceedings. Vol. 2003-05, p. 389, 2003

62 V. Kilchytska, A. Nève, L. Vancaillie, D. Levacq, S. Adriaensen, H. van Meer, K. De Meyer, S. Raynaud, M. Dehan, J.P. Raskin, D. Flandre, IEEE Transactions on Electron Devices, Vol. 50, no. 3, p. 577, 2003

63 L.T. Su, M.J. Sherony, H. Hu, J.E. Chung, and D.A. Antoniadis, IEEE Electron Device Letters, Vol. 15, no. 9, p. 85, 2001

64 F. Deng, R.A. Johnson, W.B. Dubbelday, G.A. Garcia, P.M. Asbeck, and S.S. Lau, Proceedings of the IEEE International SOI Conference, p. 78, 1996

65 H. Komatsu, H. Nakayama, K. Koyama, K. Matsumoto, T. Ohno, K. Takeshita, Proceedings of the IEEE International SOI Conference, p. 23, 2001

66 A. Ogura, H. Wakabayashi, M. Ishikawa, T. Kada, H. Machida, Y. Ohshita, Proceedings of the IEEE International SOI Conference, p. 70, 2003

67 T. Nishimura, Y. Yamaguchi, H. Miyatake, and Y. Akasaka, Proceedings of the IEEE SOS/SOI Technology Conference, p. 132, 1989

68 Y. Yamaguchi, T. Nishimura, Y. Akasaka, and K. Fujibayashi, IEEE Transactions on Electrons Devices, Vol. 39, no. 5, p.1179, 1992

69 J. Foerstner, J. Jones, M. Huang, B.Y. Hwang, M. Racanelli, J. Tsao, and N.D. Theodore, Proceedings of the IEEE International SOI Conference, p. 86, 1993

70 L.T. Su, M.J. Sherony, H. Hu, J.E. Chung and D.A. Antoniadis, Technical Digest of IEDM, p. 723, 1993

71 E. Dubois, G. Larrieu, Solid-State Electronics, Vol. 46, p. 997, 2002

72   H.I. Liu, J.A. Burns, C.L. Keast, P.W. Wyatt, IEEE Transactions on Electron Devices, Vol. 45, no. 5, p. 1099, 1998

73   T. Ichimori, N. Hirashita, 2000 IEEE International SOI Conference. Proceedings, p. 72, 2000

74   T. Ichimori, N. Hirashita, Japanese Journal of Applied Physics Part 1, Vol. 40, no. 4B, p. 2881, 2001

75   T. Ichimori, N. Hirashita, IEEE Transactions on Electron Devices, Vol. 49, no. 12, p. 2296, 2002

76   J.M. Hwang, R. Wise, E. Yee, T. Houston, and G.P. Pollack, Digest of Technical papers, Symposium on VLSI Technology, p. 33, 1994

77   M. Cao, T. Kamins, P. Vande Voorde, C. Diaz, and W. Greene, IEEE Electron Device Letters, Vol. 18, no. 6, p. 251, 1997

78   A. Vandooren, S. Egley, M. Zavala, A. Franke, A. Barr, T. White, S. Samavedam, L. Mathew, J. Schaeffer, D. Pham, J. Conner, S. Daksihna-Murthy, B.Y. Nguyen, B. White, M. Orlowski, J. Mogab, Proceedings of the IEEE International SOI Conference, p. 205, 2002

79   H. van Meer, K. De Meyer, Electrochemical Society Proceedings, Vol. 2001-3, p. 301, 2001

80   S. Bagchi, J.M. Grant, J. Chen, S. Samavedam, F. Huang, P. Tobin, J. Conner, L. Prabhu, M. Tiner, Proceedings International SOI Conference, p. 56, 2000

81   A. Vandooren, A. Barr, L. Mathew, T.R. White, S. Egley, D. Pham, M. Zavala, S. Samavedam, J. Schaeffer, J. Conner, B.Y. Nguyen, B.E. White, Jr., M.K. Orlowski, J. Mogab, IEEE Electron Device Letters, Vol. 24, no. 5, p. 342, 2003

82   S.S. Kim, T.H. Choe, H.S. Rhee, G.J. Bae, K.W. Lee, N.I. Lee, K. Fujihara, H.K. Kang, J.T. Moon, Proceedings of the IEEE International SOI Conference, p. 74, 2000

83   J.L. Egley, A. Vandooren, B. Winstead, E. Verret, B. White, B.Y. Nguyen, Electrochemical Society Proceedings, Vol. 2003-05, p. 307, 2003

84   M. Chan, F. Assaderaghi, S.A. Parke, S.S. Yuen, C. Hu, and P.K. Ko, Proceedings of the IEEE International SOI Conference, p. 172, 1993

85   O. Faynot and B. Giffard, IEEE Electron Device Letters, Vol. 15, p. 175, 1994

86   A. Toriumi, J. Koga, H. Satake, and A. Ohata, Technical Digest of IEDM, p. 847, 1995

87   C. Raynaud, J.L. Pelloie, O. Faynot, B. Dunne, and J. Hartmann, Proceedings of the IEEE International SOI Conference, p. 12, 1995

88   T. Numata, M. Noguchi, Y. Oowaki, S. Takagi, Proceedings of the IEEE International SOI Conference, p. 78, 2000

89   J.H. Lee, H.C. Shin, J.S. Lyu, B.W. Kim, and Y.J. Park, Proceedings of the IEEE International SOI Conference, p. 122, 1996

90   D. Hisamoto, K. Nakamura, M. Saito, N. Kobayashi, S. Kimura, R. Nagai, T. Nishida, and E. Takeda, Technical Digest of IEDM, p. 829, 1992

91   M. Takahashi, T. Ohno, Y. Sakakibara, K. Takayama, IEEE Transactions on Electron Devices, Vol. 48, no. 7, p. 1380, 2001

92   M. Tao, D. Udeshi, S. Agarwal, N. Basit, E. Madonaldo, W.P. Kirk, Material Research Society Symposium Proceedings, Vol. 765, p. D7.10.1, 2003

93   J. Kedzierski, M.K. Ieong, P. Xuan, J. Bokor, T.J. King, C.Hu, Proceedings of the IEEE International SOI Conference, p. 21, 2001

94   A. Itoh, M. Saitoh, M. Asada, Device Research Conference Conference Digest, p. 77, 2000

95   J. Kedzierski, P. Xuan, V. Subramanian, J. Bokor, T.J. King, C. Hu, Superlattices and Microstructures, Vol. 28, no. 5/6, p. 445, 2000

96 J. Kedzierski, P. Xuan, E.H. Anderson, J. Bokor, T.J. King, C. Hu, Technical Digest of the International Electron Device Meeting, p. 57, 2000

97 J. Guo, M.S. Lundstrom, IEEE Transactions on Electron Devices, Vol. 49, no. 11, p. 1897, 2002

98 E. Dubois, G. Larrieu, Solid-State Electronics, Vol. 46, p. 997, 2002

99 X. Tang, J. Katcki, E. Dubois, J. Ratajczak, G. Larrieu, P. Loumaye, O. Nisole, V. Bayot, Electrochemical Society Proceedings, Vol. 2003-05, p. 99, 2003

100 M.J. Kumar, D.V. Rao, IEEE Transactions on Electron Devices, Vol. 59, no. 6, p. 1070, 2002

101 S. Bagchi, J.M. Grant, J. Chen, S. Samavedam, F. Huang, P. Tobin, J. Conner, L. Prabhu, M. Tiner, Proceedings International SOI Conference, p. 56, 2000

102 H. Shimada, Y. Hirano, T. Ushiki, K. Ino, T. Ohmi, IEEE Transactions on Electron Devices, Vo. 44, no. 11, p. 1903, 1997

103 A. Vandooren, A. Barr, L. Mathew, T.R. White, S. Egley, D. Pham, M. Zavala, S. Samavedam, J. Schaeffer, J. Conner, B.Y. Nguyen, B.E. White, Jr., M.K. Orlowski, J. Mogab, IEEE Electron Device Letters, Vol. 24, no. 5, p. 342, 2003

104 P.E. Hellberg, S.L. Zhang, C.S. Petersson, IEEE Electron Device Letters, Vol. 18, no. 9, p. 456, 1997

105 Y.K. Choi, N. Lindert, P. Xuan, S. Tang, D. Ha, E. Anderson E, T.J. King, J. Bokor, C. Hu, International Electron Devices Meeting Technical Digest, p. 19.1.1, 2001

106 Z. Krivokapic, W. Maszara, K. Achutan, P. King, J. Gray, M. Sidorow, E. Shao, J. Zhang, J. Chan, A. Marathe, M.R. Lin, Technical Digest of the International Electron Device Meeting, p. 271, 2002

107 P. Ranade, Y.K. Choi, D. Ha, A. Agarwal, M. Ameen, T.J. King, International Electron Devices Meeting Technical Digest, p. 363, 2002

108 H. Wakabayashi, Y. Saito, K. Takeuchi, T. Mogami, T. Kunio, Technical Digest of the International Electron Device Meeting, p. 253, 1999

109 B. Cheng, B. Maiti, S. Samavedam, J. Grant, B. Taylor, P. Tobin, J. Mogab, Proceedings of the IEEE International SOI Conference, p. 91, 2001

110 S. Nakashima, J. Kodate, Proceedings of the IEEE International SOI Conference, p. 119, 1999

111 R. Chau, J. Kavalieros, B. Doyle, A. Murthy, N. Paulsen, D. Lionberger, D. Barlage, R. Arghavani, B. Roberds, M. Doczy, Technical Digest of IEDM, p. 29.1.1, 2001

112 B. Doyle, R. Arghavani, D. Barlage, S. Datta, M. Doczy, J. Kavalieros, A. Murthy, R. Chau, Intel Technology Journal, Vol. 06, no. 2, p. 42, 2002

113 A. Vandooren, S. Egley, M. Zavala, A. Franke, A. Barr, T. White, S. Samavedam, L. Mathew, J. Schaeffer, D. Pham, J. Conner, S. Daksihna-Murthy, B.Y. Nguyen, B. White, M. Orlowski, J. Mogab, Proceedings of the IEEE International SOI Conference, p. 205, 2002

114 A. Yagishita, T. Saito, S. Inumiya, K. Matsuo, Y. Tsunashima, K. Suguro, IEEE Transactions on Electron Devices, Vol. 49, no. 3, p. 422, 2002

115 J.P. McVittie, 1st International Symposium on Plasma-induced Damage Proceedings, American Vacuum Society, p. 7, 1996

116 A.O. Adan, T. Naka, A. Kagisawa, H. Shimizu, Proceedings of the IEEE International SOI Conference, p. 9, 1998

117 R. Giacomini, J.A. Martino, Electrochemical Society Proceedings, Vol. 2003-09, p. 68, 2003

118 N.K. Annamalai and M.C. Biwer, IEEE Transactions on Nuclear Science, Vol. 35, p. 1372, 1988

[119] H. Lee, K.S. Chun, J.H. Lee, Y.J. Park, H.S. Min, Extended Abstracts of the International Conference on Solid-State Devices and Materials, p. 276, 2001

[120] Y. Omura and K. Izumi, IEEE Transactions on Electron Devices, Vol. 35, p. 1391, 1988

[121] W. Chen, Y. Taur, D. Sadana, K.A. Jenkins, J. Sun, S. Cohen, Symposium on VLSI Technology Digest of Technical Papers, p. 92, 1996

[122] Y. Hirano, S. Maeda, T. Matsumoto, K. Nii, T. Iwamatsu, Y. Yamaguchi, T. Ipposhi, H. Kawashima, S. Maegawa, M. Inuishi, T. Nishimura, Proceedings of the IEEE International SOI Conference, p. 131, 1999

[123] K. Verhaege, G. Groeseneken, J.P. Colinge, H.E. Maes, IEEE Electron Device Letters, Vol. 14, no. 7, p. 326, 1993

[124] K. Verhaege, G. Groeseneken, J.P. Colinge, H.E. Maes, Microelectronics Reliability, Vol. 35, no.3, p. 555, 1995

[125] J.S.T. Huang, Proceedings of the IEEE International SOI Conference, p. 122, 1993

[126] J.S.T. Huang, H.J. Chen, S.J. Kueng, IEEE Transactions on Electron Devices, Vol. 39, p. 1170, 1992

[127] H.K. Yu, J.S. Lyu, S.W. Kang, C.K. Kim, IEEE Transactions on Electron Devices, Vol. 41, no. 5, p. 726, 1994

[128] N. Kistler J. Woo, IEEE Transactions on Electron Devices, Vol. 41, no. 7, p. 1217, 1994

[129] J.C. Smith, Microelectronics Reliability, Vol. 38, p. 1669, 1998

[130] J.C. Smith, M. Lien, S. Veeraraghavan, Proceedings of the IEEE International SOI Conference, p. 170, 1996

[131] M. Chan, J.C. King, P.K. Ko, C. Hu, IEEE Electron Device Letters, Vol. 16, no. 1, p. 11, 1995

[132] A.O. Adan, T. Naka, S. Kaneko, D. Urabe, K. Higashi, A. Kasigawa, Proceedings of the IEEE International SOI Conference, p. 116, 1996

[133] S. Voldman, D. Hui, L. Warriner, D. Young, J. Howard, F. Assaderaghi, G. Shahidi, Journal of Electrostatics, Vol. 49, p. 151, 2000

[134] Ming-Dou Ker, Kei-Kang Hung, H.T.H. Tang, S.C. Huang, S.S. Chen, M.C. Wang, Proceedings of the 2001 8th International Symposium on the Physical Failure Analysis of Integrated Circuits (IPFA 2001), p. 91, 2001

[135] Y. Omura, S. Wakita, Electrochemical Society Proceedings, Vol. 2003-05, p. 313, 2003

[136] C. Putnam, R. Gauthier, M. Muhammad, K. Chatty, M. Woo, Proceedings of the IEEE International SOI Conference, p. 23, 2003

[137] S. Voldman, D. Hui, L. Warriner, D. Young, R. Williams, J. Howard, V. Gross, W. Rausch, E. Leobangdung, M. Sherony, N. Rohrer, C. Akrout, F. Assaderaghi, G. Shahidi, Proceedings of the IEEE International SOI Conference, p. 68, 1999

Chapter 5

THE SOI MOSFET

The Metal-Oxide-Semiconductor Field-Effect Transistor (MOSFET) is the most widely used SOI device. The present chapter is devoted to single-gate SOI MOSFETs. Other types of SOI MOSFETs devices, such as multiple-gate SOI MOSFETs, dynamic-threshold MOSFETs, power MOS devices, etc., will be described in Chapter 6. SOI MOSFETs exhibit interesting properties that make them particularly attractive for applications such as radiation-hard circuits and high-temperature electronics. The properties of the SOI MOSFET operating in a harsh environment will be described in Chapter 7.

5.1 CAPACITANCES

Contrary to bulk CMOS where devices are isolated by reverse-biased junctions, CMOS SOI devices are dielectrically isolated from one another. This rules out latch-up between devices, as indicated in Chapter 1. Similarly, there is no leakage path between devices, while field transistor action may occur in bulk technologies. Full dielectric isolation can be interesting for monolithic integration of both high-voltage devices and low-voltage CMOS on a single chip.

5.1.1 Source and drain capacitance

In bulk MOS devices, the parasitic drain (or source)-to-substrate (or well) capacitance consists of two components: the capacitance between the drain and the substrate, and the capacitance between the drain and the channel-stop implant under the field oxide (Figure 5-1.A). As devices shrink to smaller geometries, higher substrate doping concentrations are used, and the junction capacitance increases. In SOI devices with reach-through junctions, the junction capacitance has only one component: the capacitance of the MOS structure that consists of the comprising the junction (gate electrode of the MOS structure under consideration), the buried oxide (gate oxide of the MOS structure), and the underlying silicon substrate (substrate of the MOS structure). This capacitance is always smaller than the capacitance of the buried oxide (Figure 5-1.B), and is typically lower than the junction capacitance of a bulk MOSFET. This reduction of parasitic capacitances contributes to the excellent speed performances observed in CMOS/SOI circuits. In addition, the buried oxide thickness does not necessarily have to be scaled down as device dimensions are reduced. This reinforces the capacitance advantage of SOI over bulk as technologies evolve toward deep submicron dimensions.

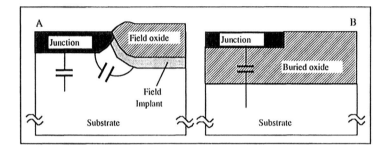

Figure 5-1. Junction capacitances. A: Capacitance between a junction and the substrate and between the junction and the field (channel-stop) implant in a bulk device. B: Capacitance between a junction and the substrate, across the buried oxide, in an SOI device.

The presence of a buried oxide underneath the devices reduces not only the junction capacitances, but other capacitances as well. Table 5-1. presents typical capacitances of a one-micrometer CMOS process in bulk and SOI.[1] The capacitances are given in fF/μm^2.

Table 5-1. Capacitances (fF/μm2) found in typical bulk and SOI 1-μm CMOS processes.

Type of capacitance	SOI (fF/μm^2)	Bulk (fF/μm^2)	Bulk to SOI capacitance ratio
Gate capacitance	1.3	1.3	1
Source (drain) to substrate	0.057	0.2 ... 0.35	4 ... 7
Polysilicon to substrate	0.04	0.1	2.5
Metal 1 to substrate	0.027	0.05	1.85
Metal 2 to substrate	0.018	0.021	1.16

The reduction of capacitance is, of course, most noticeable between the junctions and the substrate, but one can also observe that even the Metal 1-to-substrate capacitance can be reduced 40% by using SOI substrates rather than bulk silicon wafers.

The source or drain capacitance can be calculated as a function of the supply voltage. In the case of a bulk device, it is given by the classical expression for the capacitance of a PN junction:

$$C = \sqrt{\frac{q\varepsilon_s}{2}\frac{N_a N_d}{(N_a + N_d)}}\frac{1}{\sqrt{\Phi_o - V_D}} \qquad (5.1.1)$$

where V_D is the voltage across the junction (*i.e.*, the drain-substrate voltage), Φ_o is the built-in junction potential, and N_d and N_a are the doping concentrations in the N- and P-type regions, respectively. In the case of an SOI MOSFET, the drain capacitance is easily obtained from the MOS capacitor theory:

$$C = \frac{C_{BOX}}{\sqrt{1 + \dfrac{2C_{BOX}^2 V_D}{q N_a \varepsilon_{si}}}} \qquad (5.1.2)$$

where C_{BOX} is the capacitance of the buried oxide, V_d is the drain-substrate voltage, and N_a is the substrate doping concentration. Typical values of capacitances are presented in Figure 5-2.

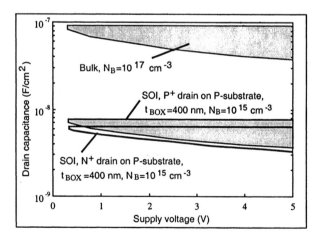

Figure 5-2. Parasitic junction capacitance per unit area as a function of supply voltage in bulk CMOS with constant substrate doping concentration (10^{17} cm^{-3}) and in standard SIMOX with a substrate doping concentration of 10^{15} cm^{-3}.[2]

5.1.2 Gate capacitance

Anomalous intrinsic gate capacitance characteristics (when compared to bulk) may be obtained, in some modes of operation of SOI MOSFETs. These characteristics result from the possibility of multiple conduction paths (*e.g.*: a back channel), source (or drain) coupling with the gate through the floating body, combinations of front and back oxide capacitances, as well as floating substrate effects and impact-ionization related phenomena [3,4]:

- The subthreshold front gate capacitance for low V_D values and for inverted back interface exceeds the conventional (bulk) capacitance value;
- The saturation intrinsic source-to-gate capacitance is equal to 0.72, instead of 0.66 times the total gate oxide capacitance (WLC_{ox1}) when the device is operated in partial depletion, as a result of a unique capacitive coupling through the source junction, floating body, and pinch-off depleted region;
- The activation of the parasitic bipolar transistor induces a decrease in the drain-to-gate capacitance, reducing it down to unusual negative values for very large V_D values;
- Impact ionization causes a steady increase in the source-to-gate capacitance with drain voltage in partial as well as in full depletion operation.

In accumulation-mode devices, the gate capacitance exhibits the following features [5]:

- A two-step dependence of the gate capacitance on the gate voltage that tends to disappear as a positive back-gate bias is applied. This effect is due to the presence of "body" conduction within the device;
- A larger (than in bulk) gate capacitance below threshold;
- A lower (than in bulk) saturation value for C_{GS} in moderate accumulation and below threshold;
- A kink in the $C_{GD}(V_{G1})$ curves in the transition region between the triode and the saturation regimes of operation.

5.2   FULLY AND PARTIALLY DEPLETED DEVICES

The physics of SOI MOSFETs is highly dependent on the thickness and the doping concentration of the silicon film in which they are made. Two types of devices can be distinguished: devices in which the silicon film in the channel region is never completely depleted ("*partially depleted device*" or "*PD device*"), and devices where the silicon film can be completely depleted ("*fully depleted device*" or "*FD device*"). Note that some authors call PD devices "*non-fully depleted devices*". Figure 5-3 shows the cross section of an SOI MOSFET and Figure 5-4 presents the energy band diagrams of bulk, partially depleted, and fully depleted n-channel SOI device at threshold.

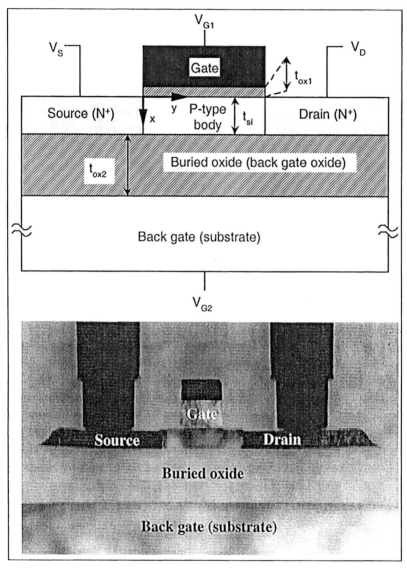

Figure 5-3. Top: schematic cross section of an n-channel SOI MOSFET illustrating some of the notations used in this chapter. Bottom: SEM cross-section of an actual SOI MOSFET (Courtesy G.K. Celler and S. Cristoloveanu).[6]

•   In a *bulk* device (Figure 5-4.A), the depletion zone extends from the Si-SiO$_2$ interface to the maximum depletion width, x_{dmax}, which is classically given by: $x_{dmax} = \sqrt{\dfrac{4\varepsilon_{si}.\Phi_F}{q\,N_a}}$, where Φ_F is the Fermi potential, which is equal to $\Phi_F = \dfrac{kT}{q}\,ln\left(\dfrac{N_a}{n_i}\right)$.

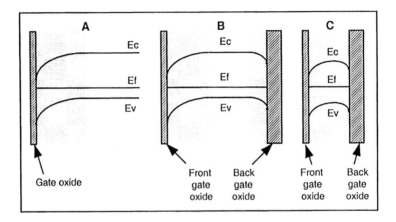

Figure 5-4. Energy band diagrams in bulk (A), partially depleted SOI (B), and fully depleted SOI (C). All devices are represented at threshold (front gate voltage = threshold voltage). The shaded areas represent the depleted zones. SOI devices are represented for a condition of weak inversion (below threshold) at the back interface.

•   In a *partially depleted* SOI (PDSOI) device (Figure 5-4.B), the silicon film thickness, t_{si}, is larger than twice the value of x_{dmax}. In such a case, there is no interaction between the depletion zones arising from the front and the back interfaces. A neutral region exixts beneath the depletion regions. If this neutral piece of silicon, called the "body", is connected to ground by a "body contact", the characteristics of the device will exactly be those of a bulk device. If, however, the body is left electrically floating, the device will present some effects called the "floating-body effects", such as the "kink effect" (see Section 5.9.1). In addition, the presence of a parasitic open-base NPN bipolar transistor between source and drain influences the device properties (see Section 5.10).

•   In a *fully depleted* SOI (FDSOI) device (Figure 5-4.C), the silicon film thickness, t_{si}, is smaller than x_{dmax}. In this case, the silicon film is fully depleted at threshold, irrespective of the bias applied to the back gate (with the exception of the possible presence of thin accumulation or inversion layers at the back interface, if a large negative or positive bias is applied to the back gate, respectively). Fully depleted SOI devices are virtually free of kink effect, if their back interface is not in accumulation. Among all types of SOI devices, fully depleted devices with depleted back interface exhibit the most attractive properties, such as low electric fields, high transconductance, excellent short-channel behavior, and quasi-ideal subthreshold slope characteristics.

• *Near-fully depleted* SOI (NFDSOI) devices are an intermediate case between fully depleted and partially depleted devices, and are obtained in those cases where $x_{dmax} < t_{si} < 2x_{dmax}$. If the back-gate bias is such that the front and back depletion zones do not touch each other, or if the back interface is neutral or accumulated, the transistor will behave as a partially depleted device. If, on the other hand, the presence of a back-gate bias induces an overlap between the front and back depletion zones, the device will be fully depleted.

Because the front and back interface can either be in accumulation, depletion or inversion, there are nine possible modes of operation in a fully depleted SOI transistor, as a function of V_{G1} and V_{G2} (Figure 5-5).[7] Most of these operation modes are not of practical use, however. Usual operating gate voltages are indicated by the shaded area of Figure 5-5.

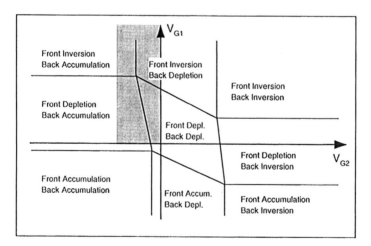

Figure 5-5. Different operation modes of a fully depleted, n-channel SOI MOS transistor as a function of front-gate bias (V_{G1}) and back-gate bias (V_{G2}) - (linear regime, low drain voltage). The shaded area represents the usual operating conditions.

To be more general, it should be mentioned that the presence of accumulation, depletion or inversion layers is also a function of the drain voltage, and that the back interface can, for instance, be accumulated near the source and depleted near the drain. Such a mode of operation will be analyzed later. It is also worthwhile noting that most of the "attractive properties" of fully depleted devices, with depleted back interface (high transconductance, etc.), can be disabled if an accumulation layer is present at the back interface.

The previous remarks are valid for enhancement-mode MOSFETs, which are the most common SOI devices (at least in the n-channel case). However, another type of devices can be realized: the accumulation-mode (or "deep-depletion") MOSFET. The operation mode of a MOSFET (enhancement or accumulation) depends on fabrication parameters, the most important being the type of gate material used (Table 5-2).[8] We will, however, consider that SOI devices are enhancement-mode devices, unless otherwise specified.

Table 5-2. Different operation modes of an SOI MOSFET as a function of gate material.

	N$^+$ poly gate	P$^+$ poly gate	Midgap gate material
N-channel device	Inversion-mode	Accumulation-mode	Either mode
P-channel device	Accumulation-mode	Inversion-mode	Either mode

The merits of the different types of SOI MOSFETs are reported in Table 5-3, where some electrical properties of the devices are compared. Bulk silicon devices are taken as a reference.

Table 5-3. Comparison of some of the electrical properties of partially depleted and fully depleted SOI devices. The bulk device is given as a reference. 0, + and - mean "similar to bulk", "better than bulk" and "worse than bulk", respectively.

	PDSOI	FDSOI
Mobility	0	0/+
Body effect	0	+
Short-channel effects	0	-/0/+
S&D capacitance	+	+
Hot-carrier degradation	0	+
Subthreshold slope	0	+
V$_{TH}$ sensitivity on t$_{si}$	0	-
Kink effect	-/+	0
Parasitic bipolar effect	0/-	-
Total-dose hardness	0/+	-
SEU/soft error hardness	+	+
Gamma-dot hardness	+	+
High-temperature operation	0/+	+

One can see that fully depleted devices offer the most attractive properties for ULSI applications. The popularity of partially depleted devices is due to the independence of their threshold voltage on silicon film thickness and on charges in the buried oxide. The kink effect is a drawback of PDSOI, but it does increase the current drive of the devices, which can be taken advantage of to increase circuit speed. As a result the kink effect in PDSOI devices receives both "+" and "-" mark in Table 5-3. Similarly, FDSOI can have better of worse short-channel effects than bulk silicon, depending on the SOI film thickness (the thinner the film, the smaller the short-channel effects). Short-channel effects are avoided if the SOI film thickness is less than a quarter of the channel length (see section 6.1.2.2).[9,10]

5.3 THRESHOLD VOLTAGE

- The threshold voltage of an enhancement-mode *bulk* n-channel MOSFET is classically given by:[11]

$$V_{TH} = V_{FB} + 2\,\Phi_F + \frac{q\,N_a\,x_{dmax}}{C_{ox}} \tag{5.3.1}$$

where V_{FB} is the flatband voltage, equal to $\Phi_{MS} - \dfrac{Q_{ox}}{C_{ox}}$, Φ_F is the Fermi potential, equal to $\dfrac{kT}{q}\,ln\left(\dfrac{N_a}{n_i}\right)$, and x_{dmax} is the maximum depletion width, which is equal to $\sqrt{\dfrac{4\varepsilon_{si}\,\Phi_F}{q\,N_a}}$.

- In a *partially depleted* SOI device where, there can be no interaction between the front and back depletion zones because $t_{si} > 2\,x_{dmax}$. In that case, the threshold voltage is the same as in a bulk transistor and is given by Equation (5.3.1).

- The threshold voltage of a *fully depleted*, enhancement-mode, n-channel SOI device can be obtained by solving the Poisson equation, using the depletion approximation: $\dfrac{d^2\Phi}{dx^2} = \dfrac{q\,N_a}{\varepsilon_{si}}$, which when integrated twice yields the potential as a function of depth in the silicon film, x (Lim and Fossum model) [12]:

$$\Phi(x) = \frac{q\,N_a}{2\,\varepsilon_{si}}\,x^2 + \left(\frac{\Phi_{s2}-\Phi_{s1}}{t_{si}} - \frac{q\,N_a\,t_{si}}{2\,\varepsilon_{si}}\right)x + \Phi_{s1} \tag{5.3.2}$$

where Φ_{s1} and Φ_{s2} is the potential at the front and back silicon/oxide interface, respectively (Figure 5-4). The doping concentration, N_a, is assumed to be uniform.

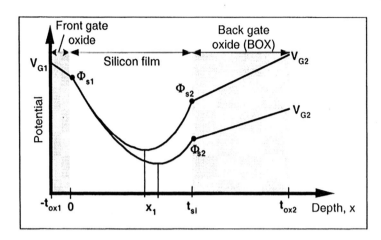

Figure 5-4. Potential in the silicon film and front and back oxides at $V_{G1} = V_{TH1}$ and for two back-gate bias conditions. x_1 is the point of minimum potential. The depletion zone between $x=0$ and $x=x_1$ is controlled by the front gate, and the depletion zone between $x=x_1$ and the back Si-SiO$_2$ interface is controlled by the back gate.

The electric field in the silicon film is given by:

$$E(x) = \frac{-q\,N_a}{\varepsilon_{si}}\,x - \left(\frac{\Phi_{s2} - \Phi_{s1}}{t_{si}} - \frac{q\,N_a\,t_{si}}{2\,\varepsilon_{si}}\right) \qquad (5.3.3)$$

The front surface electric field, E_{s1} (at $x=0$), can be calculated from 5.3.3 and is given by:

$$E_{s1} = \left(\frac{\Phi_{s1} - \Phi_{s2}}{t_{si}} + \frac{q\,N_a\,t_{si}}{2\,\varepsilon_{si}}\right) \qquad (5.3.4)$$

Applying the Gauss theorem at the front interface, one obtains the potential drop across the gate oxide, Φ_{ox1}:

$$\Phi_{ox1} = \frac{\varepsilon_{si}\,E_{s1} - Q_{ox1} - Q_{inv1}}{C_{ox1}} \qquad (5.3.5)$$

where Q_{ox1} is the fixed charge density at the front Si-SiO$_2$ interface, Q_{inv1} is the front channel inversion charge ($Q_{inv1} < 0$), and C_{ox1} is the front gate oxide capacitance. Similarly, applying the Gauss theorem at the back interface and using 5.3.4 yields the potential drop across the buried oxide, Φ_{ox2}:

$$\Phi_{ox2} = - \frac{\varepsilon_{si} E_{s1} - q N_a t_{si} + Q_{ox2} + Q_{s2}}{C_{ox2}} \qquad (5.3.6)$$

where Q_{s2} is the charge in a possible back inversion ($Q_{s2} < 0$) or accumulation ($Q_{s2} > 0$) layer.

The front and back gate voltages, V_{G1} and V_{G2}, are given by:

$$V_{G1} = \Phi_{s1} + \Phi_{ox1} + \Phi_{MS1} \quad \text{and} \quad V_{G2} = \Phi_{s2} + \Phi_{ox2} + \Phi_{MS2} \qquad (5.3.7)$$

where Φ_{MS1} and Φ_{MS2} are the front and back work function differences, respectively.

Combining 5.3.4, 5.3.5 and 5.3.7, we obtain a relationship between the front gate voltage and the surface potentials:

$$V_{G1} = \Phi_{MS1} - \frac{Q_{ox1}}{C_{ox1}} + \left(1 + \frac{C_{si}}{C_{ox1}}\right) \Phi_{s1} - \frac{C_{si}}{C_{ox1}} \Phi_{s2} - \frac{\frac{1}{2} Q_{depl} + Q_{inv1}}{C_{ox1}} \qquad (5.3.8)$$

where $C_{si} = \varepsilon_{si}/t_{si}$ and $Q_{depl} = - q N_a t_{si}$ is the total depletion charge in the silicon film.

One finds a similar relationship between the back gate voltage and the surface potentials:

$$V_{G2} = \Phi_{MS2} - \frac{Q_{ox2}}{C_{ox2}} - \frac{C_{si}}{C_{ox2}} \Phi_{s1} + \left(1 + \frac{C_{si}}{C_{ox2}}\right) \Phi_{s2} - \frac{\frac{1}{2} Q_{depl} + Q_{s2}}{C_{ox2}} \qquad (5.3.9)$$

Equations 5.3.8 and 5.3.9 are the key relationships describing the charge coupling between the front and back gates in a fully depleted SOI MOSFET. Combining them yields the dependence of the (front) threshold voltage on back-gate bias and device parameters.

We will now detail the expression of the threshold voltage of the fully depleted SOI MOSFET as a function of the different possible steady-state charge conditions at the back interface.

If the back surface is *accumulated*, Φ_{s2} is pinned to approximately 0V, and the front-side threshold voltage *with backside accumulation*, $V_{TH1,acc2}$, can be derived from equation 5.3.8 since $V_{TH1,acc2}$ is equal to V_{G1} when $\Phi_{s2} = 0$, $Q_{inv1}=0$, and $\Phi_{s1} = 2\Phi_F$. The result is:

$$V_{TH1,acc2} = \Phi_{MS1} - \frac{Q_{ox1}}{C_{ox1}} + \left(1 + \frac{C_{si}}{C_{ox1}}\right) 2\Phi_F - \frac{Q_{depl}}{2C_{ox1}} \qquad (5.3.10)$$

If the back surface is *inverted*, Φ_{s2} is pinned to approximately $2\Phi_F$. The front-side threshold voltage *with backside inversion*, $V_{TH1,inv2}$ can be derived from equation 5.3.8 by writing that $V_{TH1,inv2}$ is equal to V_{G1} when $\Phi_{s2} = 2\Phi_F$, $Q_{inv1}=0$, and $\Phi_{s1} = 2\Phi_F$. The result is:

$$V_{TH1,inv2} = \Phi_{MS1} - \frac{Q_{ox1}}{C_{ox1}} + 2\Phi_F - \frac{Q_{depl}}{2C_{ox1}} \qquad (5.3.11)$$

For this case, the device is turned ON even if $V_{G1} < V_{TH1,inv2}$ since there is an inversion channel connecting source to drain at the bottom of the silicon film. Such a device would be useless for any practical circuit application.

If the back surface is depleted, Φ_{s2} depends on the back-gate voltage, V_{G2}, and its value can range between 0 and $2\Phi_F$. The value of back-gate voltage for which the back interface reaches accumulation (the front interface being at threshold), $V_{G2,acc}$, is given by equation 5.3.9 where $\Phi_{s1} = 2\Phi_F$, $\Phi_{s2} = 0$, and $Q_{s2}=0$. Similarly, the value of back-gate voltage for which the back interface reaches inversion, $V_{G2,inv}$, is given by the same equation where $\Phi_{s1} = 2\Phi_F$, $\Phi_{s2} = 2\Phi_F$, and $Q_{s2}=0$. When $V_{G2,acc} < V_{G2} < V_{G2,inv}$ the front threshold voltage is obtained by combining equations 5.3.8 and 5.3.9 with $\Phi_{s1} = 2\Phi_F$ and $Q_{inv1} = Q_{s2} =0$. The result is:

$$V_{TH1,depl2} = V_{TH1,acc2} - \frac{C_{si}\, C_{ox2}}{C_{ox1}\,(C_{si} + C_{ox2})}\,(V_{G2} - V_{G2,acc}) \qquad (5.3.12)$$

The above relationships are valid if the thickness of the inversion or accumulation layers are small with respect to the silicon film thickness. This may no longer be the case in ultra-thin-film devices, in which case the width of the accumulation/inversion zones must be subtracted from the silicon film thickness to obtain an effective silicon thickness. The effective thickness is equal to the effective width of the depleted layer and it replaces used in t_{si} in the above relationships. In very thin films (t_{si} < 10 nm) complex interaction can take place between the front and back interfaces. This interaction has been reported to give rise to a decrease of carrier mobility and is analyzed in more detail in section 5.8.

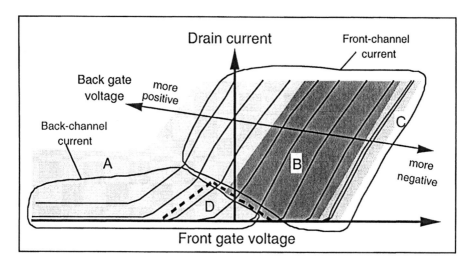

Figure 5-5. Linear $I_D(V_{G1})$ characteristics of a fully depleted SOI n-channel MOSFET for different V_{G2} values. Different back interface conditions are outlined by the shaded areas: inversion (A), depletion (B), accumulation (C), and back depletion/inversion depending on the front gate voltage (D).

The influence of the back-gate bias on the $I_D(V_{GS})$ characteristics of a fully depleted SOI n-channel MOSFET with low V_{DS} is shown in Figure 5-5. In the left portion of the graph (region A), the back interface is inverted. As a consequence, current flows in the device even if the front-gate voltage is negative. Φ_{s2} is pinned to $2\Phi_F$, and the front threshold voltage is fixed at a constant value. The apparent shift of the curves to the left when the back-gate bias is increased is actually a shift of the curves upwards due to the increase of the back channel current. In the right portion of the graph (region C), the back interface is accumulated, the front threshold voltage is constant, and no shift of the curves to the right can be obtained by further negative increase of the back bias. In region B, the back interface is depleted, and the front threshold voltage depends linearly on the back bias. The anomalous slope found in region D is explained by the following effect. As the back and front interfaces are depleted, and as the back interface is close to inversion, any increase of V_{G1} will push the point of minimum potential deeper into the silicon film, and therefore reduce the back threshold voltage. This can lead to creation of an inversion channel at the back interface. Thus, one observes a situation where a variation of the bias in the front gate creates and modulates an inversion channel at the back interface [13], which nicely illustrates the importance of the interaction between front and back gate in thin SOI devices.

5.3.1 Body effect

•   In a *bulk* device, the body effect is defined as the dependence of the threshold voltage on the substrate bias. In an SOI transistor, it is similarly defined as the dependence of the threshold voltage on the back-gate bias. In a *bulk* n-channel transistor, the threshold voltage can be written as [14]:

$$V_{TH} = \Phi_{MS} + 2\Phi_F - \frac{Q_{ox}}{C_{ox}} + \frac{Q_b}{C_{ox}} \quad \text{with} \quad Q_b = \sqrt{2\varepsilon_{si} \, q \, N_a \, (2\Phi_F - V_B)}$$

which can be rewritten:

$$V_{TH} = \Phi_{MS} + 2\Phi_F - \frac{Q_{ox}}{C_{ox}} + \frac{\sqrt{2\varepsilon_{si} \, q \, N_a \, (2\Phi_F - V_B)}}{C_{ox}} \qquad (5.3.13)$$

where the source is taken as voltage reference, and V_B is the substrate bias.

If we define

$$\gamma = \frac{\sqrt{2\varepsilon_{si} \, q \, N_a}}{C_{ox}} \qquad (5.3.14)$$

we obtain:

$$V_{TH} = \Phi_{MS} + 2\Phi_F - \frac{Q_{ox}}{C_{ox}} + \gamma\sqrt{2\Phi_F} + \gamma\left(\sqrt{2\Phi_F - V_B} - \sqrt{2\Phi_F}\right)$$

The last term of the equation describes the dependence of threshold voltage on substrate bias. When a negative bias is applied to the substrate (with respect to the source), the threshold voltage increases as a square-root function of the substrate bias. If the threshold voltage with zero substrate bias is referred to as V_{THo}, one can write:

$$V_{TH}(V_B) = V_{THo} + \gamma\left(\sqrt{2\Phi_F - V_B} - \sqrt{2\Phi_F}\right) \qquad (5.3.15)$$

The variation of threshold voltage with substrate bias is shown in Figure 5-6. If the substrate is grounded and taken as voltage reference, and if the source voltage, V_S, is different from zero, 5.3.15 can be rewritten as follows:

$$\begin{aligned} V_{TH}(V_S) &= V_S + V_{FB} + 2\Phi_F + \gamma\sqrt{2\Phi_F + V_S} \\ &= V_S + V_{FB} + 2\Phi_F + \gamma\sqrt{2\Phi_F + V_S} + \sqrt{2\Phi_F} - \sqrt{2\Phi_F} \\ &= V_{THo} + V_S + \gamma\left(\sqrt{2\Phi_F + V_S} - \sqrt{2\Phi_F}\right) \quad (5.3.16) \end{aligned}$$

Equation 5.3.16 is often linearized to simplify device modeling, in which case one can write: [15]

$$V_{TH} = V_{THo} + (1+\alpha)V_S = V_{THo} + nV_S \qquad (5.3.17)$$

Where n is called the "body-effect coefficient" or the "body factor" of the device.

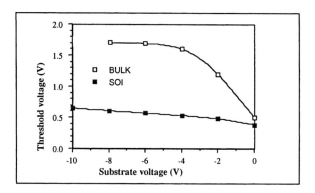

Figure 5-6. Dependence of threshold voltage on substrate bias in bulk and fully depleted SOI MOSFETs.[16]

• In a *partially depleted* device with a body contact the body effect is identical to that in a bulk MOSFET. If the body is floating the body potential, V_B, is determined by capacitive effects and by currents in the PN junctions. As a result the threshold voltage varies as a function of these parameters. This phenomenon gives rise, among others, to the kink effect. In a *partially depleted* device the threshold voltage does not vary with back-gate bias because there is no coupling between front and back gate.

• In a *fully depleted* SOI device, the variation of threshold voltage can be obtained from Equations 5.3.10 – 5.3.12:

$$\frac{dV_{TH1,acc2}}{dV_{G2}} = 0 \qquad (5.3.18)$$

$$\frac{dV_{TH1,inv2}}{dV_{G2}} = 0 \qquad (5.3.19)$$

$$\frac{dV_{TH1,depl2}}{dV_{G2}} = -\frac{C_{si}\,C_{ox2}}{C_{ox1}\,(C_{si} + C_{ox2})} = \frac{-\varepsilon_{si}\,C_{ox2}}{C_{ox1}\,(t_{si}\,C_{ox2} + \varepsilon_{si})} \equiv -\alpha \qquad (5.3.20)$$

The symbol α is chosen by analogy with the case of a bulk device. It should be noted that in a bulk device, α is the result of a linearization of a more complex expression 5.3.16, while in a fully depleted device, α is the exact expression of the variation of threshold voltage with back-gate bias. In most cases, the following approximation can be made: $\alpha \cong -\dfrac{t_{ox1}}{t_{ox2}}$.

Expression 5.3.20 is valid only when the film is fully depleted. In a first order approximation, one can consider that Φ_{S2} is pinned at $2\Phi_F$ when the back interface is inverted, and that further increase of the back-gate bias will no longer modify the front threshold voltage. Similarly, when a large negative back-gate bias is applied, the back interface is accumulated, Φ_{S2} is pinned at 0 V, and further negative increase of the back-gate bias will not modify the front threshold voltage. Variation of the front gate threshold voltage with back-gate bias, based on these assumptions, is represented in Figure 5-7. In an actual device, however, the back surface potential can exceed $2\Phi_F$ (back inversion) or become smaller than 0V (back accumulation), these excursions being limited to a few $\dfrac{kT}{q}$. As a result, the front threshold voltage slightly increases (decreases) when the back-gate voltage is increased beyond back-gate accumulation (inversion).

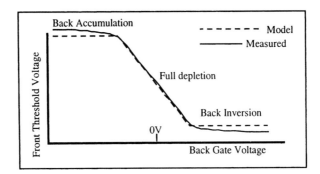

Figure 5-7. Variation of the front threshold voltage with back-gate bias in a fully depleted SOI MOSFET.

Relationship 5.3.20 is independent of the doping concentration, N_a. If C_{ox1} and C_{ox2} are known, 5.3.20 can be used to determine the thickness of the silicon film (see section 2.3).[17] The dependence of the front-gate threshold voltage on back-gate bias decreases with increasing t_{ox2}. When t_{ox2} is very thick ($C_{ox2} \cong 0$), the front threshold voltage is virtually independent of the back-gate bias. In real SOI devices, the back-gate material is not a metal, as considered in this first-order model, but a silicon substrate, the surface of which can become inverted, depleted or accumulated as a function of the back-gate bias conditions. This variation of substrate surface potential has some influence on the device threshold voltage, but this influence is small, and can be neglected as long as the thickness of the buried oxide layer is large compared to that of the front gate oxide.[18,19]

From Figure 5-6, it is clear that the back-gate effect of SOI devices is much smaller than the body effect of bulk MOSFETs. For instance, when a back (substrate) bias of –8V is applied to the bulk device, a 1.2V increase of threshold voltage is observed. Under the same bias conditions, the threshold voltage of the SOI device only increases by 0.2V.

The reduced body effect is an important feature of SOI devices. A large body effect reduces the current drive capability of transistors whose source is not directly connected to ground, such as transfer gates, nMOS load devices, and differential input pairs. Higher performance can, therefore, be expected from SOI gates than from their bulk counterparts.

5.3.2 Short-channel effects

There are numerous effects caused by the reduction of channel length in MOSFETs.[20] In this Section we will specifically deal with the so-called "short-channel effect" which results in a roll-off of the threshold voltage. It is due to the loss of control by the gate of a part of the depletion zone below it. In other words, the depletion charge controlled by the gate is no longer equal to $Q_{depl} = q\,N_a\,x_{dmax}$ (bulk MOSFET case), but to a fraction of it, which we will call Q_{d1}. This reduction of the depletion charge, due to the encroachment from the source and drain, becomes significant in short-channel devices, and brings about a lowering of threshold voltage obtained by substituting Q_{d1} for $Q_{depl=qN_ax_{dmax}}$ in equation 5.3.1. In a bulk MOSFET, Q_{d1} can be represented by the area of a trapezoid, as shown in Figure 5-8. In a long-channel device, the lengths of the upper and the lower base of the trapezoid are almost equal to L, the channel length. In a short-channel device, the upper base length is still equal to L, but the lower base is significantly shorter (it can even disappear, as in Figure 5-8). In a bulk device the value of Q_{d1} can be approximated by: [21]

$$Q_{d1} = Q_{depl}\left(1 - \frac{r_j}{L}\left(\sqrt{1 + \frac{2x_{dmax}}{r_j}} - 1\right)\right) \qquad (5.3.21)$$

where r_j is the source and drain junction depth.

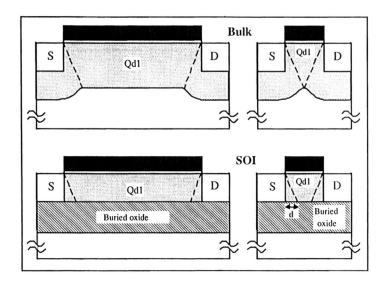

Figure 5-8. Distribution of depletion charges in long-channel (left) and short-channel (right) bulk and fully depleted SOI MOSFETs. Q_{d1} is the depletion charge controlled by the gate.

In a fully depleted SOI device, the depletion charge controlled by the gate is given by: $Q_{d1} = Q_{depl} \left(1 - \dfrac{d}{L} \right)$ where d is a distance defined by Figure 5-8 and $Q_{depl} = qN_a t_{si}$. Calculating the value of d is rather complex and necessitates iterative calculation of the back-surface potential. A description of the calculation method is given in [22,23,24].

Figure 5-9 shows the threshold voltage roll-off in bulk and fully depleted MOSFETs. It starts to occur at significantly smaller gate lengths in thin SOI transistors than in bulk devices (this can also be seen in Figure 5-8: Q_{d1} retains a reasonable trapezoid shape in the short-channel SOI device, while it is a triangular shape ($Q_{d1} \cong \dfrac{Q_{depl}}{2}$) in the bulk case).

The short-channel effect is reduced further in fully depleted devices with accumulation at the backside over fully depleted devices with depleted back interface. However, both devices exhibit better short-channel properties than bulk devices.[25,26] The short-channel effect can also be minimized by the use of ultra-thin SOI films: a threshold voltage roll-off of only 70 mV has been observed in devices with L = 40nm (as compared to long-channel devices) when a silicon film thickness of 4nm is used.[27] It seems that an optimum control of the space charge in the silicon film by the gate (which would further minimize the short-channel effect) could be obtained by using double-gate devices (one gate below the active silicon film and one above it).[28] Making such devices is a real technological challenge, but some practical solutions have already been proposed.[29] Detailed modeling of the threshold voltage in short-channel SOI MOSFETs can be found in the literature.[30, 31, 32]

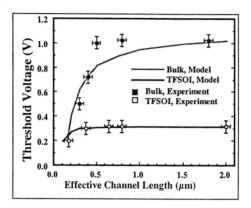

Figure 5-9. Threshold voltage as a function of gate length in bulk and fully depleted SOI n-channel MOSFETs. [33]

Another short-channel effect, called drain-induced barrier lowering (DIBL), is also due to charge sharing between the gate and the junctions.[34] DIBL is caused by the reduction of the depletion charge controlled by the gate due to size increase of the drain junction-related depletion zone, which increases with V_{DS}. In order to model the DIBL effect in a MOSFET, an analysis involving the solution of the two-dimensional Poisson equation must be carried out. It can be shown that DIBL is lower in fully depleted SOI devices than in bulk devices.[35] In SOI devices short-channel effects can be further minimized by heavily doping the top of the substrate under the BOX to form ground-plane electrode.[36] Such an electrode, however, does increase the source and drain capacitance to the substrate and may degrade cross-talk characteristics. It is possible to realize a localized ground-plane electrode that is self-aligned to the gate and does not overlap with source and drain.[37]

5.4 CURRENT-VOLTAGE CHARACTERISTICS

- Models for the current-voltage characteristics of the *bulk* MOSFET can be found in many textbooks. When the transistor is non-saturated the expression of the current is [38]:

$$I_D = \mu_{eff} C_{ox} \frac{W}{L}\left[\left(V_G - V_{FB} - 2\Phi_F - V_S - \frac{V_{DS}}{2}\right)V_{DS} - \frac{2}{3}\gamma\left((2\Phi_F + V_D)^{\frac{3}{2}} - (2\Phi_F + V_S)^{\frac{3}{2}}\right)\right]$$

$$(5.4.1)$$

where V_S, V_G, V_D, V_{FB}, Φ_F are the source voltage, gate voltage, drain voltage, flat-band voltage, and Fermi potential, respectively. V_{DS} is equal to V_D-V_S, and γ is defined by relationship 5.3.14. Most of the time, expression 5.4.1 is linearized, which yields a much more familiar expression for nonsaturation:

$$I_D = \mu_{eff} C_{ox} \frac{W}{L}\left[(V_G - V_{THS})V_{DS} - \frac{1}{2}nV_{DS}^2\right] \qquad (5.4.2)$$

and

$$I_{Dsat} = \frac{1}{2n}\mu_{eff} C_{ox} \frac{W}{L}(V_G - V_{THS})^2 \qquad (5.4.3)$$

in saturation. The drain saturation voltage is given by

$$V_{DSAT} = \frac{V_G - V_{THS}}{n} + V_S \qquad (5.4.4)$$

where

$$V_{THS} = V_{THo} + nV_S. \qquad (5.4.5)$$

The body factor, n, is equal to $1 + \dfrac{C_D}{C_{ox}}$, where $C_D = \varepsilon_{si}/x_{d\,max}$ is the depletion capacitance. Comparing 5.3.17 with 5.4.5 one finds:

$$n = 1 + \alpha = 1 + \frac{\varepsilon_{si}/x_{dmax}}{C_{ox}} \qquad (5.4.5)$$

- The current-voltage characteristics of a *partially depleted* SOI MOSFET with grounded body are identical to those of a bulk transistor. If the body is not grounded a series of effects, called "floating-body effects" appear. These will be described in section 5.10.

- The current-voltage characteristics of a *fully depleted* SOI MOSFET are described by the Lim & Fossum model.

5.4.1 Lim & Fossum model

The current characteristics of a fully depleted SOI device were derived using the classical gradual-channel approximation by H.K. Lim and J.G. Fossum in 1984.[39] The model assumes constant mobility as a function of y, uniform doping of the silicon film in the channel region, and negligible diffusion currents. It considers the case of an n-channel device. More elaborate models including short-channel effects can be found in later references.[40,41,42,43]

Using Ohm's law in an elemental section of the inversion channel, we can write:

$$I_D = - W \, \mu_n \, Q_{inv1}(y) \frac{d\Phi_{s1}(y)}{dy} \qquad (5.4.6)$$

where W is the width of the transistor, and μ_n is the mobility of the electrons in the inversion layer. Integration from source ($y=0$) to drain ($y=L$) yields:

$$I_D = - \frac{W}{L} \, \mu_n \int\limits_{2\Phi_F}^{2\Phi_F + V_{DS}} Q_{inv1}(y) \, d\Phi_{s1}(y) \qquad (5.4.7)$$

Assuming full depletion in the silicon film and assuming that the thickness of the inversion and accumulation layers at the interfaces is equal to zero, one obtains the inversion charge density in the front channel, $Q_{inv1}(y)$, from equation 5.3.8:

$$-Q_{inv1}(y) = C_{ox1}\left(V_{G1} - \Phi_{MS1} + \frac{Q_{ox1}}{C_{ox1}} - \left(1 + \frac{C_{si}}{C_{ox1}}\right)\Phi_{s1}(y) + \frac{C_{si}}{C_{ox1}}\Phi_{s2}(y) + \frac{Q_{depl}}{2C_{ox1}}\right) \qquad (5.4.8)$$

where the back surface potential, $\Phi_{s2}(y)$, can be extracted from 5.3.9:

$$\Phi_{s2}(y) = \frac{C_{ox2}}{C_{ox2} + C_{si}}\left(V_{G2} - \Phi_{MS2} + \frac{Q_{ox2}}{C_{ox2}} + \frac{C_{si}}{C_{ox2}}\Phi_{s1}(y) + \frac{Q_{depl}}{2C_{ox2}} - \frac{Q_{s2}(y)}{C_{ox2}}\right) \qquad (5.4.9)$$

Different cases can now be distinguished for the <u>back interface</u>: the film can be fully depleted from source to drain (DS+DD = "<u>d</u>epleted <u>s</u>ource + depleted <u>d</u>rain"), accumulated from source to drain (AS+AD), accumulated near the source and depleted near the drain (AS+DD), inverted from source to drain (IS+ID), and inverted near the source and depleted near the drain (IS+DD). Because back-channel inversion is generally undesirable and avoided as an operating region, we will describe only the (DS+DD), (AS+AD) and (AS+DD) cases.

1.   The back interface is *accumulated* from source to drain (AS+AD) under the condition $V_{G2} < V_{G2,acc}(L)$

where

$$V_{G2,acc}(L) = V_{G2,acc} - \frac{C_{si}}{C_{ox2}} V_{DS} \qquad (5.4.10)$$

and

$$V_{G2,acc} = \Phi_{MS2} - \frac{Q_{ox2}}{C_{ox2}} - \frac{C_{si}}{C_{ox2}} 2\Phi_F - \frac{Q_{depl}}{2C_{ox2}} \qquad (5.4.11)$$

which is obtained from (5.3.9) assuming $\Phi_{s1} = 2\Phi_F$, $\Phi_{s2} = 0$, and $Q_{s2}=0$. Equation (5.3.18.a) is also obtained from (5.3.9) with $\Phi_{s1}(L) = 2\Phi_F + V_{DS}$. Note that $V_{G2,acc}(L) < V_{G2,acc}(0)$.

Combining 5.4.7 and 5.4.8 one obtains:

$$I_{D,acc2} = \frac{W}{L}\,\mu_n\,C_{ox1}\left((V_{G1}-V_{TH1,acc2})V_{DS}-\left(1+\frac{C_{si}}{C_{ox1}}\right)\frac{V_{DS}^2}{2}\right) \tag{5.4.12}$$

where $V_{TH1,acc2}$ is given by 5.3.10.

The drain saturation voltage is obtained from (5.4.12) under the condition $dI_D/dV_{DS} = 0$:

$$V_{Dsat,acc2} = \frac{V_{G1}-V_{TH1.acc2}}{1+\dfrac{C_{si}}{C_{ox1}}} \tag{5.4.13}$$

and the saturation current is given by:

$$I_{Dsat,acc2} = \frac{1}{2}\frac{W}{L}\frac{\mu_n\,C_{ox1}}{1+C_{si}/C_{ox1}}(V_{G1}-V_{TH1,acc2})^2 \tag{5.4.14}$$

2.   The back interface is *depleted* from source to drain (DS+DD) under the condition $V_{G2,acc} < V_{G2} < V_{G2,inv}$

where $V_{G2,inv} = \varPhi_{MS2} - \dfrac{Q_{ox2}}{C_{ox2}} + 2\varPhi_F - \dfrac{Q_{depl}}{2C_{ox2}}$ \hfill (5.4.15)

which is obtained from 5.3.9 assuming $\varPhi_{s1} = \varPhi_{s2} = 2\varPhi_F$ and $Q_{s2}=0$.

For a back interface depleted from source to drain we have $Q_{s2}(y)=0$. Using equations 5.3.15, 5.3.16 and 5.3.17, one obtains:

$$I_{D,depl2} = \frac{W}{L}\,\mu_n\,C_{ox1}\left((V_{G1}-V_{TH1,depl2})V_{DS}-\left(1+\frac{C_{si}C_{ox2}}{C_{ox1}(C_{si}+C_{ox2})}\right)\frac{V_{DS}^2}{2}\right) \tag{5.4.16}$$

where the front threshold voltage under depleted back interface conditions, $V_{TH1,depl2}$, is given by 5.3.12.

The drain saturation voltage is obtained from (5.4.16) under the condition $dI_D/dV_{DS} = 0$:

$$V_{Dsat,depl2} = \frac{V_{G1}-V_{TH1,depl2}}{1+\dfrac{C_{si}C_{ox2}}{C_{ox1}(C_{si}+C_{ox2})}} \tag{5.4.17}$$

and the saturation current is given by:

$$I_{Dsat,depl2} = \frac{1}{2}\frac{W}{L}\frac{\mu_n\,C_{ox1}}{1+\dfrac{C_{si}C_{ox2}}{C_{ox1}(C_{si}+C_{ox2})}}(V_{G1}-V_{TH1,depl2})^2 \tag{5.4.18}$$

3.   If the back surface is *accumulated near the source and depleted near the drain* (AS+DD), the accumulation layer occurs at the back interface from $y=0$ to $y=y_t$, where y_t is the point where $\Phi_{s2}(y_t)=0$ and $Q_{s2}(y_t)=0$.

The drain current is then given by the following expression:

$$I_{D,AS+DD}= -\frac{W}{L}\ \mu_n\left(\int_{2\Phi_F}^{\Phi_{s1}(y_t)} Q_{inv1}(y)\,d\Phi_{s1}(y) + \int_{\Phi_{s1}(y_t)}^{2\Phi_F+V_{DS}} Q_{inv1}(y)\,d\Phi_{s1}(y)\right) \qquad (5.4.20)$$

where $\Phi_{s1}(y_t)$ is given by combining 5.3.9, where $\Phi_{s2}(y_t)=0$ and $Q_{s2}(y_t)=0$, and 5.4.11:

$$\Phi_{s1}(y_t)= 2\,\Phi_F + \frac{C_{ox2}}{C_{si}}(V_{G2,acc} - V_{G2}) \qquad (5.4.21)$$

Combining 5.4.20 and 5.4.21 one obtains:

$$I_{D,AS+DD} = \frac{W}{L}\ \mu_n\ C_{ox1}\left\{(V_{G1}-V_{TH1,acc2})V_{DS}-\left(1+\frac{C_{si}\,C_{ox2}}{C_{ox1}\,(C_{si}+C_{ox2})}\right)\frac{V_{DS}^2}{2}\right.$$
$$-\frac{C_{si}\,C_{ox2}}{C_{ox1}\,(C_{si}+C_{ox2})}\ V_{DS}\,(V_{G2,acc} - V_{G2})$$
$$\left.+\frac{C_{si}\,C_{ox2}}{C_{ox1}\,(C_{si}+C_{ox2})}\frac{C_{ox2}}{C_{si}}(V_{G2,acc} - V_{G2})^2\right\} \qquad (5.4.22)$$

where $V_{TH1,acc2}$ is given by 5.3.10, and $V_{G2,acc}$ is given by 5.4.11.

The drain saturation voltage is obtained from 5.4.22 under the condition $dI_D/dV_{DS} = 0$:

$$V_{Dsat,AS+DD} = \frac{V_{G1} - V_{TH1,acc2} - \dfrac{C_{si}C_{ox2}}{C_{ox1}(C_{si} + C_{ox2})}(V_{G2,acc} - V_{G2})}{1 - \dfrac{C_{si}C_{ox2}}{C_{ox1}(C_{si} + C_{ox2})}} \qquad (5.4.23)$$

and the drain saturation current is given by:

$$I_{Dsat,AS+DD} = \frac{W}{L}\frac{\mu_n\,C_{ox1}}{1+\dfrac{C_{si}C_{ox2}}{C_{ox1}(C_{si+}C_{ox2})}}\ \left\{(V_{G1}-V_{TH1,acc2})^2\right.$$
$$-\frac{2C_{si}C_{ox2}}{C_{ox1}(C_{si+}C_{ox2})}\left(V_{G1} - V_{TH1,acc2}\right)\left(V_{G2,acc} - V_{G2}\right)$$
$$\left.+\frac{C_{ox2}^2\,(C_{si+}C_{ox1})}{C_{ox1}^2\,(C_{si+}C_{ox2})}(V_{G2,acc}-V_{G2})^2\right\} \qquad (5.4.24)$$

The saturation current in an SOI MOSFET (fully depleted with depleted back interface or fully depleted with accumulated back interface) is given by equations 5.4.14 and 5.4.18 and can be written in the general form:

$$I_{Dsat} = \frac{1}{2n} \frac{W}{L} \mu_n C_{ox1} [V_{G1} - V_{TH}]^2 \qquad (5.4.25)$$

where n is the body-effect coefficient or body factor defined in 5.3.17. We have $n = 1 + \alpha = 1 + C_{si}/C_{ox1}$ in a fully depleted device with accumulation at the back interface, and $n = 1 + \alpha = 1 + \dfrac{C_{si} C_{ox2}}{C_{ox1}(C_{si} + C_{ox2})}$ in a fully depleted transistor with depleted back interface. The AS+DD case basically follows similar rules, but it is more complicated and will not be described here. Expression 5.4.25 is also valid in a bulk transistor [44] if we write $n = 1 + \alpha$ $= 1 + C_D/C_{ox}$ where C_D is the depletion capacitance and is equal to $\dfrac{\varepsilon_{si}}{x_{dmax}}$ (see relationship 5.4.5).

The numerical values of n are in the following order:

$$n_{\text{fully depleted SOI}} < n_{\text{bulk}} < n_{\text{back accum SOI}}$$

As a result, the drain saturation current is highest in the fully depleted device, lower in the bulk device, and even lower in the fully depleted device with back accumulation. It is also interesting to note that α represents the ratio, C_b/C_{ox1}, of two capacitors: C_{ox1} is the front gate oxide capacitance, and C_b is the capacitance between the inversion channel and the grounded back-gate electrode or grounded substrate. The value of C_b is either that of the depletion capacitance $C_D = \varepsilon_{si}/x_{dmax}$ in a bulk device, the silicon film capacitance $C_{si} = \varepsilon_{si}/t_{si}$ in a fully depleted device with back interface accumulation, or the value given by the series association of C_{si} and C_{ox2} ($C_b = \dfrac{C_{si} C_{ox2}}{(C_{si} + C_{ox2})}$) in a fully depleted device with back interface depleted.

The low body-effect coefficient of fully depleted SOI MOSFETs brings about an increase of current drive (compared to bulk devices), which largely contributes to the excellent speed performances of fully depleted (FD) SOI CMOS circuits. Figure 5-10 compares the drain saturation current in a bulk and a FD SOI device, as a function of $V_{G1} - V_{TH}$. The larger current drive of the SOI device is evident and it is actually 20-30% higher that the drive of the bulk device.[45] The superior current drive capability of FD SOI devices is somewhat degraded as short gate lengths are considered, as a result of velocity saturation effects. It has been shown, however, that FD SOI devices still present a 25% current drive improvement over bulk devices for a gate length of 0.2 μm.[46]

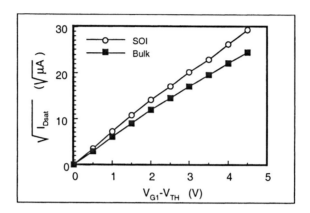

Figure 5-10. $\sqrt{I_{Dsat}}$ as a function of V_{G1}-V_{TH} in a bulk and a fully depleted SOI device with same technological parameters.

The SOI MOSFET, which was so far considered as a four-terminal device (source, drain, front gate and back gate), is actually a five-terminal device, the fifth terminal being the body. In many applications, the body is left unconnected, but its presence has considerable influence on the dynamic characteristics of the device. This influence is more pronounced in thick-film devices than in thin, fully depleted, devices. A charge-based model of the five-terminal SOI MOSFET can be found in Ref. [47]. Detailed analysis of the short-channel effects is described in Ref. [48].

5.4.2 C_∞-continuous model

In some cases, such as the simulation of analog circuits, it is necessary to have a model which is continuous in all regimes of operation (weak or strong inversion), and whose derivatives are also continuous (C_∞-continuous model). The EKV model and the ENSERG model belong to that class of models.[49,50,51,52]

We will describe here a C_∞-continuous model developed for fully depleted SOI MOSFETs and keep the notations used in Figure 5-3.[53] According to charge-sheet models the relationship between the inversion charge density at the front interface, Q_n, and the front surface potential, Ψ_{s1}, is linear and can be written as: [54]

$$Q_n = -C_{ox1}\left(V_{G1} - V_{FB1} + \frac{Q_b}{2C_{ox1}} - \alpha\left(V_{G2} - V_{FB2} + \frac{Q_b}{2C_{ox2}}\right) - n\Psi_{s1}\right) \quad (5.4.26)$$

where $Q_b = -qN_at_{si}$, $\alpha = \dfrac{C_{si}C_{ox2}}{C_{ox1}(C_{si}+C_{ox2})}$ and where $n=1+\alpha$ is the body factor.

If there is no inversion at the back interface the drain current is written as the sum of the drift and diffusion terms of Q_n:

$$I_D = -W\mu_n Q_n \frac{dV_{ch}}{dy} = -W\mu_n\left(Q_n\frac{d\Psi_{s1}}{dy} - u_T\frac{dQ_n}{dy}\right) \quad (5.4.27)$$

where μ_n is the electron mobility, V_{ch} is the potential in the channel and $u_T = kT/q$. Using Equations 5.5.26 and 5.5.27 and integrating along the channel one obtains:

$$\frac{d\Psi_{s1}}{dy} = \frac{1}{nC_{ox1}}\frac{dQ_n}{dy} \Rightarrow I_D\int_0^L dy = W\mu_n\int_{Q_{ns}}^{Q_{nd}}\left(u_T - \frac{Q_n}{nC_{ox1}}\right)dQ_n$$

$$\Downarrow$$

$$I_D = \mu_n\frac{W}{L}\left(u_T(Q_{nd} - Q_{ns}) - \frac{Q_{nd}^2 - Q_{ns}^2}{2nC_{ox1}}\right) \quad (5.4.28)$$

where Q_{ns} and Q_{nd} are the inversion charge densities at the source and drain edges, calculated for V_{ch} being equal to V_S and V_D, respectively. Following the EKV model one can develop an explicit model using an approximate expression for Q_n that tends to the desired limits in weak and strong inversion: [55]

$$Q_n = -C_{ox}nu_T S_{NT}\ln\left[1 + \frac{-Q_0/C_{ox}}{nu_T S_{NT}}\exp\left(\frac{V_G - V_{Ti} - nV_{ch}}{nu_T}\right) + \exp\left(\frac{V_G - V_T - nV_{ch}}{S_{NT}nu_T}\right)\right] \quad (5.4.29)$$

In the latter expression the second term of the logarithm dominates in weak inversion and the third term dominates in strong inversion. The parameter S_{NT} (<1) controls the transition from weak to strong inversion. Q_0 is the inversion charge density at $V_{G1}=V_{Ti}$, V_{Ti} being the threshold voltage which corresponds to a surface potential of $2\Phi_F+V_{ch}$, i.e., $V_{Ti}=V_{T0i}-\alpha V_{G2}$ where V_{T0i} is the threshold voltage when $V_{G2}=0$. This threshold value appears in the weak inversion term of (5.4.29), and one has to introduce a different threshold voltage value in the strong inversion term, $V_T = V_{T0} - \alpha V_{G2}$, to take into account the fact that in strong inversion the surface potential is always larger than $2\Phi_F + V_{ch}$ by a few kT/q. The use of two threshold voltages (a higher threshold in strong than in weak inversion) may seem awkward, but it is physically more correct than other approaches that, for instance, overestimate field mobility degradation in order to fit the model with experimental data. V_{T0} and V_{T0i} can be considered as fitting parameters obtained from parameter extraction, or as threshold voltage parameters that can be derived from the device's physical characteristics in the following manner:

$$V_{T0i} = V_{FB1} - \frac{Q_b}{2C_{ox1}} - \alpha\left(\frac{Q_b}{2C_{ox2}} - V_{FB2}\right) + 2\,n\,\Phi_F \tag{5.4.30}$$

and

$$V_{T0} = V_{FB1} - \frac{Q_{Db}}{2C_{ox1}} - \alpha\left(\frac{Q_b}{2C_{ox2}} - V_{FB2}\right) + n\,\Phi_B \tag{5.4.31}$$

where Φ_B is $2\Phi_F$ plus a few kT/q.

Equation 5.4.28 can also be obtained under an equivalent way by using, in 5.4.27, an expression of dV_{ch} expressed in terms of Q_n. Q_n can be derived from the unified charge-control model adapted to the SOI MOSFET: [56]

$$V_{G1} - V_T - n\,V_{ch} = n\,u_T\,ln\left(\frac{Q_n}{Q_0}\right) - \frac{Q_n - Q_0}{C_{ox1}} \tag{5.4.32}$$

The latter equation allows one to extract simple but accurate analytical expressions for the conductances. For instance, by differentiating 5.4.32 with respect to V_{G1}, one finds:

$$\frac{dQ_n}{dV_{G1}} = \frac{Q_n}{nu_T - Q_n/C_{ox1}} \tag{5.4.33}$$

By differentiating 5.4.28 with respect to the front gate voltage and by replacing dQ_n/dV_{G1} by the right-hand term of 5.4.33 one obtains the front-gate transconductance, g_{m1}:

$$g_{m1} = \frac{dI_D}{dV_{G1}} = \frac{\mu_n W}{nL}(Q_{nd} - Q_{ns}) \tag{5.4.34}$$

Similarly, by differentiating 5.4.28 with respect to the drain voltage, the back-gate voltage and the source voltage, one obtains the drain conductance, g_d, the back-gate transconductance, g_{m2}, and the source conductance, g_s, respectively:

$$g_d = \frac{dI_D}{dV_D} = -\mu_n \frac{W}{L} Q_{nd} \qquad (5.4.35)$$

$$g_{m2} = \frac{dI_D}{dV_{G2}} = \alpha \frac{\mu_n W}{nL} (Q_{nd} - Q_{ns}) \qquad (5.4.36)$$

$$g_s = \frac{dI_D}{dV_s} = \mu_n \frac{W}{L} Q_{ns} \qquad (5.4.37)$$

The above expressions become explicit functions of the applied voltage, by expressing Q_{ns} and Q_{nd} as a function of voltage values using equation 5.4.29 with $V_{ch}=V_S$ and $V_{ch}=V_D$, respectively. More complete expressions, including mobility degradation, DIBL (drain-induced barrier lowering), and other short-channel effects, can be found in Ref. [57].

The charge densities are obtained by integrating the charge densities along the channel. Using (5.4.26) and (5.4.27), one finds a relationship between dQ_n and dy which can be integrated to yield the total electron charge in the channel:

$$dy = -\frac{W\mu_n}{I_D} (Q_n \, d\Psi_{s1} - u_T \, dQ_n) \quad \text{and} \quad d\Psi_{s1} = \frac{dQ_n}{nC_{ox1}}$$

$$\Downarrow$$

$$dy = -\frac{W\mu_n}{I_D} \left(\frac{Q_n}{nC_{ox1}} - u_T \right) dQ_n$$

$$\Downarrow$$

$$W \int_0^L Q_n dy = \frac{W^2 \mu_n}{I_D} \int_0^L \left(\frac{Q_n^2}{nC_{ox1}} - u_T Q_n \right) dQ_n$$

Using (5.4.28) one obtains:

$$Q_N = \frac{WL\left(u_T \frac{Q_{nd}+Q_{ns}}{2} - \frac{Q_{ns}^2+Q_{ns}Q_{nd}+Q_{nd}^2}{3nC_{ox1}}\right)}{u_T - \frac{Q_{ns}+Q_{nd}}{2nC_{ox1}}}$$ (5.4.38)

The drain and source charges are given by:

$$Q_D = W\int_0^L \frac{y}{L} Q_n dy$$

$$= WL\left\{\frac{C_{ox1}u_T n}{2} + \frac{n^2 C_{ox1}^2 \mu_n W}{3LI_D}\left(u_T - \frac{Q_{nd}}{nC_{ox1}}\right)^3 - \frac{n^3 C_{ox1}^3 \mu_n^2 W^2}{15L^2 I_D^2}\left[\left(u_T - \frac{Q_{ns}}{nC_{ox1}}\right)^5 - \left(u_T - \frac{Q_{nd}}{nC_{ox1}}\right)^5\right]\right\}$$ (5.4.39)

and $$Q_S = Q_N - Q_D$$ (5.4.40)

The front and back gate charges, Q_{G1} and Q_{G2}, are obtained by integrating the front and back charge densities $Q_{g1} = C_{ox1}[V_{G1}-V_{FB1}-\Psi_{s1}]$ and $Q_{g2} = C_{ox2}[V_{G2}-V_{FB2}-\Psi_{s2}]$ along the channel, respectively:

$$Q_{G1} = WLC_{ox1}\left[\frac{\alpha}{n}V_{g1} - \frac{Q_b}{2nC_{ox1}} - \frac{\alpha}{n}\left(V_{g2} + \frac{Q_b}{2C_{ox2}}\right)\right] - \frac{Q_N}{n} - WLQ_{ox1}$$ (5.4.41)

and

$$Q_{G2} = WLC_{ox2}\left[V_{g2}\left(1 - \frac{C_{ox2}}{C_{ox2}+C_{si}} - \frac{\alpha C_{si}}{n(C_{ox2}+C_{si})}\right) - \frac{Q_b}{2(C_{ox2}+C_{si})}\left\{1 + \frac{C_{si}}{n}\left(\frac{1}{C_{ox1}} + \frac{\alpha}{C_{ox2}}\right)\right\} - V_{g1}\frac{C_{si}}{n(C_{si}+C_{ox2})}\right]$$
$$- Q_N \frac{C_{si}C_{ox2}}{nC_{ox1}(C_{si}+C_{ox2})} - WLQ_{ox2}$$ (5.4.42)

where $V_{g1}=V_{G1}-V_{FB1}$ and $V_{g2}=V_{G2}-V_{FB2}$ and where Q_{ox1} and Q_{ox2} are the fixed charge densities in the front and back oxide, respectively. The total depletion charge in the silicon film is equal to

$$-Q_{G1}-Q_{G2}-Q_{ox1}-Q_{ox2}=W L Q_b$$ (5.4.43)

The intrinsic capacitances of the transistor are given by:

$$C_{ij} = \chi_{ij}\frac{dQ_i}{dV_j}$$ (5.4.44)

where $\chi_{ij} = 1$ if $i=j$ and $\chi_{ij} = -1$ if $i\neq j$. These expressions are valid up to a frequency limit of $\mu_n(V_{G1}-V_{FB1}-nV_S)/(3nL^2)$, which is over 1 GHz for a 1 µm-long device with a front gate oxide thickness of 30 nm. A similar C_∞-continuous model has been developed for accumulation-mode MOSFETs.[58]

5.5 TRANSCONDUCTANCE

The transconductance of a MOSFET, g_m, is a measure of the effectiveness of the control of the drain current by the gate voltage. The transconductance of a bulk or partially depleted SOI MOSFET in saturation can easily be derived from equation 5.4.3:

$$g_m = dI_{Dsat}/dV_G \qquad (for \ V_{DS} > V_{Dsat})$$

$$g_m = \frac{\mu_n C_{ox1}}{(1+\alpha)} \frac{W}{L}(V_{G1} - V_{TH}) \quad \text{with} \quad \alpha = \frac{\varepsilon_{si}}{x_{dmax} C_{ox}} \qquad (5.5.1)$$

The transconductance of a fully depleted SOI MOSFET can be obtained from (5.4.14) and (5.4.18):

$$g_m = dI_{Dsat}/dV_{G1} = \frac{\mu_n C_{ox1}}{(1+\alpha)} \frac{W}{L}(V_{G1} - V_{TH}) \qquad (5.5.2)$$

for $V_{DS} > V_{Dsat}$ and with $\alpha = C_{si}/C_{ox1}$ in a fully depleted device with accumulation at the back interface, and $\alpha = \dfrac{C_{si} C_{ox2}}{C_{ox1} (C_{si}+C_{ox2})}$ in a fully depleted transistor with depleted back interface. [59]

As in the case of the analysis of the saturation current (Section 5.4.1), α represents the ratio C_b/C_{ox1} of two capacitors: C_{ox1} is the front gate oxide capacitance, and C_b is the capacitance between the inversion channel and ground.

- In a *bulk device* with grounded substrate, C_b is equal to the depletion capacitance $C_D = \varepsilon_{si}/x_{dmax}$.

- In a *fully depleted SOI MOSFET with back accumulation* the potential of the accumulation layer is pinned at zero volt and acts as a virtual ground. The capacitance between the channel and this grounded back accumulation layer is equal to the silicon film capacitance $C_{si} = \varepsilon_{si}/t_{si}$.

- In a *fully depleted device with back depletion*, the capacitance between the channel and the grounded back gate is given by the series association of C_{si} and C_{ox2}, and therefore, $C_b = \dfrac{C_{si}C_{ox2}}{(C_{si}+C_{ox2})}$. It is once again worth noting that the values of α are typically in the following order:

$$\alpha_{\text{fully depleted SOI}} < \alpha_{\text{bulk}} < \alpha_{\text{back accum SOI}} \qquad (5.5.3)$$

As a result, the transconductance is highest in the fully depleted device, lower in the bulk device, and even lower in the fully depleted device with back accumulation.

Because of electrostatic coupling between the front and back gate the variation of transconductance with front gate voltage in a fully depleted SOI MOSFET can look very different from that in a bulk device. Since the cuurent depends on both the front-gate bias and the back-gate voltage, it is possible to plot g_m as a function of V_{G1} for different values of V_{G2}. Such a set of curves is obtained by simple derivation of the current characteristics in Figure 5-5 and is shown in Figure 5-11.

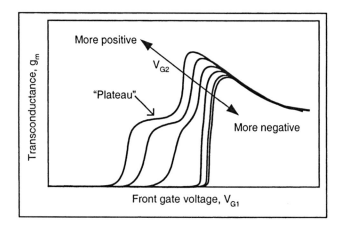

Figure 5-11. Transconductance of a fully depleted n-channel SOI MOSFET for different values of back-gate voltage.[60,61]

If a negative back-gate bias is applied, an accumulation layer is formed at the bottom of the device and the transconductance curve is similar to that of a bulk device. If V_{G2} is increased, the back interface becomes depleted and the g_m curve shifts to the left as the front threshold voltage is reduced by the back-gate bias, according to relationship 5.3.20. For more negative V_{G2} values, inversion appears at the back interface. The inversion charge depends not only on the back-gate bias, but also on the front-gate voltage, since the back threshold voltage decreases when the front gate voltage is increased. This variation of current in the back channel with front gate voltage gives rise to a plateau in the $g_m(V_{G1})$ characteristics. This plateau is unique to FDSOI devices and is not found in bulk or PDSOI MOSFETs.

5.5.1 g_m/I_D ratio

The voltage gain of a MOSFET is given by:

$$\frac{\Delta V_{out}}{\Delta V_{in}} = \frac{\Delta I_d}{g_D} \frac{1}{\Delta V_{in}} = \frac{\Delta V_{in} g_m}{g_D} \frac{1}{\Delta V_{in}} = \frac{g_m}{g_D} = \frac{g_m}{I_D} V_A \qquad (5.5.4)$$

where g_D is the output conductance, and V_A is the Early voltage.[62] The output conductance of a FDSOI transistor is basically identical to that of a bulk device, but in SOI it can be significantly improved (decreased) by varying the doping concentration in the channel from source to drain (graded-channel MOSFET).[63,64] The g_m/I_D ratio is a direct measure of the efficiency of the transistor, since it represents the amplification (g_m) obtained from the device, divided by the energy supplied to achieve this amplification (I_D). In a MOS transistor g_m/I_D is maximum when in weak inversion.[65] Its value is:

$$\frac{g_m}{I_D} = \frac{dI_D}{I_D \, dV_G} = \frac{ln(10)}{S} = \frac{q}{(1+\alpha)kT} = \frac{q}{nkT} \qquad (5.5.5)$$

where S is the subthreshold swing and n is the body factor (see Section 5.4). In strong inversion, g_m/I_D decreases with drain current and is equal to: [66]

$$\frac{g_m}{I_D} = \sqrt{\frac{2 \, \mu \, Cox \, W/L}{(1+\alpha)I_D}} = \sqrt{\frac{2 \, \mu \, Cox \, W/L}{n \, I_D}} \qquad (5.5.6)$$

Because of the lower body factor, n, fully depleted SOI MOSFETs have values of g_m/I_D significantly higher than bulk devices. The g_m/I_D ratio can reach a value of 35 V^{-1} in FDSOI MOSFETs, while the maximum value is 25 to 30 V^{-1} in bulk devices.

5.5.2 Mobility

The mobility of the electrons in the inversion layer of an n-channel MOSFET has so far been considered constant. It is actually a function of the vertical electric field below the gate oxide. It can be approximated by:

$$\mu_n(y) = \mu_{max} \, [E_c/E_{eff}(y)]^c \quad \text{for } E_{eff}(y) > E_c \qquad (5.5.6)$$

where μ_{max}, E_c, and c are fitting parameters depending on the gate oxidation process and the device properties [67,68], and

$$E_{eff}(y) = E_{s1}(y) - \frac{Q_{inv1}(y)}{2\varepsilon_{si}} \qquad (5.5.7)$$

The vertical electric field below the gate oxide is given by (5.3.4):

$$E_{s1}(y) = (\frac{\Phi_{s1}(y) - \Phi_{s2}(y)}{t_{si}} + \frac{q \, N_a \, t_{si}}{2 \, \varepsilon_{si}}) \qquad (5.5.8)$$

The inversion charge in the channel, $Q_{inv1}(y)$, can be obtained from (5.4.8):

$$- Q_{inv1}(y) = C_{ox1}\left(V_{G1} - \Phi_{MS1} + \frac{Q_{ox1}}{C_{ox1}} - \left(1 + \frac{C_{si}}{C_{ox1}}\right)\Phi_{s1}(y) + \frac{C_{si}}{C_{ox1}} \, \Phi_{s2}(y) + \frac{Q_{depl}}{2C_{ox1}}\right) \qquad (5.5.9)$$

The back surface potential, $\Phi_{s2}(y)$, is given by (5.4.9):

$$\Phi_{s2}(y) = \frac{C_{ox2}}{C_{ox2} + C_{si}}\left(V_{G2} - \Phi_{MS2} + \frac{Q_{ox2}}{C_{ox2}} + \frac{C_{si}}{C_{ox2}}\Phi_{s1}(y) + \frac{Q_{depl}}{2C_{ox2}} - \frac{Q_{s2}(y)}{C_{ox2}}\right) \qquad (5.5.10)$$

The expression of the surface electric field can be simplified by considering the case of a fully depleted device operating with a low drain voltage $(V_{DS} \cong 0)$, so that the front and back surface potentials are independent of y. If the back interface is depleted and close to inversion, $\Phi_{s1} - \Phi_{s2} \cong 0$, and the front surface electric field is equal to $\dfrac{q \, N_a \, t_{si}}{2 \, \varepsilon_{si}}$. In this case, the surface electric field is lower than the surface electric field in the corresponding bulk device, which is equal to $\dfrac{qN_a x_{dmax}}{\varepsilon_{si}}$, since $t_{si} < x_{dmax}$. If

the film is fully depleted, yet not close to inversion, then $E_{s1} \cong \dfrac{q N_a x_1}{\varepsilon_{si}}$ is a

good approximation, where x_1 is the point of minimum potential in the film in Figure 5-4, and the surface electric field is still lower than in a bulk devices, since $x_1 < t_{si} < x_{dmax}$.

Figure 5-12 presents the electric field, E, as a function of depth, x, in a bulk and a fully depleted, SOI device. The slope of the $E(x)$ is the same if both devices have the same doping concentration, N_a. The (front) surface electric field, $E_{s1}=E(x=0)$, is lower in the SOI device than in the bulk device.

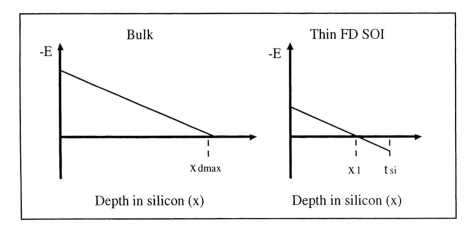

Figure 5-12. Electric field distribution in a bulk and a fully depleted (FD) SOI device with depleted back interface.

An increase of surface mobility in fully depleted SOI devices has been reported by several authors.[69,70] This high mobility is obtained right above threshold, but it rapidly disappears when the gate voltage is increased due to the rapid increase of Q_{inv1} with V_{G1} when $V_{G1}>V_{TH1}$ in 5.5.9. In general SOI MOSFETs obey the same empirical mobility reduction law as bulk MOSFETs. The mobility degradation can be taken into account in the equation for the MOSFET operating in non-saturation: [71]

$$I_D = \dfrac{\mu_o}{1+\dfrac{\theta_1}{n}\left(V_G - V_{TH}\right)} C_{ox} \dfrac{W}{L}\left[\left(V_G - V_{TH}\right)V_D - \dfrac{1}{2n}V_D^2\right] \quad (5.5.11)$$

and operating in saturation:

$$I_{Dsat} = \frac{1}{2n} \frac{\mu_o}{1 + \frac{\theta}{n}(V_{GS} - V_{TH})} C_{ox} \frac{W}{L_{eff}}(V_{GS} - V_{TH})^2 \qquad (5.5.12)$$

where Θ is the mobility reduction factor and n is the body factor. Carrier mobility is dependent on film thickness in very thin SOI devices. In general a mobility increase is observed when the film thickness is reduced below 20 nm. If the film thickness is reduced below 10 nm, however, severe mobility reduction occurs (see section 5.8.2)

5.6 BASIC PARAMETER EXTRACTION

The extraction of MOSFET parameters such as threshold voltage, mobility and source and drain resistance, is an important part of the device modeling and characterization process. This section describes parameter extraction techniques that have been developed for SOI MOSFETs.

5.6.1 Threshold voltage and mobility

The non-saturated current in a MOS transistor is given by 5.5.11:

$$I_D = \frac{\mu_o}{1 + \frac{\theta_1}{n}(V_G - V_{TH})} C_{ox} \frac{W}{L}\left[(V_G - V_{TH})V_D - \frac{1}{2n}V_D^2\right] \qquad (5.6.1)$$

Using a small V_D value (*e.g.*: V_D = 50 mV) and neglecting the body factor ($n=1$), we can write:[72]

$$I_D = \frac{\mu_o}{1 + \theta_1(V_G - V_{TH})} C_{ox} \frac{W}{L}(V_G - V_{TH})V_D \qquad (5.6.2)$$

or

$$I_D = \frac{A(V_G - V_{TH})}{1 + \theta_1(V_G - V_{TH})} \quad \text{with} \quad A = \mu_o C_{ox} \frac{W}{L}V_D \qquad (5.6.3)$$

Taking the derivative of the inverse of the current with respect to gate voltage we find:

$$F_1(V_G) = -\frac{d\left(\frac{1}{I_D}\right)}{dV_G} = \frac{1}{A(V_G - V_{TH})^2} = \frac{\frac{dI_D}{dV_G}}{I_D^2} = \frac{g_m}{I_D^2} \qquad (5.6.4)$$

or: $\qquad \sqrt{A}\left(V_G - V_{TH}\right) = \dfrac{I_D}{\sqrt{g_m}} \equiv F_1(V_G)$ $\qquad$ (5.6.5)

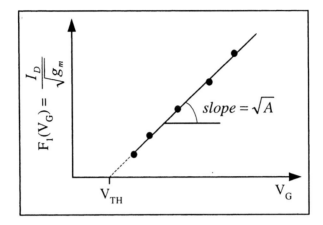

Figure 5-13. Measurement of threshold voltage and mobility.[72]

Plotting $F_1(V_G) = \dfrac{I_D}{\sqrt{g_m}}$ yields a straight line whose intercept with the x-axis is the threshold voltage and whose slope can be used to find the mobility using $\mu_o = \dfrac{A L}{W C_{ox} V_D}$ (Figure 5-13). One advantage of this measurement technique is that it uses the derivative of $I_D(V_G)$. Source and drain series resistance, which act as additive factors to equation 5.6.2, are naturally eliminated from the measurement by the differentiation process.

A more refined version of this measurement technique has been developed, which accounts for more complex mobility reduction mechanisms and effective channel length.[73,74] If V_D is small (*e.g.*: $V_D = 50$ mV) and if we neglect the body effect ($n=1$), we can write:

$$I_D = \frac{\mu_o}{1 + \theta_1\left(V_G - V_{TH}\right) + \theta_2\left(V_G - V_{TH}\right)^2} \, C_{ox} \frac{W}{L_{eff}}\left(V_G - V_{TH}\right)V_D \qquad (5.6.6)$$

or

$$I_D = \frac{A\left(V_G - V_{TH}\right)}{1 + \theta_1\left(V_G - V_{TH}\right) + \theta_2\left(V_G - V_{TH}\right)^2} \qquad (5.6.7)$$

where

$$A = \mu_o \, C_{ox} \frac{W}{L_{eff}} V_D \qquad (5.6.8)$$

and where $L_{eff} = L - \Delta L$ is the effective channel length. The use of two mobility reduction factors, θ_1 and θ_2, is necessary to accurately represent the current dependence on gate voltage when thin gate oxides (< 7 nm, typically) are used. Taking the first and second derivative of the inverse of the current we can define:

$$F_2(V_G) = \left[\frac{d^2\left(\frac{1}{I_D}\right)}{dV_G^2} \right]^{-\frac{1}{3}} = \sqrt[3]{A/2} \, (V_G - V_{TH}) \qquad (5.6.9)$$

and

$$F_3(V_G) = -\frac{d\left(\frac{1}{I_D}\right)}{dV_G} = \frac{g_m}{I_D^2} = \frac{1}{A}\left(\theta_2 - \frac{1}{(V_G - V_{TH})^2} \right) \qquad (5.6.10)$$

Plotting F_2 as a function of V_G yields V_{TH} and A (Figure 5-14A). Plotting $1/A$ for different channel lengths allows the extraction of both ΔL and μ_o (Figure 5-14B). The mobility reduction factor θ_2 can be obtained by plotting F_3 as a function of $(V_G - V_{th})^{-2}$ (Figure 5-14C).

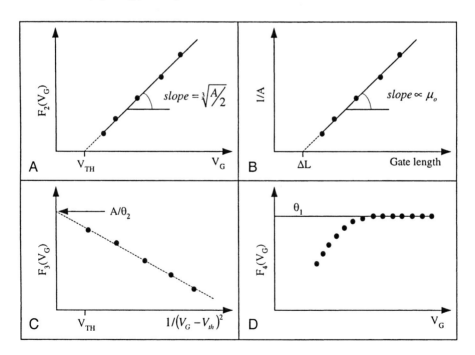

Figure 5-14. Measurement of threshold voltage, mobility and mobility reduction factors. [74]

From relationship 5.6.7 we can define $F_4(V_G)$ as:

$$F_4(V_G) = \theta_1(V_G) = -A\,\theta_2\,(V_G - V_{TH})\frac{1}{I_D} - \frac{1}{V_G - V_{TH}} \qquad (5.6.11)$$

F_4 increases with V_G until it reaches a plateau, at which point $F_4 = \theta_1$. Figure 5-14.D shows a plot of F_4 as a function of V_G which allows for the extraction of the primary mobility reduction factor, θ_1.

5.6.2 Source and drain resistance

The saturation current of a MOS transistor is given by 5.5.12:

$$I_{Dsat} = \frac{K}{2}(V_{GS} - V_{TH})^2 \quad \text{with } K = \frac{K_o}{1 + \theta_1(V_{GS} - V_{TH})} \qquad (5.6.12)$$

and

$$K_o = \frac{W\,\mu_o\,C_{ox}}{n\,L_{eff}} \qquad (5.6.13)$$

where V_G is the voltage applied to the gate (Figure 5-15), $L_{eff} = L - \Delta L$ is the effective channel length and n is the body factor. If the source resistance is not negligible and if the source is grounded, the *intrinsic* gate-to-source voltage, V_{GS}, is given by:

$$V_{GS} = V_G - I_{Dsat}\,R_s \qquad (5.6.14)$$

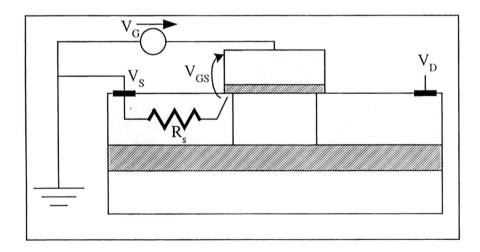

Figure 5-15. Set-up for source and drain resistance measurement.

Combining 5.6.12, 5.6.13 and 5.6.14 we find:[75]

$$V_G = V_{TH} + R_t I_{Dsat} + \sqrt{\frac{2 I_{Dsat}}{K_o} + R_\theta^2 I_{Dsat}^2} \qquad (5.6.15)$$

with $\qquad R_\theta \equiv \frac{\theta}{K_o}$ and $R_t = R_s + R_\theta \qquad (5.6.16)$

where R_s is the resistance of the source and R_θ is an effective resistance due to mobility degradation in the channel.[75]

In a MOSFET in saturation $\dfrac{2 I_{Dsat}}{K_o}$ is typically much larger than $R_\theta^2 I_{Dsat}^2$,

such that $R_\theta^2 I_{Dsat}^2$ can be neglected and equation 5.6.15 becomes:

$$V_G = V_{TH} + R_t I_{Dsat} + \sqrt{\frac{2 I_{Dsat}}{K_o}} \qquad (5.6.17)$$

Let us now define the following function:

$$G_1(V_G, I_{Dsat}) = V_G - \frac{2}{I_{Dsat}} \int_0^{V_G} I_{Dsat}(V_G) dV_G \qquad (5.6.18)$$

The function $G_1(V_G, I_{Dsat})$ can be extracted numerically from device measurements. Using integration by parts $\int u\, dv = uv - \int v\, du$, G_1 can be rewritten in the following form:

$$G_1(V_G, I_{Dsat}) = \frac{2}{I_{Dsat}} \int_0^{I_{Dsat}} V_G(I_{Dsat}) dI_{Dsat} - V_G \qquad (5.6.19)$$

Replacing V_G in 5.6.19 by its value from equation 5.6.17 we find:

$$G_1(I_{Dsat}) = V_{TH} + \frac{1}{3}\sqrt{\frac{2 I_{Dsat}}{K_o}} \qquad (5.6.20)$$

Thus by plotting G_1 vs. $\sqrt{I_{Dsat}}$ we can extract both V_{TH} and K_o (Figure 5-16). Once V_{TH} and K_o are known the resistance R_t can then be evaluated using 5.6.17:

$$R_t = \frac{V_G - V_{TH} - \sqrt{\dfrac{2 I_{Dsat}}{K_o}}}{I_{Dsat}} \qquad (5.6.21)$$

At this point the source (or drain) resistance, R_s, can be calculated using 5.6.16 if $R_s \gg R_\theta$ (*i.e.*, if θ is small, t_{ox} is small or W/L is large). If devices with different channel lengths are available, additional measurement can be performed to calculate R_s and θ_l with higher accuracy. Combining 5.6.13 and 5.6.16 we find:

$$R_t = R_s + \theta_1 \frac{n L_{eff}}{\mu_o C_{ox}} = R_s + \theta_1 K_o^{-1} \qquad (5.6.22)$$

Thus, a plot of R_t vs. K_o^{-1} will yield θ_l and R_s, as shown in Figure 5-17. Most MOSFETs are symmetrical, in which case the drain resistance is equal to the source resistance ($R_D = R_S$).

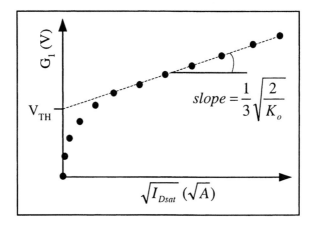

Figure 5-16. Measurement of V_{TH} and K_o.[75]

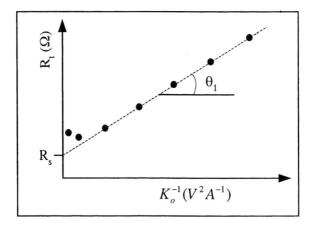

Figure 5-17. Extraction of source (drain) resistance and mobility reduction factor.[75]

5.7 SUBTHRESHOLD SLOPE

The inverse subthreshold slope (or, in short, the subthreshold slope, or subthreshold swing) is defined as the inverse of the slope of the $I_D(V_G)$ curve in the subthreshold regime, presented on a semilogarithmic plot (Figure 5-18).

$$S = \frac{d\,V_G}{d\,(log\ I_D)}$$ (5.7.1)

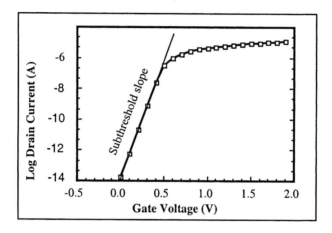

Figure 5-18. Semilogarithmic plot of $I_D(V_G)$ of an nMOS device.

The subthreshold current of an MOS transistor is independent of drain voltage, which suggests it is due to a diffusion mechanism rather than to a drift mechanism.[76,77] We can thus write:

$$I_D = -q\,A\,D_n \frac{dn}{dy} = q\,A\,D_n \frac{n(0)-n(L)}{L}$$ (5.7.2)

where A is the cross-sectional area of a vertical section of the channel region through which the electrons flow, D_n is the diffusion coefficient for electrons, and $n(0)$ and $n(L)$ are the electron concentrations at the edge of the source and drain junction, respectively. The latter can be expressed as follows:

$$n(0) = n_{po}\ exp\left(\frac{q\,\Phi_S}{kT}\right)\quad and\quad n(L) = n_{po}\ exp\left(\frac{q(\,\Phi_S\text{-}V_D)}{kT}\right)$$ (5.7.3)

where the source is considered at ground and $n_{po} = n_i^2/N_a$.

We know that the electron concentration varies as $exp(q\Phi(x)/kT)$. To simplify calculations we will approximate the exponential electron profile by a constant electron density extending to a depth d below the surface. The depth d is defined as the depth at which the potential has decreased by kT/q below the surface potential value. Therefore, one can write:

$$d = \frac{kT/q}{E_s} \quad where \ E_s = - \ \frac{d\Phi(x)}{dx}\bigg|_{x=0} \tag{5.7.4}$$

The area of the section through which the electrons flow is thus equal to $A = Wd$, where W is the transistor width. Using Equations 5.7.2 to 5.7.4 and Einstein's relationship $D_n = \frac{kT}{q} \ \mu_n$ we obtain the subthreshold current:

$$I_D = \mu_n \frac{W}{L} \ q \left(\frac{kT}{q}\right)^2 \frac{n_i^2}{N_a} \ [1 - exp \ (-qV_D/kT)] \ \frac{exp \ (q\Phi_S/kT)}{-\frac{d\Phi_S}{dx}} \tag{5.7.6}$$

On a log plot such as Figure 5-18 the subthreshold current appears as a straight line. The inverse of the slope of that line is called "inverse subthreshold slope", "subthreshold swing", or more simply, "subthreshold slope". It is expressed in volts or millivolts per decade. The lower the value of the subthreshold slope, S, the more efficient and rapid the switching of the device from the off state to the on state.

By definition the subthreshold slope is given by:

$$S = \frac{dV_G}{d \ log(I_D)} \tag{5.7.7}$$

or, if we change the logarithm base to the natural logarithm base:

$$S = \frac{ln(10)}{\frac{d(ln \ (I_D))}{dV_G}} \tag{5.7.8}$$

Noting that $\qquad \frac{d(ln(I_D))}{dV_G} = \frac{1}{I_D} \ \frac{d(ln(I_D))}{d\Phi_S} \ \frac{d\Phi_S}{dV_G} \tag{5.7.9}$

the log of the subthreshold current can be differentiated, which yields:

$$\frac{d(\ln(I_D))}{dV_G} = \left[\frac{q}{kT} - \frac{\frac{d}{d\Phi_s}\left(-\frac{d\Phi_s}{dx}\right)}{-\frac{d\Phi_s}{dx}} \right] \frac{d\Phi_s}{dV_G} \qquad (5.7.10)$$

- In the case of a *bulk* or a *partially depleted* SOI device, the electric field at the silicon surface ($x=0$) is equal to:

$$E_s = -\frac{d\Phi_s}{dx} = \sqrt{\frac{2\,q\,N_a\,\Phi_s}{\varepsilon_{si}}} = \frac{q\,N_a}{C_D}$$

where C_D is the depletion capacitance. We also have: $\frac{d}{d\Phi_s}(-\frac{d\Phi_s}{dx}) = \frac{C_D}{\varepsilon_{si}}$. The second term in between brackets in 5.7.10 is thus equal to:

$$\frac{\frac{d}{d\Phi_s}(-\frac{d\Phi_s}{dx})}{-\frac{d\Phi_s}{dx}} = \frac{C_D^2}{q\,N_a\,\varepsilon_{si}} = \frac{1}{2\,\Phi_s} \qquad (5.7.11)$$

Usually, this term is small compared to kT/q (approximately 4% of it when $\Phi_s = \frac{2}{3}\Phi_F$, and will be neglected here.

Using the relationship between gate voltage and charges in the silicon and at the interfaces:

$$V_G = \Phi_{MS} + \Phi_s + \frac{-Q_D - Q_{ox} + C_{it}\,\Phi_s}{C_{ox}}$$

we find: $\qquad \frac{d\Phi_s}{dV_G} = \frac{C_{ox}}{C_{ox} + C_D + C_{it}} \qquad (5.7.12)$

where $C_D = dQ_D/d\Phi_s$, $Q_D = q\,N_a\,x_{dmax}$, and $C_{it} = q\,N_{it}$.

Usually, the second term in between brackets in 5.7.10 is small compared to kT/q (approximately 4% of it when $\Phi_s = \frac{2}{3}\Phi_F$), and will be neglected here. Thus, using 5.7.8, 5.7.9, 5.7.10 and 5.7.12, we find:

$$S = \frac{kT}{q}\,ln\,(10)\,(1 + \frac{C_D + C_{it}}{C_{ox}}) \qquad (5.7.13)$$

or if the interface traps are neglected:

$$S = \frac{kT}{q}\,ln\,(10)\,(1 + \frac{C_D}{C_{ox}}) \qquad (5.7.14)$$

The right-side term of 5.7.12 (with $C_{it}=0$) can be represented by the capacitor network of Figure 5-19, in which we can observe that

$$C_{ox}\,d(V_G\text{-}\Phi_s) = I_G = C_D\,d\Phi_s \text{ and, hence, } \frac{d\,V_G}{d\,\Phi_s} = 1 + \frac{C_D}{C_{ox}} \, .$$

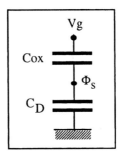

Figure 5-19. Equivalent capacitor network (bulk case).

- In the case of a *fully depleted* SOI device with depleted back interface, we use equations 5.3.8 and 5.3.9:

$$V_{G1} = \Phi_{MS1} - \frac{Q_{ox1}}{C_{ox1}} + \frac{q\,N_a\,t_{si}}{2\,C_{ox1}} + \Phi_{s1}(\frac{\varepsilon_{si}}{t_{si}\,C_{ox1}} + 1) + \Phi_{s2}(\frac{-\varepsilon_{si}}{t_{si}\,C_{ox1}}) \qquad (5.7.15)$$

$$V_{G2} = \Phi_{MS2} - \frac{Q_{ox2}}{C_{ox2}} + \frac{q\,N_a\,t_{si}}{2\,C_{ox2}} + \Phi_{s1}(\frac{-\varepsilon_{si}}{t_{si}\,C_{ox2}}) + \Phi_{s2}(\frac{\varepsilon_{si}}{t_{si}\,C_{ox2}} + 1) \qquad (5.7.16)$$

If we define $\varepsilon_{si}/t_{si}=C_{si}$ as the capacitance of the depleted silicon film, and $qN_at_{si} = Q_{depl}$ as the depletion charge in the silicon film, we obtain:

$$V_{G1} = \Phi_{MS1} - \frac{Q_{ox1}}{C_{ox1}} + \frac{Q_{depl}}{2\,C_{ox1}} + \Phi_{s1}(\frac{C_{si}}{C_{ox1}} + 1) + \Phi_{s2}(\frac{-C_{si}}{C_{ox1}}) \qquad (5.7.17)$$

$$V_{G2} = \Phi_{MS2} - \frac{Q_{ox2}}{C_{ox2}} + \frac{Q_{depl}}{2\,C_{ox2}} + \Phi_{s1}(\frac{-C_{si}}{C_{ox2}}) + \Phi_{s2}(\frac{C_{si}}{C_{ox2}} + 1) \qquad (5.7.18)$$

Eliminating Φ_{s2} between these two equations, we can express Φ_{s1} as a function of V_{G1} and V_{G2}:

$$\Phi_{s1}\left((1 + \frac{C_{si}}{C_{ox1}})\frac{C_{ox1}}{C_{si}}(1 + \frac{C_{si}}{C_{ox2}}) - \frac{C_{si}}{C_{ox2}} \right) =$$
$$V_{G2} - \Phi_{MS2} + \frac{Q_{ox2}}{C_{ox2}} - \left(\frac{Q_{depl}}{2\,C_{ox1}} + \Phi_{MS1} - \frac{Q_{ox1}}{C_{ox1}} - V_{G1}\right)\frac{C_{ox1}}{C_{si}}(1 + \frac{C_{si}}{C_{ox2}}) + \frac{Q_{depl}}{2\,C_{ox2}} \qquad (5.7.19)$$

From which we derive:

$$\frac{d\Phi_{sl}}{dV_{Gl}} = \frac{\frac{C_{ox1}}{C_{si}}(1+\frac{C_{si}}{C_{ox2}})}{(1+\frac{C_{si}}{C_{ox1}})\frac{C_{ox1}}{C_{si}}(1+\frac{C_{si}}{C_{ox2}})-\frac{C_{si}}{C_{ox2}}} = \frac{\frac{1}{C_{si}}+\frac{1}{C_{ox2}}}{\frac{1}{C_{ox1}}+\frac{1}{C_{si}}+\frac{1}{C_{ox2}}} \tag{5.7.20}$$

and thus
$$\frac{dV_{Gl}}{d\Phi_{sl}} = \left(\frac{d\Phi_{sl}}{dV_{Gl}}\right)^{-1} = 1+\frac{1}{C_{ox1}}\left(\frac{C_{si}C_{ox2}}{C_{si}+C_{ox2}}\right) \tag{5.7.21}$$

Using 5.7.8, 5.7.9, 5.7.10 and 5.7.21, we finally obtain the expression of the subthreshold slope in a fully depleted device:

$$S = \frac{kT}{q}\ln(10)\left[1+\frac{1}{C_{ox1}}\left(\frac{C_{si}C_{ox2}}{C_{si}+C_{ox2}}\right)\right] \tag{5.7.22}$$

The right-hand side term of 5.7.20 can be represented by the capacitor network of Figure 5-20.

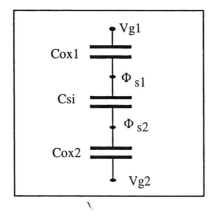

Figure 5-20. Equivalent capacitor network (fully depleted SOI device).

In fully depleted SOI MOSFETs we usually have $C_{ox2} << C_{ox1}$ and $C_{ox2} << C_{si}$. As a result, $\frac{d\Phi_{sl}}{dV_{Gl}}$ tends to unity, and the expression of the subthreshold slope becomes:

$$S \cong \frac{kT}{q}\ln(10) \tag{5.7.23}$$

The complete expression of the subthreshold slope, including the presence of interface states, is given by:

$$S=\frac{dV_{G1}}{d\log(I_D)} = \frac{kT}{q} \ln(10) \left[\left(1+\frac{C_{it1}}{C_{ox1}}+\frac{C_{si}}{C_{ox1}}\right) - \frac{\dfrac{C_{si}}{C_{ox2}}\dfrac{C_{si}}{C_{ox1}}}{1+\dfrac{C_{it2}}{C_{ox2}}+\dfrac{C_{si}}{C_{ox2}}} \right] \qquad (5.7.24)$$

where V_{G1}, $C_{it1}=qN_{it1}$, $C_{it2}=qN_{it2}$, C_{ox1} and C_{ox2} are the front gate voltage, the top interface trap capacitance, the bottom interface trap capacitance, the front gate oxide capacitance, and the bottom gate oxide capacitance, respectively. C_{si} is the silicon film capacitance, and is defined by $C_{si}=\varepsilon_{si}/t_{si}$, where t_{si} is the silicon film thickness.

The expression between brackets in Equation (5.7.24) is usually close to unity. For instance, it is equal to 1.07, when t_{si}=90 nm, t_{ox1}=15 nm, t_{ox2} = 400 nm, N_{it1}=10^{10} cm^{-2}V^{-1}, and N_{it2} =5×10^{10} cm^{-2}V^{-1}.

A mere comparison between figures 5-19 and 5-20 (corresponding to equations 5.7.14 and 5.7.22) shows that the inverse subthreshold slope of a fully depleted SOI MOSFET is lower than that of a bulk or thick-film SOI device having the same parameters. The theoretical minimum value of S is 60 mV/decade at room temperature (*i.e.,* a 60 mV increase of gate voltage results in a tenfold increase of subthreshold drain current). This also means that, in the subthreshold regime, any increase of gate bias ΔV_G will give rise to an increase of surface potential $\Delta\Phi_{s1}$ equal to ΔV_G (perfect coupling between V_G and Φ_{s1}).

The subthreshold slopes of a fully depleted SOI MOSFET and a partially depleted device can be compared in Figure 5-21, which shows the $I_D(V_G)$ characteristics of two n-channel devices (a fully depleted device with t_{si}=100nm and a partially depleted with t_{si}=200nm).[78] The subthreshold slope of the thicker-film device is identical to that of a bulk device with the same channel doping concentration. It is interesting to note that the thinner device has a slightly lower leakage current at V_G=$0V$ than the thicker device, although its threshold voltage is lower.

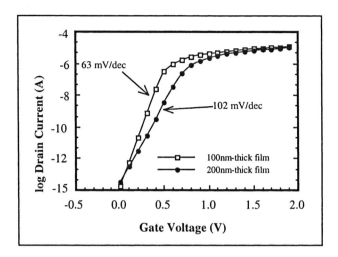

Figure 5-21. Simulated $I_D(V_G)$ characteristics of a 200 nm-thick, partially depleted, n-channel device and of a 100 nm-thick, fully depleted MOSFET.

In general, neglecting the presence of interface states, the subthreshold slope can be written as:

$$S = \frac{kT}{q} \ln(10)(1+\alpha) = n\frac{kT}{q} \ln(10) \qquad (5.7.25)$$

where, α represents the ratio C_b/C_{ox1} of two capacitors. C_{ox1} is the gate oxide capacitance, C_b is the capacitance between the inversion channel and the back-gate electrode, and n is the body factor. The value of C_b is either that of the depletion capacitance $C_D = \varepsilon_{si}/x_{dmax}$ in a bulk device, the silicon film capacitance $C_{si} = \varepsilon_{si}/t_{si}$ in a fully depleted device with back interface accumulation, or the value given by the series association of C_{si} and C_{ox2} $(C_b = \frac{C_{si}C_{ox2}}{(C_{si}+C_{ox2})})$ in a fully depleted device. The values of α are typically in the following sequence:

$$\alpha_{\text{fully depleted SOI}} < \alpha_{\text{bulk}} < \alpha_{\text{back accum SOI}}$$

As a result, the inverse subthreshold slope has the lowest (*i.e.* best) value in the fully depleted device, it is larger in the bulk device, and even larger in the fully depleted device with back accumulation.

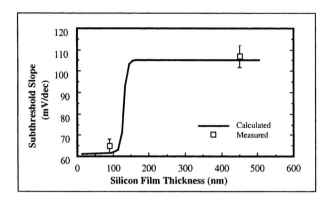

Figure 5-22. Simulated and measured subthreshold slope as a function of silicon film thickness. $N_a = 8 \times 10^{16}$ cm^{-3}, gate oxide thickness is 25 nm.[79]

The minimum theoretical value of 60 mV/decade at room temperature is never reached because of the presence of traps at the Si-SiO$_2$ interfaces and because of the finite value of C_{ox2}, but values as low as 65 mV/dec can easily be obtained. The transition between the fully depleted and the partially depleted regimes is quite sharp, as illustrated in Figure 5-22.

The excellent value of the subthreshold slope in fully-depleted SOI devices allows one to use smaller values of threshold voltage than in bulk (or thick SOI) devices without increasing the leakage current at $V_{G1}=0V$. As a result, better speed performances can be obtained, especially at low supply voltage. [80] As in bulk devices, an increase of the subthreshold slope is observed in short-channel devices, but thin SOI MOSFETs show less degradation than bulk transistors.[81,82]

5.8 ULTRA-THIN SOI MOSFETS

To reduce short-channel effects such as DIBL (drain-induced barrier lowering) in SOI MOSFETs one has to reduce the silicon film thickness to 20 nm or below (see section 6.1.2.2). Extremely thin SOI devices are often referred to as "ultrathin SOI" (UT SOI) or "ultrathin body" (UTB SOI) devices. Several effects related to the modification of the band structure do appear when such thin films are used. When the silicon film becomes thin enough the energy bands split into subbands. The minimum energy of the lowest conduction subband is obtained by posing $n=1$ in equation 6.1.16:

$$E_C = E_{C0} + \frac{\hbar^2}{2m}\left(\frac{\pi}{t_{si}}\right)^2 \qquad (5.8.1)$$

where m^* is the electron effective mass, $\hbar$ is the reduced Planck constant and E_{co} is the "original" conduction band minimum of bulk silicon.[83] One can easily see that $E_C > E_{C0}$, and as a result the bandgap of silicon is larger in ultrathin silicon films than in bulk silicon. A more detailed analysis shows that the density of states becomes a staircase function of the energy in the bands.[84,85,86,87]

5.8.1 Threshold voltage

Classical theory predicts that the threshold voltage decreases in a fully depleted SOI MOSFET when silicon film thickness is decreased, assuming the doping concentration N_a is held constant (equation 5.3.12). This is due to the reduction of depletion charge $qN_a t_{si}$ as the film thickness is decreased (Figure 5-23). When the film thickness is below 10nm, however, the conduction band splits into subbands and the mimimum energy of the conduction band increases as the film thickness is decreased, according to relationship 5.8.1. This causes the threshold voltage to increase, as shown in Figure 5-23. This quantum phenomenon was first reported in 1993 by Omura *et al.* [88] and has since been confirmed and measured by several authors.[89,90,91,92]

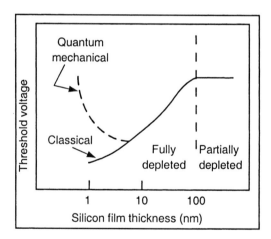

Figure 5-23. Variation of threshold voltage with film thickness according to classical and quantum-mechanical models.

An increase of short-channel effects is predicted for very thin silicon films (<10 nm) because of the increase of bandgap energy and the decrease of effective density of states in the conduction band due to 2D quantization.[93,94]

5.8.2 Mobility

Several factors influence mobility in thin SOI films. Surface roughness scattering plays a role in very thin films, and so does scattering by interface charges.[95,96] Both scattering mechanisms become important in extremely thin (a few nanometers) films where electrons are tightly squeezed between two Si-SiO$_2$ interfaces. The most interesting behavior, however, is that of the phonon-limited low-field mobility. When the silicon thickness is reduced below 20 nm the energy bands split into subbands and the electrons are redistributed in these several subbands. Electrons in the lower subband have a lower mass, and therefore, a higher mobility than the other subbands (Figure 5-24A).

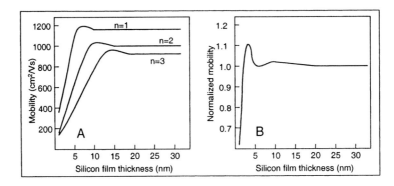

Figure 5-24. A: Mobility in the three first subbands and B: mobility (normalized to bulk mobility) in thin SOI films as a function of silicon film thickness.[97,98]

When the film thickness becomes smaller than 15 nm the mobility slightly increases in the subbands and then plummets as the film thickness is further reduced. The slight increase is due to a reduction of the effective mass of the confined electrons when film thickness is decreased, while the plummeting is caused by an increase of scattering rate due to the overlap of the electron wavefunctions in the different subbands and Coulomb scattering due to charged centers at the Si-SiO$_2$ interfaces.[99] The latter scattering rate is inversely proportional to the silicon film thickness. When the film thickness is decreased below 20nm the total mobility, taking into account all subbands, increases slightly below 15nm (Figure 5-24B). Then, as the film thickness is further reduced below 5nm, the population of the higher energy subbands decreases abruptly to the profit of the lowest-energy subband (*n=1*). Since this subband exhibits the highest mobility the overall macroscopic mobility increases and it can reach a value 10% larger than bulk mobility. Finally, as the thickness is reduced below 3nm, the mobility drops sharply [100] because all the electrons are confined within a very limited space and inter- and intra-subband scattering rates become very high.[101,102,103,104,]

5.9 IMPACT IONIZATION AND HIGH-FIELD EFFECTS

Several parasitic phenomena related to impact ionization in the high electric field region near the drain appear in SOI MOSFETs. Some of them, such as the reduced drain breakdown voltage, are related to the parasitic NPN bipolar transistor found in the n-channel SOI MOSFET, and will be dealt with in Section 5.10. The present section will neglect body current multiplication by bipolar effect and focus on two phenomena. The first of these is the kink effect, and the second is hot-electron degradation. The kink effect is normally not found in bulk devices operating at room temperature when substrate or well contacts are provided. Hot-electron degradation takes place in bulk MOSFETs having a short channel and constitutes a major reliability hazard in submicron devices. A comparison of hot-electron degradation phenomena between bulk and SOI devices will be made in Section 5.9.2.

5.9.1 Kink effect

The kink effect is characterized by the appearance of a "kink" in the output characteristics of an SOI MOSFET, as illustrated in Figure 5-25. The kink appears above a given drain voltage. It can be very strong in n-channel transistors, but is usually absent from p-channel devices. The kink effect is not observed in bulk devices at room temperature when substrate or well contacts are provided, but it can be observed in bulk MOSFETs operating at low temperatures [105] or in devices realized in floating p-wells. The kink effect was already observed in early SOS devices.[106]

The kink effect can be explained as follows. Let us consider a *partially depleted* SOI n-channel transistor. When the drain voltage is high enough, the channel electrons can acquire sufficient energy in the high electric field zone near the drain to create electron-hole pairs, due to an impact ionization mechanism. The generated electrons rapidly move into the channel and the drain, while the holes (which are majority carriers in the p-type body) migrate towards the place of lowest potential, *i.e.*, the floating body (Figure 5.6.2.C).

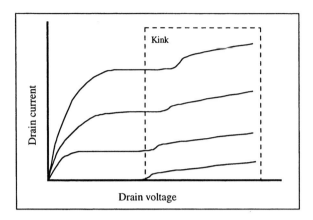

Figure 5-25. Kink in the output characteristics of an n-channel SOI MOSFET.

The injection of holes into the floating body forward biases the source-body diode. The floating body reaches a positive potential, which can be calculated using the following equation:

$$I_{holes,gen} = I_{so}\left(exp\left(\frac{q\,V_{BS}}{n\,k\,T}\right) - 1\right)$$ (5.9.1)

where $I_{holes\,gen}$ is the hole current generated near the drain (Figure 5.6.2.A), I_{so} is the saturation current of the source-body diode, V_{BS} is the potential of the floating body, and n is the ideality factor of the diode. The value of $I_{holes,gen}$ depends on different parameters, including the body potential. Its calculation requires iterative solving of a set of complex equations and will not be described here.[107]

The increase of body potential (up to 700 mV) gives rise to a decrease of the threshold voltage. This is again a manifestation of the body effect (equation 5.3.15). The threshold voltage shift can be calculated using the expression of the body effect in a *bulk* transistor where the substrate bias is replaced by the floating body potential. This decrease of threshold voltage induces an increase of the drain current as a function of drain voltage which can be observed in the output characteristics of the device (Figure 5-25), and which is called "kink effect". If the minority carrier lifetime in the silicon film is high enough, the kink effect can be reinforced by the NPN bipolar transistor structure present in the device (the "base" hole current is amplified by the bipolar gain, which gives rise to an increased net drain current, sometimes called "second kink").

Let us now consider the case of a *fully depleted* SOI n-channel transistor. Comparison between Figures 5-26A and 5-26B reveals that the electric field near the drain is lower in the fully depleted device than in the partially depleted case (the density of isopotential lines is lower). As a result, less electron-hole pair generation will take place in the fully depleted device.[108] As in the case of the partially depleted device, the generated electrons rapidly move into the channel and the drain, while the holes migrate towards the place of lowest potential, *i.e.*, near the source junction. However, contrary to the case of the partially depleted transistor, the source-to-body diode is "already forward biased" (the body-source potential barrier is very small), due to the full depletion of the film (Figure 5-26D), and the holes can readily recombine in the source without increasing the body potential (there is no significant potential barrier between the body and the source). As a result, the body potential remains unchanged, the body effect is virtually equal to zero, and there is no threshold voltage decrease as a function of drain voltage. This explains why fully depleted n-channel SOI MOSFETs are free of the kink effect. If a negative back-gate bias is used, however, to induce an accumulation layer at the back interface, the device behaves as a partially depleted device, and the kink reappears.

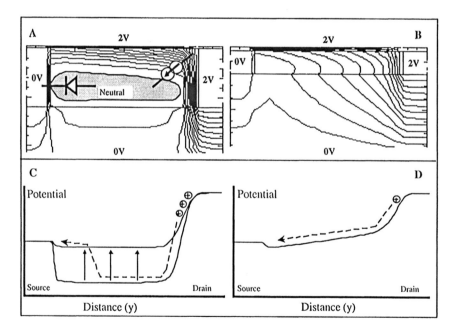

Figure 5-26. A: Isopotential lines in a partially depleted SOI n-channel MOSFET (one curve equals 200 mV). The grey area is the neutral, floating body; B:Isopotential lines in a fully depleted device; C: Potential in the neutral region from source to drain in the PD device before and after the onset of the kink effect (lower and upper curve, respectively); D: Potential from source to drain in the FD device.

For completeness, one can mention that p-channel SOI transistors are usually free of the kink effect, because the coefficient of pair generation by energetic holes is much lower than that of pair generation by energetic electrons. The kink effect is not observed in bulk devices if the majority carriers generated by impact ionization can escape into the substrate or to a well contact. If the well is left floating, or if the silicon is not conducting (*e.g.* at such low temperatures that the carriers are frozen out), the kink effect appears. Finally, the kink effect can be eliminated from partially depleted SOI MOSFETs if a substrate contact is provided for the removal of excess majority carriers from the device body. This technique is, unfortunately, not 100% effective because of the relatively high resistance of the body (see also section 5.10.1).

The presence of a floating substrate can also give rise to anomalous subthreshold slope effects [109]: when the body is floating, the relatively weak impact ionization that can occur near the drain in the subthreshold regime (in addition to the other carrier generation mechanisms) can result in holes being injected into the neutral body. The injection of holes in the body produces a forward bias on the body-source junction that reduces the threshold voltage while the device operates in the subthreshold region. As a result, the subthreshold current "jumps" from a high-V_{TH} characteristic to a low-V_{TH} $I_D(V_{G1})$ curve, and a subthreshold slope lower than 60 mV/decade can be observed at room temperature.[110]

5.9.2 Hot-carrier degradation

The horizontal electric field in the channel of a MOS transistor is roughly proportional to the ratio of the supply voltage to the gate length. Thus it increases as device dimensions are reduced, and, in modern MOSFETs, the electric field near the drain can reach values high enough to affect the reliability of the device. Let us take the example of an n-channel device. When the transistor operates in the saturation mode, a high electric field can develop between the channel pinchoff and the drain junction. This electric field gives rise to high-energy "hot" electrons which can be injected into the gate oxide, thereby damaging the oxide-silicon interface [111]. At very high injection levels, gate current can even be measured [112]. The high-energy channel electrons can also create electron-hole pairs by impact ionization. In a bulk device, the generated holes escape into the substrate and give rise to a substrate current. There exists a relationship between the substrate current and the gate current [113], and the lifetime of the device can be related to the magnitude of the hot-electron injection into the gate oxide. The device lifetime, τ, associated with the (static) hot-carrier degradation of the gate oxide, can be correlated to the impact-ionization current according to the following relationship: [114]

$$\tau \propto \frac{W}{I_D} (M\text{-}1)^{-m} \qquad (5.9.2)$$

where $m \cong 3$ [115], and M is the multiplication factor due to impact ionization, defined by $I_{body} = (M\text{-}1) I_{Dsat}$, where I_{body} is the hole current generated by impact ionization. In a fully depleted SOI MOSFET, the multiplication factor can be obtained by integrating the ionization coefficient over the high field region near the drain, and depends on both the drain and gate voltages [116]:

$$M\text{-}1 \cong \frac{A_i}{B_i} [V_{DS}\text{-}V_{Dsat}] \exp\left(-\frac{B_i l_c}{V_{DS}\text{-}V_{Dsat}}\right) \qquad (5.9.3)$$

where l_c is a characteristic length defined by: $l_c = t_{si} \sqrt{\dfrac{C_{si}\,\beta}{2 C_{ox1}\,(1+\alpha)}}$ with

$\beta = 1$ if the back interface is accumulated, and $\beta = 1 + \dfrac{C_{si}}{C_{si}+C_{ox2}}$ if the

back interface is depleted. A_i and B_i are impact ionization constants for electrons, assumed to be equal to 1.4×10^6 cm^{-1} and 2.6×10^6 V/cm, respectively. As in the case of the analysis of the saturation current, the transconductance and the subthreshold slope, α represents the ratio C_b/C_{ox1} of two capacitors: C_{ox1} is the gate oxide capacitance, and C_b is the capacitance between the inversion channel and the back-gate electrode. The value of C_b is either that of the depletion capacitance $C_D = \varepsilon_{si}/x_{dmax}$ in a bulk device, the silicon film capacitance $C_{si} = \varepsilon_{si}/t_{si}$ in a fully depleted device with back interface accumulation, or the value given by the series association of C_{si} and C_{ox2} ($C_b = \dfrac{C_{si}C_{ox2}}{(C_{si}+C_{ox2})}$) in a fully depleted device. Since the values of α are typically in the following sequence:

$$\alpha_{\text{fully depleted SOI}} < \alpha_{\text{bulk}} < \alpha_{\text{back accum SOI}}$$

The multiplication factor is lowest in the fully depleted device, higher in the bulk device, and the highest in the fully depleted device with back accumulation.

Figure 5-27 presents the multiplication factor *M-1* in bulk and fully depleted SOI MOSFETs of similar geometries. The multiplication factor is related to the lifetime of the device through expression (5.9.2). Clearly, a reduction in carrier-electron degradation can be expected from the fully depleted SOI MOSFETs. Numerical modeling of bulk and FDSOI devices shows that the electric field near the drain is minimized if fully depleted SOI devices are used. [117]

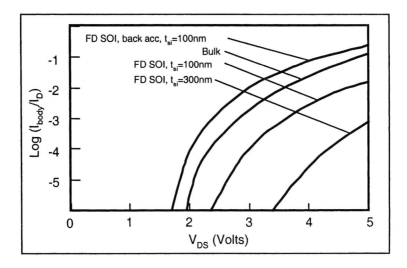

Figure 5-27. Multiplication factor $(M-1)=I_{body}/I_D$ as function of drain voltage in bulk and fully-depleted SOI nMOSFETs. The SOI MOSFETs have a thickness of either 100 or 300 nm, with or without back accumulation. L=1μm, V_G-V_{TH}=1.5 V.

Devices with back accumulation can present higher drain electric fields than bulk devices and can, therefore, be submitted to greater hot-carrier degradation. Finally, the use of thicker SOI films permits one to further minimize hot-electron degradation, provided the device remains fully depleted. Confirmation of the reduced hot-electron degradation in fully depleted SOI devices has been experimentally obtained. Both threshold voltage shift and transconductance degradation as a function of stress time at high V_{DS} have been found to be significantly smaller in fully depleted devices with back depletion than in either bulk devices or fully depleted SOI devices with back-interface accumulation.

It seems that there is no consensus on the hot-carrier degradation issues in SOI devices. Several papers have been published on the topic, leading to contradictory conclusions. Several publications support the idea that SOI devices are less prone to hot-carrier degradation than bulk devices, while others draw the opposite conclusion. These discrepancies are partly due to the fact that it is almost impossible to make one-to-one comparisons between fully depleted SOI and bulk devices. Indeed, if the same channel doping concentration is used in both types of devices, the threshold voltages and saturation currents will be different. If the threshold voltages are similar, the doping profiles are different, etc. In addition, the drain breakdown characteristics are different as well. There is some agreement, however, on the following point: the peak drain electric field is lower in fully depleted SOI devices than in bulk devices, leading to less hot-carrier generation.[118,119,120,121]

Because the degradation mechanism is not only a function of the hot-carrier generation rate, but includes on many other different parameters (the quality of the buried oxide, the direction and magnitude of the vertical electric field, etc.) the hot carriers degrade SOI devices in a different way than they do in bulk devices.[122,123] One difference from bulk is the carrier injection in the buried oxide, although some authors report degradation of the front oxide only.[124]

Figure 5-28 presents the threshold voltage shift and the transconductance degradation in SOI and bulk transistors. Much larger initial transconductance degradation, a more pronounced initial threshold voltage shift, and a weaker time dependence are observed in the SOI devices. The most affected parameter in the SOI devices is the threshold voltage. The degradation depends on both front- and back-gate bias.[125,126] Depending on the applied front and back gate bias either holes or electrons can be injected in the buried oxide. A t^n time dependence of the degradation is observed, with $n=0.25$ in the case of pure hole or pure electron injection, and $n=0.5$ when both holes and electrons are injected in the buried oxide.[127,128] The complex interaction between front and back interface degradation due to the combined injection of hot electrons and holes, combined with the influence of floating-body and bipolar effects, is described in the literature.[129]

It has also been observed that the substrate current generated by hot-carrier effects decreases in SOI devices once the channel length is shorter than 0.2 μm (Figure 5-29). This is caused by velocity saturation overshoot of the carriers in the channel. These can then reach the drain without loosing energy via an impact ionization process. This observation further supports the use of SOI for very short channel devices. Finally, it appears that devices made in lightly doped, ultrathin SOI layers have excellent hot-carrier reliability characteristics.[130]

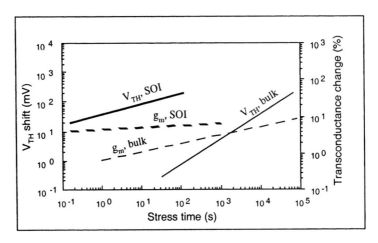

Figure 5-28. Threshold voltage shift and transconductance degradation in SOI and bulk n-channel transistors as a function of stress time. L=0.5 μm, V_{GS}=2 V, and V_{DS}=6 V.[131]

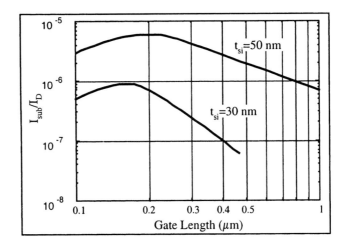

Figure 5-29. I_{sub}/I_D versus gate length in FD SOI MOSFETs for two different values of the silicon film thickness.[132]

5.10 FLOATING-BODY AND PARASITIC BJT EFFECTS

The neutral body of a partially depleted SOI MOSFETs is electrically floating if no connection is made to it. The potential of the body is determined by currents flowing from and to the body, as well as by the capacitive coupling between the body and the source, drain, front and back gate. Since the body potential influences the threshold voltage, body voltage transients result in threshold voltage and drain current instabilities. Figure 5-

30A shows the capacitive coupling of the body potential to other terminals, as well as the two junction diodes and the current source associated with impact ionization (see section 5.9.1). Furthermore, there exists a parasitic bipolar transistor in the MOS structure. If we consider an n-channel device, the N⁺ source, the P-type body and the N⁺ drain also form the emitter, the base, and the collector of an NPN bipolar transistor (BJT) (Figure 5-30B). In a bulk device, the base of the bipolar transistor is usually grounded by means of a substrate contact. In an SOI device, however, the body (= the base of the bipolar transistor) is floating if no body contact is provided. This parasitic bipolar transistor can amplify impact ionization current and become the origin of several undesirable effects in SOI devices, and can trigger an extremely low inverse subthreshold slope (S is very small), and reduce the drain breakdown voltage.

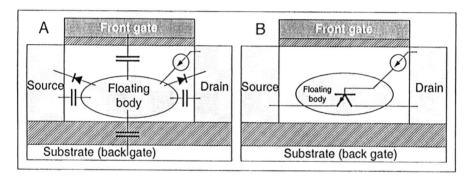

Figure 5-30. Partially depleted SOI MOSFET. A: Capacitors and diodes connecting the floating body to other terminals. B: Parasitic bipolar transistor.

5.10.1 Anomalous subthreshold slope

The generation of majority carriers (holes in the case of an n-channel transistor) by impact ionization near the drain can give rise to an increase of the body potential and decrease of the threshold voltage (see section 5.9.1). Sometimes, a similar effect can occur at gate voltages lower than the threshold voltage. If the drain voltage is high enough, impact ionization can occur in the subthreshold region, even though the drain current is very small. This effect is observed in partially depleted devices and in fully depleted devices with back accumulation, and can be explained as follows. When the device is turned off, there is no impact ionization, and the body potential is equal to zero, since there is no base-to-source current. When the gate voltage is increased the weak inversion current can induce impact ionization in the high electric field region near the drain, holes are generated, the body potential increases, and the threshold voltage is reduced. Consequently, the whole $I_D(V_{GS})$ characteristic shifts to the left, and the current can increase with gate voltage with a slope larger than 60 mV/decade. In other words, inverse subthreshold slopes lower than the theoretical limit (in absence of the multiplication effect) of 60 mV/decade can be observed.[133,134,135]

If the minority carrier lifetime in the silicon film is high enough, the parasitic bipolar transistor present in the NPN structure of the MOS device can amplify the base current (*i.e.* amplify the hole current generated by impact ionization near the drain). The base current is given by:

$$I_{body} = (M-1) I_{Dsat} + (M-1) I_{ch} \qquad (5.10.1)$$

where I_{ch} is the channel current, M is the multiplication factor, and the resulting increase of drain current is given by $\Delta I_D = \beta_F I_{body} = \beta_F (M-1) I_{ch}$ where β_F is the common-emitter current gain of the BJT (Figure 5-31).

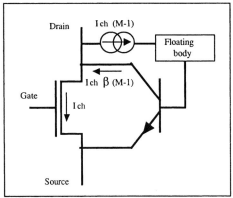

Figure 5-31. Parasitic bipolar transistor of the SOI MOSFET. I_{ch} is the channel current.

This increase of drain current constitutes a positive feedback loop on the current flowing through the device: the drain current suddenly increases, and a subthreshold slope equal to zero millivolt per decade is observed (Figure 5-32). This phenomenon is known as the "single-transistor latchup".[136,137,138,139] It can be explained as follows (Figure 5-32). At low drain bias (curve a), a normal subthreshold slope is obtained, under forward as well as reverse gate voltage conditions. If the drain voltage is increased (curve b), the impact ionization near the drain raises the body potential (forward gate voltage case). This reduces the threshold voltage and leads to an increase of the drain current, which in turn increases impact ionization near the drain. When the gain of the positive feedback loop, $\beta_F(M-1)$, reaches unity, the current suddenly increases. The positive feedback is self-limiting; increased body bias also increases the drain saturation voltage, which results in a lower electric field near the drain and a smaller impact ionization current. During a descending gate voltage scan (curve b), the impact ionization current under a large drain voltage keeps the body voltage high, which in turn keeps the threshold voltage of the device low, and a high drain current is observed until the positive feedback can no longer be maintained, so that $\beta_F(M-1) < 1$, and the drain current suddenly drops. It can be noticed that the gate voltage at which the current suddenly raises, during a forward scan, is higher than the voltage at which the current suddenly falls, during a reverse scan. As a consequence, hysteresis is observed in the $I_D(V_G)$ curve (curve b). If the drain bias is large enough, the positive feedback loop cannot be turned off once it has been triggered (since $\beta_F(M-1)$ remains larger or equal to 1), and the device does not turn off (curve c). It is worth noting that this parasitic phenomenon was not observed in devices made in early SOI material, where the recombination lifetime of minority carriers in the base was low and where the gain of the bipolar transistor, β_F, was small. The single-transistor latch effect is not observed if the body of the device is grounded, and it is reduced if the device is fully depleted.

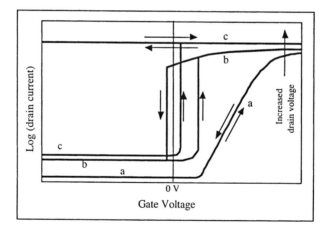

Figure 5-32. Illustration of the single-transistor latch. "Normal" subthreshold slope at low drain voltage (a), infinite subthreshold slope and hysteresis (b), and device "latch-up" (c).

5.10.2 Reduced drain breakdown voltage

It has been demonstrated that the peak electrical field at the drain junction of SOI devices is lower than what is commonly found in bulk devices if the junction abuts the buried oxide [140]. From this result, a higher junction breakdown voltage could be expected in SOI transistors compared to bulk. Unfortunately, the SOI MOSFET includes a parasitic bipolar transistor with floating base. From bipolar transistor theory [141], the collector (drain) breakdown voltage with open base, BV_{CE0}, is smaller than when the base is grounded, BV_{CB0}. Both breakdown voltages relate as follows:

$$BV_{CE0} = \frac{BV_{CB0}}{\sqrt[n]{\beta_F}} \qquad (5.10.2)$$

where β_F is the gain of the bipolar device, and n typically ranges between 3 to 6. The above relationship is quite a simplification of the breakdown mechanisms occurring in SOI MOSFETs, since both β_F and M-1 (the multiplication factor) depend on the drain voltage in a highly nonlinear fashion.[142] From bipolar transistor theory, one can also write:

$$\beta_F \cong 2 \, (L_n/L_B)^2 - 1 \qquad (5.10.3)$$

where L_B is the base width, which can be assumed, in a first approximation, to be equal to the effective channel length, L, and L_n is the electron diffusion length: $L_n^2 = D_n \, \tau_n$, where D_n and τ_n are the diffusion coefficient and the lifetime of the minority carriers in the base, respectively. From (5.7.1) and (5.7.2), and taking BV_{CE0} as the drain breakdown voltage with floating body equal to BV_{DS}, one obtains:

$$B_{VDS} = \frac{BV_{CB0}}{\sqrt[n]{\dfrac{2D_n\tau_n}{L^2} - 1}} \qquad (5.10.4)$$

Using Einstein's relationship $D_n = \dfrac{kT}{q} \, \mu_n$, an family of curves relating BV_{DS}/BV_{CB0} to channel length can be plotted, with the minority carrier lifetime in the SOI material as parameter (Figure 5-33) [143]. It should, however, be noted that equations 5.10.3 and 5.10.4, and hence Figure 5-33 are valid only if $L_n > L$, *i.e.* if β_F is significantly larger than unity. This implies that the device has a short channel and that the lifetime is relatively long. A closed-form model for breakdown voltage in partially depleted SOI MOSFETs can be found in the literature.[144]

This reduction of the drain breakdown voltage was not observed in devices made early on with higly defective SOI material, where the minority carrier lifetime was low, and its control constitutes a challenges for current SOI research [145], especially when deep-submicron devices are considered.

Possible solutions of the problem include the use of lightly-doped source and drain extensions (LDD) [146], the introduction of lifetime killers by *e.g.* argon implantation [147,148,149,150], growth of silicides on source and drain [151,152], the use of SiGe source and drain [153,154,155], and the use of a body contact.[156,157]

Finally, it should be noted that the above model stems from the presence of a neutral base region, and is, therefore, valid only for partially depleted devices. The understanding of parasitic bipolar effects in fully depleted devices necessitates the development of a model for a bipolar transistor with fully depleted base.[158,159] Numerical simulations show that, even though no kink is discernible in fully depleted SOI MOSFETs, these devices are subjected to significant body charging by impact ionization, which results in significant parasitic bipolar effects. The drain breakdown voltage is controlled by the common-emitter bipolar breakdown voltage BV_{CE0} which occurs under the condition that $\beta_F(M-1)=1$. Both β_F and M are highly nonlinear. M increases strongly with V_{DS} due to the increasing drain electric field, and β_F decreases strongly with V_{DS} due to the the onset of high injection in the body (base of the bipolar device).

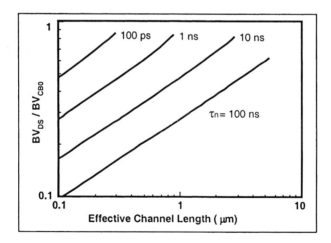

Figure 5-33. Relationship between the reduction of drain breakdown voltage, the gate length and the minority carrier lifetime (τ_n) in SOI MOSFETs.

5.10.3 Other floating-body effects

The presence of a floating body gives rise to a series of problems of transient effects, hysteresis, etc. All these effects have their origin in the charging/discharging of the floating body by currents coming from the source or the drain and in the capacitive coupling between the gate and the floating body. One can list the following effects:

- Transient leakage current in pass transistors, which can affect SOI SRAM and DRAM circuitry.[160,161,162,163,164]

- Drain current transient or overshoot when a voltage step is applied to the gate (or drain) due to capacitive coupling between the gate (or drain) and the floating body – Figure 5-34.[165,166,167,168]

- Variations of drain current and propagation delays depending on device switching history (history effect) [169,170,171,172,173,174] or clock frequency.[175]

- Irregular signal propagation (frequency-dependent delay times, pulse stretching).[176,177,178] High-frequency operation may not allow the device enough time to return to an equilibrium state between pulses (Figure 5-34).

- Degradation of logic states in dynamic circuits.[179]

- Double snapback of the $I_D(V_D)$ characteristics.[180,181,182]

- Dependence of characteristics on gate current. This happens when ultrathin gate oxides are used and tunneling current occurs between gate and body.[183,184,185,186]

- Occurrence of a transconductance peak in the linear mode of operation due to the tunnel current between the gate and the floating body (linear kink effect, LKE).[187,188]

The classical remedy to floating-body problems is the use of a body contact (or body tie) in partially depleted devices.[189] In fully depleted devices there are, usually, no floating body problems, and if there are, these are much weaker than in partially depleted transistors.

Circuit design using partially depleted SOI MOSFETs without body contacts requires careful modeling of floating-body effects. Advantage can be taken of current overshoot effects and increased current drive due to the kink effect to boost switching speed and, therefore, circuit performance.[190] However, elaborate circuit models need to be used to avoid unexpected device behavior due to floating body effects.[191,192]

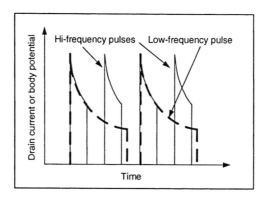

Figure 5-34. Drain current or body potential in a partially depleted SOI transistor clocked with low- or high-frequency pulses.[193,194]

The presence of an electrically floating *back gate* underneath an SOI MOSFET can influence the high-frequency response of the output conductance. This effect, called the floating effective back-gate (FEBG) effect depends strongly on substrate doping and is caused by the variation of potential at the substrate-buried oxide interface. A lower doping concentration in the substrate results in a more pronounced FEBG effect.[195]

5.11 SELF HEATING

SOI transistors are thermally insulated from the substrate by the buried insulator. As a result, removal of excess heat generated within the device is less efficient than in bulk, which results in substantial elevation of device temperature.[196,197] The conduction paths for excess heat are multiple: heat diffuses mostly vertically through the buried oxide but some of it is conducted laterally through the silicon island into the contacts and the metallization.[198,199]

Due to self-heating, a negative resistance can be seen in the output characteristics of SOI MOSFETs.[200] It is due to a mobility reduction effect caused by the elevation of temperature.[201] This effect is clearly visible on the output curves if sufficient power is dissipated in the device. Because of the relatively low thermal conductivity of the buried oxide, devices can heat up by 50 to 150°C and a mobility reduction is observed.[202]. This effect has been included in device and circuit simulators.[203]

One should not forget, however, that this effect takes place as power is dissipated in the device. This is the case when the device is measured in a quasi-dc mode with a curve tracer or an HP4145, but not in an operating CMOS circuit. Indeed, in an operating CMOS circuit, there is virtually no current flowing through the devices in the standby mode, and power is dissipated in the devices only during switching for brief periods of time ($<$ 1 nanosecond). Dynamic power dissipation ($\frac{1}{2} fCV^2$) must be taken account as well, especially if the operating frequency, f, is high.[204]

It has been shown by a pulsed measurement technique [205] that the time constants involved in the self heating of SOI transistors are on the order of several tens of nanoseconds, and that no negative resistance effect is observed when the devices are measured in the pulse mode. In pulse mode the time during which power is dissipated is much shorter than the thermal time constant of the devices. A similar experiment reports that the self-heating does not influence the output characteristics of transistors, if the measurement is carried out at a slew rate higher than 20 V/μs.[206] It seems thus that the negative resistance is not a problem for digital circuits (with the possible exception of output buffers). There might, however, be an influence of the duty cycle and the frequency at which the devices are switched on the overall local temperature, which could modify the mobility.

For analog circuits, some heating effects are observed as well. For instance, the output conductance (g_D) of a transistor becomes frequency dependent because of the self-heating effect. At low frequencies the self-heating mechanism can follow the signal, and a reduction of g_D is observed. Above 100 kHz, an increase in conductance is observed. Moreover, three different conductance increases have been observed in SIMOX devices, with time constants approximately equal to 1μs, 100 ns, and 3 ns. There can be up to a factor of two difference between the high- and low-frequency output conductance values.[207] The heat can also propagate from one device to another, and thermal coupling effects can be observed in sensitive structures such as current mirrors.[208] Self-heating can have a significant effect on commonly used analog circuit blocks fabricated in SOI. Analog designers using SOI MOSFETs must take into account self-heating, and hence must have circuit simulation models in which this effect is included.[209]

5.12 ACCUMULATION-MODE MOSFET

It is possible to fabricate SOI MOSFETs in which the doping polarity of the body is the same as that of the source and drain. Such a device contains no PN junctions, but its electrical characteristics are very similar to those of a "regular", inversion-mode SOI MOSFET. The following sections describe the physics and the electrical characteristics of accumulation-mode MOSFETs.

5.12.1 I-V characteristics

We will take the example of a p-channel device. The source and drain are P+-doped and the body is P-type (Figure 5.35). The gate material is N+ polysilicon. When the device is turned OFF, the silicon film is fully depleted due to the presence of positive interface charges and to the negative value of the work function difference between the N+ polysilicon gate and the P-type body of the device. When the device is turned ON, the film is no longer fully depleted and conduction occurs in both the body of the device and at the accumulated surface channel. [210,211]

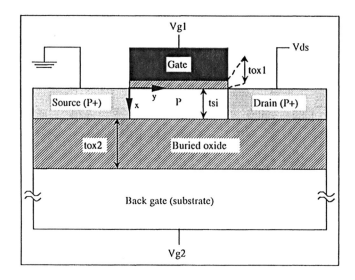

Figure 5-35. Cross-section of a p-channel SOI MOSFET illustrating some of the notations used in this section.

When a zero bias is applied to the gate, the film is fully depleted due to the use of an N+-polysilicon gate. The gate material-silicon work function difference is given by:

$$\Phi_{MS1} = -\frac{E_g}{2} - \frac{kT}{q} \, ln \frac{N_a}{n_i} \qquad (5.12.1)$$

where n_i is the intrinsic carrier concentration in silicon. When a negative voltage is applied to the gate, the hole concentration in the silicon film increases, the depletion charge decreases and eventually disappears. At that point, any further negative increase in gate voltage supports the formation of a surface accumulation channel. Threshold of accumulation is reached when the front surface potential is equal to zero ($\Phi_{s1}=0$). The gate voltage needed to reach accumulation is given by:

$$V_{TH,acc} = \Phi_{MS1} - \frac{Q_{ox1}}{C_{ox1}} = V_{fb1} \qquad (5.12.2)$$

where V_{fb1} is the front flat-band voltage, Q_{ox1} is the charge density per unit area in the gate oxide, and C_{ox1} is the gate oxide capacitance per unit area. When the gate voltage, V_{G1}, is lower (*i.e.* larger in absolute value) than the accumulation threshold voltage, the accumulation charge per unit area in the channel is given by: $Q_{acc}(y) = -[V_{G1} - V_{fb1} - V(y)] \, C_{ox1}$, where $V(y)$ is the local potential along the channel (y=0 at the source junction and y=L at the drain junction). Using the gradual channel approximation, one obtains the following expression for the current in the accumulation channel:

$$\int_0^{V_{DS}} dV = I_{acc} \int_0^L dR(y) \quad \text{with} \quad dR(y) = \frac{dy}{W \, \mu_s \, Q_{acc}} \qquad (5.12.3)$$

The integration of 5.12.3 yields

$$I_{acc} = \frac{W}{L} \, \mu_s \, C_{ox1} \left((V_{G1} - V_{fb1}) \, V_{DS} - \frac{V_{DS}^2}{2} \right) \qquad (5.12.4)$$

in the the non-saturated regime ($V_{DS} > V_{G1} - V_{fb1}$), and

$$I_{Dsat,acc} = \frac{W}{L} \, \mu_s \, \frac{C_{ox1}}{2} \, (V_{G1} - V_{fb1})^2 \qquad (5.12.5)$$

above saturation ($V_{DS} < V_{G1} - V_{fb1}$). In these expressions μ_s is the hole surface mobility, which is equal to $\dfrac{\mu_{s0}}{1+\theta(V_{fb1}-V_{G1})}$, with μ_{s0} being the zero-field surface mobility, and θ being a field mobility reduction factor.

The other current flowing from source to drain is the body current, which appears if there is a portion of non-depleted silicon below the channel. This current component is similar to the current flowing in a JFET. Conduction between source and drain occurs in the neutral part of the body. The width of this conduction path is modulated by the vertical extension of the depletion zones related to the front and the back gate. The front gate depletion depth can be found by solving the Poisson equation using the depletion approximation:

$$\frac{d^2 \Phi(x)}{dx^2} = \frac{q N_a}{\varepsilon_{si}} \qquad (5.12.6)$$

where $\Phi(x)$ is the potential in the x-direction (from the front oxide-silicon interface into the silicon film), and N_a is the acceptor doping concentration. We find:

$$-E(x) = \frac{q N_a x}{\varepsilon_{si}} + A \quad \text{and} \quad \Phi(x) = \frac{q N_a x^2}{2\varepsilon_{si}} + Ax + B$$

where A and B are integration constants. Using the following boundary conditions: $E(x_{depl})=0$ and $\Phi(x_{depl})=0$ where x_{depl} is the depletion depth we find the surface potential:

$$\Phi(0) = \frac{q N_a x_{depl}^2}{2\varepsilon_{si}}$$

Using Gauss' law at the front Si-SiO$_2$ interface, we finally obtain:

$$V_G - V_{FB} = \frac{q N_a x_{depl}^2}{2\varepsilon_{si}} + \frac{q N_a x_{depl}}{C_{ox}} \qquad (5.12.7)$$

The gate voltage for which the depletion depth equals the silicon thickness is:

$$V_G = V_{FB} + \frac{q N_a t_{si}^2}{2\varepsilon_{si}} + \frac{q N_a t_{si}}{C_{ox}} \equiv V_{FB} + V_{depl} \qquad (5.12.8)$$

V_{depl} being defined by relationship 5.12.8.

As long as there is a portion of neutral silicon in the film (at a distance y from the source), the depth of the front depletion region is given by:

$$X_{d1}(y) = \frac{-\varepsilon_{si}}{C_{ox1}} + \sqrt{\varepsilon_{si}^2/C_{ox1}^2 + 2\,\varepsilon_{si}\,(V_{G1} - V_{fb1} - V(y))/q\,N_a} \qquad (5.12.9)$$

The depth of the depletion zone at the back interface is:

$$X_{d2}(y) = \frac{-\varepsilon_{si}}{C_{ox2}} + \sqrt{\varepsilon_{si}^2/C_{ox2}^2 + 2\,\varepsilon_{si}\,(V_{G2} - V_{fb2} - V(y))/q\,N_a} \qquad (5.12.10)$$

If $X_{d1}(y) + X_{d2}(y)$ is equal to or larger than t_{si}, the silicon film thickness, the film is locally fully depleted. For the sake of simplicity, we will assume that the width of the back depletion region, X_{d2}, is independent of y and given by expression 5.12.10 where $V(y)$ is held constant and equal to $V(y=0)$. The resistance of an elementary resistor in the body channel is given by:

$$dR = \frac{dy}{W \mu_b q N_a (t_{eff} - X_{d1})} \qquad (5.12.11)$$

where t_{eff} is equal to $t_{si}-X_{d2}$ and μ_b is the bulk hole mobility. Using again the gradual channel approximation, and integrating $dV=IdR$ from source to drain for the various operation modes of the device, and using Equation (5.12.11) to find R and Equation (5.12.9) to calculate X_{d1}, one obtains the following expression:

$$I_{body} \int dy = W q N_a \mu_b \int [t_{eff} + \frac{\varepsilon_{si}}{C_{ox1}} - \sqrt{\frac{\varepsilon_{si}^2}{C_{ox1}^2} + \frac{2\varepsilon_{si}(V_{G1}-V_{fb1}-V(y))}{q N_a}}] \, dV$$

$$(5.12.12)$$

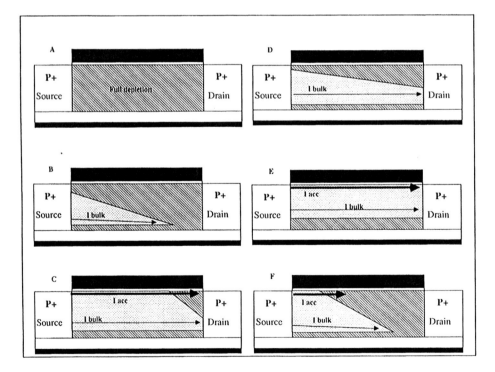

Figure 5-36. Cross-section of a p-channel accumulation-mode SOI MOSFET illustrating the possible distributions of depletion and neutral zones.

The integration of 5.12.12 from source to drain yields an expression in the form:

$$I_{body} = q\, N_a\, \mu_b \frac{W}{L}\, \xi$$

where ξ depends on the bias applied to the drain, front and back gate. The value of ξ depends on whether the film is fully depleted or not, whether the neutral zone extends all the way to the drain, etc. We will also define V'_{depl} as the V_{depl} of equation 5.12.8 where $t_{si}=t_{eff}$, or:

$$V'_{depl} \equiv \frac{q\, N_a\, t_{eff}^2}{2\varepsilon_{si}} + \frac{q\, N_a\, t_{eff}}{C_{ox}}$$

The following cases can be distinguished:

1. $\xi = 0$ when $V_{G1} > V_{fb1}+V'_{depl}$, i.e. when the film is fully depleted. V'_{depl} is the change in front gate voltage with respect to flatband necessary to obtain $X_{d1} = t_{eff}$ near the source where $V(y)=0$ using equation 5.12.9. There is no accumulation channel (Figure 5-36A).

2. $\xi = t_{eff}\, V_{DS}$ if $V_{G1} - V_{fb1} < 0$ and $V_{G1} - V_{fb1} - V_{DS} < 0$ (Figure 5-36E), i.e. when the front interface is in accumulation from source to drain.

3. In the case where $0 < V_{G1}-V_{fb1} < V'_{depl}$ and $0 < V_{G1}-V_{fb1}-V_{DS} < V'_{depl}$, i.e. when neither the accumulation channel nor the neutral body is pinched off (Figure 5-36D), equation 5.12.12 can be written as follows:

$$I_{body}\int_0^L dy = W\, q\, N_a\, \mu_b \int_0^{V_{DS}} \left[t_{eff} + \frac{\varepsilon_{si}}{C_{ox1}} - \sqrt{\frac{\varepsilon_{si}^2}{C_{ox1}^2} + \frac{2\varepsilon_{si}\,(V_{G1}-V_{fb1}-V(y))}{q\, N_a}}\, \right] dV$$

which yields:

$$\xi = (t_{eff} + \varepsilon_{si}/C_{ox1})\, V_{DS} + \frac{q N_a}{3\varepsilon_{si}}\, [(\varepsilon_{si}/C_{ox1})^2 + \frac{2\varepsilon_{si}}{q N_a}(V_{G1}-V_{fb1}-V_{DS})\,]^{3/2}$$

$$-\frac{q N_a}{3\varepsilon_{si}}\, [(\varepsilon_{si}/C_{ox1})^2 + \frac{2\varepsilon_{si}}{q N_a}(V_{G1}-V_{fb1})\,]^{3/2} \qquad (5.12.13)$$

4. In the case where $0 < V_{G1}-V_{fb1} < V'_{depl}$ and $V_{G1}-V_{fb1}-V_{DS} > V'_{depl}$, i.e. when body channel is pinched off, but there is no accumulation at the source end nor full depletion of the film (Figure 5-36B), we have:

$$I_{body} \int_0^L dy = WqN_a\mu_b \int_0^{V_{G1}-V_{FB1}-V'_{depl}} [t_{eff}+\frac{\varepsilon_{si}}{C_{ox1}}-\sqrt{\frac{\varepsilon_{si}^2}{C_{ox1}^2}+\frac{2\varepsilon_{si}(V_{G1}-V_{fb1}-V(y))}{qN_a}}] \, dV$$

which yields:

$$\xi = (t_{eff}+\varepsilon_{si}/C_{ox1})(V_{G1}-V_{fb1}-V'_{depl})+\frac{qN_a}{3\varepsilon_{si}}[(\varepsilon_{si}/C_{ox1})^2+\frac{2\varepsilon_{si}}{qN_a}V'_{depl}]^{3/2}$$

$$-\frac{qN_a}{3\varepsilon_{si}}[(\varepsilon_{si}/C_{ox1})^2+\frac{2\varepsilon_{si}}{qN_a}(V_{G1}-V_{fb1})]^{3/2} \tag{5.12.14}$$

5.   In the case where $0 > V_{G1}-V_{fb1}$ and $0 < V_{G1}-V_{fb1}-V_{DS} < V'_{depl}$, i.e. when body channel is not pinched off and there is accumulation at the source end (Figure 5-36C), we have:

$$I_{body}\int_0^L dy = WqN_a\mu_b \int_{V_{G1}-V_{FB1}}^{V_{DS}} [t_{eff}+\frac{\varepsilon_{si}}{C_{ox1}}-\sqrt{\frac{\varepsilon_{si}^2}{C_{ox1}^2}+\frac{2\varepsilon_{si}(V_{G1}-V_{fb1}-V(y))}{qNa}}] \, dV + WqN_a\mu_b \int_0^{V_{G1}-V_{FB1}} t_{eff} \, dV$$

which yields:

$$\xi = t_{eff}(V_{G1}-V_{fb1})+(t_{eff}+\varepsilon_{si}/C_{ox1})(V_{DS}-V_{G1}+V_{fb1})$$

$$+\frac{qN_a}{3\varepsilon_{si}}[(\varepsilon_{si}/C_{ox1})^2+\frac{2\varepsilon_{si}}{qN_a}(V_{G1}-V_{fb1}-V_{DS})]^{3/2}$$

$$-\frac{qN_a}{3\varepsilon_{si}}[\varepsilon_{si}/C_{ox1}]^3 \tag{5.12.15}$$

6. In the case where $0 > V_{G1}-V_{fb1}$ and $V_{G1}-V_{fb1}-V_{DS} > V'_{depl}$, i.e. when body channel is pinched off and there is accumulation at the source end. (Figure 5-36F), we have:

$$I_{body}\int_0^L dy = WqN_a\mu_b \int_{V_{G1}-V_{FB1}}^{V_{G1}-V_{FB1}-V'_{depl}} [t_{eff}+\frac{\varepsilon_{si}}{C_{ox1}}-\sqrt{\frac{\varepsilon_{si}^2}{C_{ox1}^2}+\frac{2\varepsilon_{si}(V_{G1}-V_{fb1}-V(y))}{qNa}}] \, dV + WqN_a\mu_b \int_0^{V_{G1}-V_{FB1}} t_{eff} \, dV$$

which yields:

$$\xi = t_{eff}(V_{G1}-V_{fb1})-(t_{eff}+\varepsilon_{si}/C_{ox1})(V'_{depl})$$

$$+\frac{qN_a}{3\varepsilon_{si}}[(\varepsilon_{si}/C_{ox1})^2+\frac{2\varepsilon_{si}}{qN_a}V'_{depl}]^{3/2}$$

$$-\frac{qN_a}{3\varepsilon_{si}}[\varepsilon_{si}/C_{ox1}]^3 \tag{5.12.16}$$

Finally, the total drain current in the device is given by the sum of the current in the accumulation channel and the current in the body of the device:

$$I_{DS} = I_{acc} + I_{body} \qquad (5.12.17)$$

Figure 5-37 presents the linear $I_D(V_{G1})$ characteristics of an accumulation-mode, p-channel transistor for different back-gate voltages. The *apparent* front threshold voltage of this device shows a dependence on back-gate bias similar to that observed in a fully depleted, enhancement-mode device. Actually, the front accumulation threshold voltage is independent of back-gate bias, and the shift of the $I_D(V_{G1})$ characteristics with (more negative) back-gate bias is due to the onset and increase of the body current. At large negative back biases, even an accumulation channel can be created at the back interface. This case, however, was not dealt with in the above model, since the presence of a back accumulation channel is undesirable for most practical applications. Figure 5-38 presents the output characteristics of an accumulation-mode, p-channel SOI device (t_{si}=100 nm, N_a=4x10^{16} cm^{-3}). Device characteristics such as output conductance and gate capacitance can be derived from the above model.[212]

In most applications, p-channel SOI MOSFETs are operating with a negative back-gate bias. If one considers, for instance, the case of a CMOS inverter, the source of the p-channel device is connected to $+V_{DD}$ (*e.g.* 5 volts). If the mechanical substrate (the back gate) is grounded, the p-channel transistor is then operating with a back-gate bias of $-V_{DD}$ (*e.g.* -5 V).

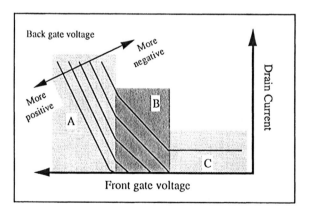

Figure 5-37. Linear $I_D(V_{Gl})$ characteristics of an accumulation-mode, p-channel transistor for different back-gate voltage biases. The front accumulation current, the body current, and the

back channel accumulation current are outlined by the shaded zones A, B, and C, respectively.

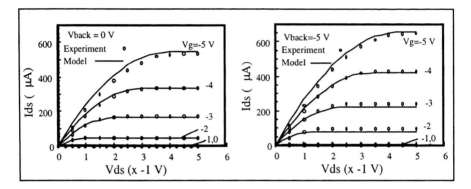

Figure 5-38. Output characteristics of an accumulation-mode, p-channel SOI device, for two values of back-gate bias.

Good performance is usually obtained from accumulation-mode p-channel MOSFETs. Because of the reduced scattering encountered by the bulk current, a higher mobility is observed in these devices compared to enhancement-mode p-channel MOSFETs [213]. Accumulation-mode ($N^+N^-N^+$) n-channel MOSFETs can also be fabricated, using either P^+ polysilicon or N^+ polysilicon as gate material.[214] In the latter case, however, devices with negative threshold voltage are obtained. The short-channel characteristics of accumulation-mode MOSFETs are are more severe than inversion-mode devices because they usually have a lower body doping concentration than their accumulation-mode counterparts. This is not the case if ultrathin silicon films are used to fabricate the devices.[215]

5.12.2 Subthreshold slope

In the subthreshold regime (when neither body nor accumulation currents are flowing), the body of the device is fully depleted because of the presence of positive oxide charges in the gate oxide and the buried oxide, and because of the work function difference between the N^+ polysilicon gate and the p-type body. When the body is fully depleted, the classical expressions developed to describe the relationship between the gate voltages and the surface potentials in fully depleted, enhancement-mode, *n-channel* MOSFETs can be employed for the accumulation-mode p-channel device (equations 5.3.8 and 5.3.9) [216,217]:

$$V_{G1} = V_{FB1} + \Phi_{s1}\left(1 + \frac{C_{it1}}{C_{ox1}} + \frac{C_{si}}{C_{ox1}}\right) - \Phi_{s2}\frac{C_{si}}{C_{ox1}} + \frac{Q_D}{2C_{ox1}}$$

and

$$V_{G2} = V_{FB2} + \Phi_{s2}\left(1 + \frac{C_{it2}}{C_{ox2}} + \frac{C_{si}}{C_{ox2}}\right) - \Phi_{s1}\frac{C_{si}}{C_{ox2}} + \frac{Q_D}{2C_{ox2}}$$

where $V_{FB1} = \Phi_{MS1} - \dfrac{qN_{ox1}}{C_{ox1}}$ is the front flatband voltage, $V_{FB2} = \Phi_{MS2} - \dfrac{qN_{ox2}}{C_{ox2}}$ is the back flatband voltage, and $Q_D = qN_a t_{si}$ is the absolute value of the total depletion charge in the silicon film. V_{G1}, V_{G2}, Φ_{s1} and Φ_{s2} are the front and back gate voltages (the source being grounded and taken as reference voltage), and the front and back surface potentials in the silicon film, respectively. Using the above equations, the surface potentials can be expressed as a function of the front and back gate voltages:

$$\Phi_{s1} = \frac{V_{G1} - V_{FB1} + \dfrac{C_{si}}{C_{ox1}}\left(\dfrac{V_{G2} - V_{FB2} - \dfrac{Q_D}{2C_{ox2}}}{1 + \dfrac{C_{it2}}{C_{ox2}} + \dfrac{C_{si}}{C_{ox2}}}\right) - \dfrac{Q_D}{2C_{ox1}}}{\left(1 + \dfrac{C_{it1}}{C_{ox1}} + \dfrac{C_{si}}{C_{ox1}}\right) - \dfrac{\dfrac{C_{si}}{C_{ox1}}\dfrac{C_{si}}{C_{ox2}}}{1 + \dfrac{C_{it2}}{C_{ox2}} + \dfrac{C_{si}}{C_{ox2}}}}$$
$$(5.12.18a)$$

and

$$\Phi_{s2} = \frac{V_{G2} - V_{FB2} + \dfrac{C_{si}}{C_{ox2}}\left(\dfrac{V_{G1} - V_{FB1} - \dfrac{Q_D}{2C_{ox1}}}{1 + \dfrac{C_{it1}}{C_{ox1}} + \dfrac{C_{si}}{C_{ox1}}}\right) - \dfrac{Q_D}{2C_{ox2}}}{\left(1 + \dfrac{C_{it2}}{C_{ox2}} + \dfrac{C_{si}}{C_{ox2}}\right) - \dfrac{\dfrac{C_{si}}{C_{ox2}}\dfrac{C_{si}}{C_{ox1}}}{1 + \dfrac{C_{it1}}{C_{ox1}} + \dfrac{C_{si}}{C_{ox1}}}}$$
$$(5.12.18b)$$

Equation 5.3.2 yields the vertical potential distribution in the film:

$$\Phi(x) = \frac{q N_a}{2 \varepsilon_{si}} x^2 + \left(\frac{\Phi_{s2} - \Phi_{s1}}{t_{si}} - \frac{q N_a t_{si}}{2 \varepsilon_{si}}\right) x + \Phi_{s1}$$

where x is the depth in the silicon film ($x=0$ at the silicon-gate oxide interface and $x=t_{si}$ at the silicon-buried oxide interface). Assuming that the drain-to-source voltage drop is small ($V_D \cong V_S$), the hole concentration in the silicon film in any section of the device between the source and the drain is given by:

$$p(y) = N_a \int\limits_0^{t_{si}} exp\left(\frac{-q\Phi(x)}{kT}\right) dx$$

$$= N_a \, exp\left(\frac{-q\Phi_{s1}}{kT}\right)\int\limits_0^{t_{si}} exp\left(\frac{-q\left(\dfrac{qN_a}{2\varepsilon_{si}} x^2 + \left(\dfrac{\Phi_{s2}-\Phi_{s1}}{t_{si}} - \dfrac{qN_a t_{si}}{2\varepsilon_{si}}\right)x\right)}{kT}\right) dx \qquad (5.12.19)$$

The above expression cannot be integrated analytically but it can readily be evaluated using numerical integration. One can write:

$$p(y) = N_a \, exp\left(\frac{-\Phi_{s1}}{kT}\right) F(t_{si}, \Phi_{s1}, \Phi_{s2}) \qquad (5.12.20)$$

where $F(t_{si}, \Phi_{s1}, \Phi_{s2})$ is the result of the numerical integration of equation 5.12.19.

Measurements [218] and numerical simulations of accumulation-mode devices show that the subthreshold current is independent of the drain voltage, as long as V_{DS} is larger than several kT/q. This suggests that the transport mechanism is due to diffusion rather than drift, as in the case of enhancement-mode MOSFETs.

The diffusion current in the transistor is given by:

$$I_D = -A\,q\,D_p\,\frac{dp(y)}{dy}$$

where D_p is the hole diffusion coefficient, and $A = W t_{si}$ is the area of the device cross section. If the absolute value of the drain voltage is small enough (with $V_{DS} < 0$), this last expression can be simplified to:

$$I_D = -A\,q\,D_p\,\frac{p(L)-p(0)}{L} = \frac{A}{L}\,q\,p(0)\,D_p\left(1 - exp\left(\frac{qV_{DS}}{kT}\right)\right)$$

$$= \frac{A}{L}\,q\,N_a\,exp\left(\frac{-q\Phi_{s1}}{kT}\right) F(t_{si}, \Phi_{s1}, \Phi_{s2})\,D_p\left(1 - exp\left(\frac{qV_{DS}}{kT}\right)\right) \qquad (5.12.21)$$

By definition, the inverse subthreshold slope is given by 5.7.1:

$$S = \frac{dV_{G1}}{d\log(I_D)} = \frac{\ln(10)}{\dfrac{d\ln(I_D)}{dV_{G1}}} \tag{5.12.22}$$

with

$$\frac{d\ln(I_D)}{dV_{G1}} = \frac{1}{I_D}\frac{dI_D}{dV_G} = \frac{1}{I_D}\frac{dI_D}{d\Phi_{S1}}\frac{d\Phi_{S1}}{dV_{G1}}$$

Using equation 5.12.21, the above expression can be rewritten:

$$\frac{d\ln(I_D)}{dV_{G1}} = \frac{d\left(\exp\left(\dfrac{-q\Phi_{s1}}{kT}\right)F(t_{si},\Phi_{s1},\Phi_{s2})\right)}{\exp\left(\dfrac{-q\Phi_{s1}}{kT}\right)F(t_{si},\Phi_{s1},\Phi_{s2})\,d\Phi_{S1}}\frac{d\Phi_{S1}}{dV_{G1}}$$

$$= \left(\frac{-q}{kT} + \frac{d}{d\Phi_{S1}}\ln(F(t_{si},\Phi_{s1},\Phi_{s2}))\right)\frac{d\Phi_{S1}}{dV_{G1}} \tag{5.12.23}$$

Using expressions 5.12.18a, 5.12.22 and 5.12.23 one finally obtains the inverse subthreshold slope of the device:

$$S = \frac{-\ln(10)}{\dfrac{q}{kT} - \dfrac{d}{d\Phi_{S1}}\ln(F(t_{si},\Phi_{s1},\Phi_{s2}))}\left[\left(1 + \frac{C_{it1}}{C_{ox1}} + \frac{C_{si}}{C_{ox1}}\right) - \frac{\dfrac{C_{si}}{C_{ox1}}\dfrac{C_{si}}{C_{ox2}}}{1 + \dfrac{C_{it2}}{C_{ox2}} + \dfrac{C_{si}}{C_{ox2}}}\right] \tag{5.12.24}$$

In most practical cases the difference $\Phi_{s2}-\Phi_{s1}$ is only weakly dependent on Φ_{s1}. In that case, $\Phi_{s2} - \Phi_{s1}$ can be considered as a constant, and the integral term $F(t_{si},\Phi_{s1},\Phi_{s2})$ can be approximated by $F(t_{si})$ which is obtained from Equation (5.12.19), where $\Phi_{s2}-\Phi_{s1}$ is held. If we use the latter approximation, $F(t_{si})$ becomes independent of Φ_{s1}, and equation 5.12.24 becomes:

$$S = -\frac{kT}{q}\ln(10)\left[\left(1 + \frac{C_{it1}}{C_{ox1}} + \frac{C_{si}}{C_{ox1}}\right) - \frac{\dfrac{C_{si}}{C_{ox2}}\dfrac{C_{si}}{C_{ox1}}}{1 + \dfrac{C_{it2}}{C_{ox2}} + \dfrac{C_{si}}{C_{ox2}}}\right] \tag{5.12.25}$$

Equation 5.12.25 is identical, in absolute value, to equation 5.7.18 and can be evaluated analytically without the need for integrating expression

5.12.19. Numerically, the exact subthreshold slope given by 5.12.2 and the approximate value found using 5.12.25 are equal within a few percent.

Figure 5-39 presents the subthreshold slope of an enhancement-mode n-channel (electrons) and an accumulation-mode p-channel device (holes).[219] Regions 1, 2, 3 and 4 correspond to a front-surface inversion channel, a back inversion channel, a front body current, and a back body current, respectively. The lower limit of the subthreshold slope (dotted line B) is given by: $S = \dfrac{kT}{q} \ln(10)(1+\alpha)$, where $\alpha = \dfrac{C_{si}C_{ox2}}{C_{ox1}(C_{si}+C_{ox2})}$, and its upper limit (dotted line A) is obtained using the same equation with $\alpha = \dfrac{C_{ox2}(C_{ox1}+C_{si})}{C_{ox1}C_{si}}$. For large positive and negative back gate biases accumulation or inversion is reached at the back interface, and high subthreshold swing values are observed.

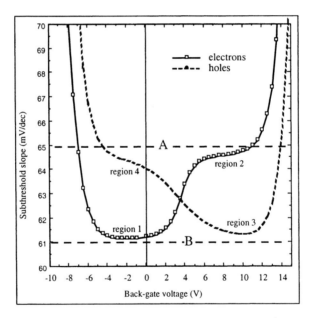

Figure 5-39. Subthreshold slope of an enhancement-mode n-channel (electrons) and an accumulation-mode p-channel device (holes). Regions 1, 2, 3 and 4 correspond to a front-surface inversion channel, a back inversion channel, a front body current, and a back body current, respectively. The meaning of dotted lines A and B is explained in section 5-13.

5.13 UNIFIED BODY-EFFECT REPRESENTATION

The body-effect coefficient (or body factor) of an MOS transistor, which is represented by the letter n, is an image of the efficiency of the coupling between the gate voltage and the channel. It is equal to $1 + \alpha$ and plays a role in each regime of operation of the device. In strong inversion, it is in the expression of the drain current (equation 5.4.2, 5.4.3, 5.4.16 and 5.4.18):

Non-saturation regime:
$$I_D = \mu C_{ox} \frac{W}{L}\left[\left(V_G - V_{TH}\right)V_D - \frac{1}{2}nV_D^2\right] \qquad (5.13.1)$$

Saturation regime:
$$I_{Dsat} = \frac{1}{2n}\mu\, C_{ox}\frac{W}{L}\left(V_G - V_{TH}\right)^2 \qquad (5.13.2)$$

It is also found in the expression of the subthreshold swing (equation 5.7.25):
$$S = n\frac{kT}{q}\ln(10) \qquad (5.13.3)$$

and the g_m/I_D ratio in weak inversion is given by equation 5.5.5:
$$\frac{g_m}{I_D} = \frac{dI_D}{I_D\, dV_G} = \frac{\ln(10)}{S} = \frac{q}{nkT} \qquad (5.13.4)$$

and in strong inversion, given by equation 5.5.6:

$$\frac{g_m}{I_D} = \sqrt{\frac{2\mu C_{ox}W/L}{nI_D}} \qquad (5.13.5)$$

In all instances the body factor, n, is given by the following expression:

$$n = 1 + \alpha \quad \text{with} \quad \alpha = \frac{C_{CH\text{-}GND}}{C_{G\text{-}CH}} \qquad (5.13.5)$$

where $C_{CH\text{-}GND}$ is the capacitance between the channel (which is a channel "under formation" in the case of the subthreshold regime) and ground (the back gate or the substrate), and $C_{G\text{-}CH}$ is the capacitance between the gate and the channel. The values of $C_{G\text{-}CH}$ and $C_{CH\text{-}GND}$ are presented in Table 5-4.

Equivalent capacitor circuits can be established for each device and are presented in Figure 5-40. These equivalent circuits illustrate well the fact that the body factor is a measure of the efficiency of the coupling between the gate voltage and the channel. Indeed, $C_{G\text{-}CH}$ represents the capacitive coupling between the gate and the channel, a surface inversion channel, except in the case of an accumulation-mode (AM) device in the subthreshold regime, where the subthreshold body current is located at a depth $x(\Phi_{min})$ within the device. The other capacitance, $C_{CH\text{-}GND}$, represents the "force" that tends to prevent the potential in the channel from following the gate voltage. The body factor in weak and strong inversion is different in a bulk device, while n remains constant for all regimes of operation in the case of a fully depleted SOI MOSFET.[220] An additional capacitor can be added below each of these equivalent circuits to represent the influence of a depletion layer in the substrate, underneath the buried oxide.[221] The variation of the value of the subthreshold slope in fully depleted and accumulation-mode SOI MOSFETs (Figure 5-39) can be best understood using the equivalent circuits of Figure 5-40. In an inversion-mode MOSFET there can only be a channel at the top or the bottom of the silicon film, such that the subthreshold slope abruptly jumps from a minimum (dotted line B in Figure 5-39) to a maximum value (dotted line A) depending on whether the channel is at the top or the bottom of the silicon film. In an accumulation-mode device, the body subthreshold current flows within the silicon film, at a depth that varies continuously from 0 to t_{si} as the back-gate voltage is decreased.

Table 5-4. Values of $C_{G\text{-}CH}$ and $C_{CH\text{-}GND}$ for the different types of MOSFETs. In the case of the accumulation-mode (AM) MOSFET, $C_{si1}=\varepsilon_{si}/x(\Phi_{min})$ and $C_{si2}=\varepsilon_{si}/(t_{si}\text{-}x(\Phi_{min}))$, where $x(\Phi_{min})$ is the depth at which the potential is minimum.

Device	$C_{G\text{-}CH}$	$C_{CH\text{-}GND}$
Bulk MOSFET (strong inversion)	C_{ox}	$\dfrac{\varepsilon_{si}}{x_{dmax}}$
Bulk MOSFET (weak inversion)	C_{ox}	$\dfrac{\varepsilon_{si}}{x_d}$
FD SOI MOSFET	C_{ox1}	$\dfrac{C_{si}\,C_{ox2}}{C_{si}+C_{ox2}}$
FD SOI MOSFET with back accumulation	C_{ox1}	$\dfrac{\varepsilon_{si}}{t_{si}}=C_{si}$
FD SOI MOSFET with back channel	$\dfrac{C_{ox1}C_{si}}{C_{ox1}+C_{si}}$	C_{ox2}
AM SOI MOSFET (subthreshold regime)	$\dfrac{C_{ox1}C_{si1}}{C_{ox1}+C_{si1}}$	$\dfrac{C_{ox2}C_{si2}}{C_{ox2}+C_{si2}}$

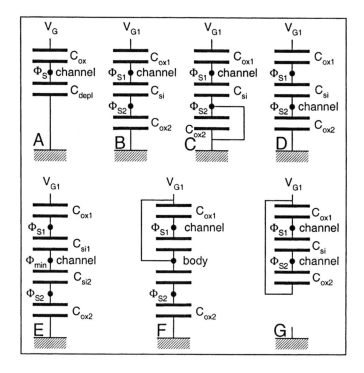

Figure 5-40. Capacitance model representing the coupling between the gate voltage and the subthreshold channel: A: bulk device; B: fully depleted SOI MOSFET; C: fully depleted SOI MOSFETwith back accumulation; D: fully depleted SOI MOSFET with back inversion channel; E: accumulation-mode SOI MOSFET in the subthreshold regime; F: MTCMOS/DTMOS (section 6.2); G: double-gate SOI MOSFET (section 6.1).

Figure 5-40 also extends the capacitor model to the MTCMOS/DTMOS device described in section 6.2 and to the double-gate SOI MOSFET (section 6.1). These capacitive models intuitively explain why these devices have $d\Phi_{s1}/dV_G \cong 1$ (and $d\Phi_{s2}/dV_G \cong 1$ in the case of a double-gate device), and therefore, $n \cong 1$ and $S = 60$ mV/decade at room temperature.

5.14 RF MOSFETS

It has been demonstrated that high-frequency n- and p-channel MOSFETs with transition frequencies above 10 GHz can be fabricated in SOI CMOS. In addition, the use of high-resistivity SOI substrates (5,000 Ω.cm or higher) allows for the fabrication of passive elements, such as strip or slot lines with relatively low losses [222,223], as well as planar inductors with a relatively good quality factor.[224] Rectangular planar inductors with quality factors of 5, 8, and 11 have been obtained when realized on 20, 4,000, and 10,000 Ω.cm substrates, respectively.[225] Moreover, the use of high-resistivity SOI substrates dramatically reduces cross-talk between devices.[226, 227, 228]

The main parameters characterizing radio-frequency (RF) operation are: the cutoff frequency, f_T, the maximum oscillation frequency, f_{max}, and the noise figure, NF. The cutoff frequency, f_T, corresponds to frequency at which the small-signal current gain is equal to unity. It is characteristic of the "intrinsic" device and does not account for resistances other than that of the channel. The maximum oscillation frequency, f_{max}, is the frequency at which the small-signal power gain is equal to unity under optimum matching conditions; it includes the effect of resistances that are in series with the channel. The cutoff frequency and the maximum oscillation frequency are given by the following expressions:[229, 230]

$$f_T = \frac{g_m}{2\pi(C_{gs} + C_{gd})} \tag{5.14.1}$$

and

$$f_{max} = \frac{g_m}{2\pi C_{gs}} \frac{1}{2\sqrt{\left(R_g + R_s + R_i\right)\left(g_d + g_m\frac{C_{gd}}{C_{gs}}\right)}} \cong \frac{f_T}{2\sqrt{(R_g + R_i)(g_d + 2\pi f_T C_{gd})}} \tag{5.14.2}$$

where g_m is the transconductance, C_{gs} is the gate-source capacitance, C_{gd} is the gate-drain capacitance, R_g is the gate resistance, R_s is the source resistance, g_d is the output conductance, and R_i is the equivalent non-quasi-static resistance of the device. The noise figure is the contribution by the device to the noise at its output. It is expressed in decibels (dB).

For correct microwave operation of a MOSFET, the gate resistance must be low. If the resistance is high, the gate behaves like a delay line for the input signal. The section of the transistor which is farthest from the gate contact does not respond to high frequencies.[231] In addition, a low gate resistance is crucial for obtaining a low noise factor.[232] Therefore, the use of silicide or metal gates is preferred. Table 5-5 presents the characteristics of SOI transistors used in low-voltage RF applications. The performances of these devices are comparable to those of GaAs MESFETs or SiGe heterojunction bipolar transistors, which enables the use of SOI CMOS for portable wireless applications.[233]

Figure 5-41 presents the current gain (h_{21}), the Maximum Available Gain (MAG) and the Unilateral Gain (ULG) as a function of frequency in an n-channel SOI transistor having a length of 0.75 μm (L_{eff}=0.65 μm) and a width of 125 μm. These parameters were measured through s-parameter extraction under a supply voltage ($V_{DS} = V_{GS}$) of 0.9 volt. The MAG is 11 dB at 2 GHz. The cutoff frequency, f_T, (found when $h_{21} = 0\ dB$) is equal to 10 GHz, and the maximum oscillation frequency, f_{max} (found when ULG = 0 dB), is equal to 11 GHz.

Table 5-5. Partially depleted (PD), fully depleted (FD) and dynamic threshold (DTMOS) RF SOI MOSFETs

PD/FD/ DTMOS	L (μm)	V_D (V)	f_T (GHz)	f_{max} (GHz)	Noise figure / Associated gain (dB) at 2GHz	Ref.
FD	0.75	0.9	12.9	30	- / 13.9 (10.4)	234
PD	0.3	2	-	24.3	0.9 / 14	235
PD	0.2	2	28.4	46	1/15.3	236
DTMOS	0.25	0.6	16	33		237
FD	0.2	1.5	76	40	0.4/18	238
PD	0.25	1.5	50	75	-/-	239
PD	0.07	1.2	114	135	-/-	240
PD	0.07	1.2	141	98	0.5/-	241
DTMOS	0.08	1.5	140	60	-/-	242
FD	0.25	1	42	70	0.8/- (6 GHz)	243
FD	0.25	1.5	30	42	0.45/-	244, 245
PD	0.13	1.2	120	150	0.8/16 (6 GHz)	246

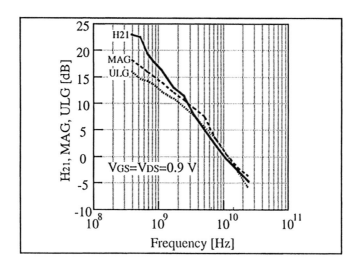

Figure 5-41. Current gain (H_{21}), the Maximum Available Gain (MAG) and the Unilateral Gain (ULG) as a function of frequency in an n-channel SOI transistor, having a length of 0.75 μm and a width of 125 μm at $V_{DS} = V_{GS} = 0.9$ V.[247]

The cutoff frequency, f_T, and the maximum oscillation frequency, f_{max}, are presented in Figure 5-42 as a function of supply voltage ($V_{DS} = V_{GS}$). These frequencies rapidly increase as the supply voltage is increased, up to 1 volt. At higher frequencies, f_T and f_{max} tend to saturate, due to electron velocity saturation in the channel. The values of f_T and f_{max} are 13 and 15.8 GHz, respectively, for a supply voltage of 3 volts. The dc power consumption as a function of supply voltage is also presented. The dc power consumed at Vdd=0.9 V, where f_T = 10 GHz and f_{max} = 11 GHz, is equal to 2.8 mW.

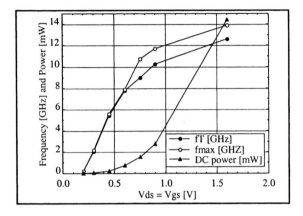

Figure 5-42. Unit-gain frequency, f_T, maximum oscillation frequency, f_{max}, and dc power consumption as a function of supply voltage. Same transistor as in Figure 5-41.[248]

5.15 CAD MODELS FOR SOI MOSFETS

Several SOI MOSFET models and SOI CAD tools can be found on the internet or in the literature. These are listed in Table 5-6.

Table 5-6. SOI models for CAD

Model name	Developer	Reference
SOISPICE	University of Florida	249
UFSOI	University of Florida	ibidem
UFDG (double-gate FETs)	University of Florida	ibidem
BSIMSOI	U.C. Berkeley	250
MINIMOS	Technical University Vienna	251
EKV	EPFL	252
ELDO	Mentor Graphics	253
LETISOI	LETI	254, 255
STAG	University of Southampton	256

REFERENCES

[1] A.J. Auberton-Hervé, Proceedings of the Fourth International Symposium on Silicon-on-Insulator Technology and Devices, Ed. by D.N. Schmidt, Vol.90-6, The Electrochemical Society, p. 455, 1990

[2] A.J. Auberton-Hervé, J.M. Lamure, T. Barge, M. Bruel, B. Aspar, and J.L. Pelloie, Semiconductor International, p. 97, October 1995

[3] D. Flandre, IEEE Transactions on Electron Devices, Vol. 40, no. 10, p. 1789, 1993

[4] D. Flandre, F. Van de Wiele, P.G.A. Jespers, and M. Haond, IEEE Electron Device Letters, Vol. 11, no. 7, p. 291, 1990

[5] B. Gentinne, D. Flandre, and J.P. Colinge, Solid-State Electronics, Vol. 39, no. 7, p. 1071, 1996

[6] G.K. Celler, S. Cristoloveanu, Journal of Applied Physics, Vol. 93, no. 9, p. 4955, 2003

[7] D. Flandre and F. Van de Wiele, Proceedings of the IEEE SOS/SOI Technology Conference, p. 27, 1989

[8] N.J. Thomas and J.R. Davis, Proc. of the IEEE SOS/SOI Technology Conference, p. 130, 1989

[9] L. Dreeskornfeld, J. Hartwich, E. Landgraf, R.J. Luyken, W. Rösner, T. Schulz, M. Städele, D. Schmitt-Landsiedel, L. Risch, Electrochemical Society Proceedings, Vol. 2003-05, p. 361, 2003

[10] E. Suzuki, K. Ishii, S. Kanemaru, T. Maeda, T. Tsutsumi, T. Sekigawa, K. Nagai, H. Hiroshima, IEEE Transactions on Electron Devices, Vol. 47, no. 2, p. 354, 2000

[11] S.M. Sze, *Physics of Semiconductor Devices*, 2nd Ed., J. Wiley & Sons, 1981

[12] H.K. Lim and J.G. Fossum, IEEE Transactions on Electron Devices, Vol. 30, p. 1244, 1983

[13] J.P. Colinge, IEEE Electron Device Lett., Vol. 6, p. 573, 1985

[14] R.S. Muller and T.I. Kamins, *Device Electronics for Integrated Circuits*, J. Wiley & Sons, p. 436, 1986

[15] J.P. Colinge and C.A. Colinge, *Physics of Semiconductor Devices*, pp. 192-196, Kluwer Academic Publishers, 2002

[16] J.P. Colinge, Microelectronic Engineering, Vol. 8, p. 127, 1988

[17] J. Witfield and S. Thomas, IEEE Electron Dev. Lett., Vol. 7, p. 347, 1986

[18] J.A. Martino, L. Lauwers, J.P. Colinge, and K. De Meyer, Electronics Letters, Vol. 26, p. 1462, 1990

[19] M.A. Pavanello, J.A. Martino, and J.P. Colinge, Solid-State Electronics, Vol. 41, no. 9, p. 1241, 1997

[20] D. Hisamoto, T. Kaga, Y. Kawamoto, and E. Takeda, Technical Digest of International Electron Devices Meeting, p. 833, 1989

[21] R.S. Muller and T.I. Kamins, *Device Electronics for Integrated Circuits*, J. Wiley & Sons, p. 487, 1986

[22] S.Veeraraghavan and J.G. Fossum, IEEE Transactions on Electron Devices, Vol. 35, p. 1866, 1988

[23] T.W. MacElwee and D.I. Calder, Proceedings of the Second International Symposium on Ultra Large Scale Integration Science and Technology, Ed. by C.M. Osburn and J.M. Andrews, Vol.89-9, The Electrochemical Society, p. 693, 1989

[24] T.W. MacElwee and I.D. Calder, Proceedings of the IEEE SOS/SOI Technology Conference, p. 171, 1989

[25] J.G. Fossum, Proceedings of the Fourth International Symposium on Silicon-on-Insulator Technology and Devices, ed. by D.N. Schmidt, Vol. 90-6, The Electrochemical Society, p. 491, 1990

[26] S.Veeraraghavan and J.G. Fossum, IEEE Transactions on Electron Devices, Vol. 36, p. 522, 1989

[27] E. Suzuki, K. Ishii, S. Kanemaru, T. Maeda, T. Tsutsumi, T. Sekigawa, K. Nagai, H. Hiroshima, IEEE Transactions on Electron Devices, Vol. 47, no. 2, p. 354, 2000

[28] T. Sekigawa and Y. Hayashi, Solid-State Electronics, Vol. 27, p. 827, 1984

[29] D. Hisamoto, T. Kaga, Y. Kawamoto, and E. Takeda, Technical Digest of International Electron Devices Meeting (IEDM), p. 833, 1989

[30] K.K. Young, IEEE Transactions on Electron Devices, Vol. 36, p. 399, 1989

[31] S.R. Banna, P.C.H Chan, P.K. Ko, C.T. Nguyen, M. Chan, IEEE Transactions on Electron Devices, Vol. 42, p. 1949, 1995

[32] T.K. Chiang, Y.H. Wang, M.P. Houng, Solid-State Electronics, Vol. 43, p. 123, 1999

[33] J.P. Colinge, Technical Digest of International Electron Devices Meeting (IEDM), p. 817, 1989

[34] S.Veeraraghavan and J.G. Fossum, IEEE Transactions on Electron Devices, Vol. 35, p. 1866, 1988

[35] S.Veeraraghavan and J.G. Fossum, IEEE Transactions on Electron Devices, Vol. 36, p. 522, 1989

[36] T. Ernst, S. Cristoloveanu, Electrochemical Society Proceedings, Vol. 99-3, p. 329, 1999

[37] W. Xiong, J.P. Colinge, Electronics Letters, Vol. 35, p. 2059, 2000

[38] J.P. Colinge and C.A. Colinge, *Physics of Semiconductor Devices*, pp. 187-196, Kluwer Academic Publishers, 2002

[39] H.K. Lim and J.G. Fossum, IEEE Transactions on Electron Devices, Vol. 31, p. 401, 1984

[40] S.Veeraraghavan and J.G. Fossum, IEEE Transactions on Electron Devices, Vol. 35, p. 1866, 1988

[41] K.K. Young, IEEE Transactions on Electron Devices, Vol. 36, p. 504, 1989

[42] S.L. Jang, B.R. Huang, J.J. Ju, IEEE Transactions on Electron Devices, Vol. 46, p. 1872, 1998

[43] K. Suzuki, S. Pidin, IEEE Transactions on Electron Devices, Vol. 50, p. 1297, 2003

[44] J.G. Fossum, J.Y. Choi, and R. Sundaresan, IEEE Transactions on Electron Devices, Vol. 37, p. 724, 1990

[45] J.C. Sturm and K. Tokunaga, Electronics Letters, Vol. 25, p. 1233, 1989

[46] J.G. Fossum and S. Krishnan, IEEE Transactions on Electron Devices, Vol. 39, p. 457, 1993

[47] H.K. Lim and J.G. Fossum, IEEE Transactions on Electron Devices, Vol. 32, p. 446, 1985

[48] S.Veeraraghavan and J.G. Fossum, IEEE Transactions on Electron Devices, Vol. 35, p. 1866, 1988

[49] C.C. Enz, in *Low-power HF microelectronics: a unified approach*, edited by G.A.S. Machado, IEE Circuits and Systems Series 8, the Institution of Electrical Engineers, p. 247, 1996

[50] C. Enz, F. Krummenacher, and E.A. Vittoz, Analog Integrated Circuit and Signal Processing, Vol. 8, No1, p. 83, 1995

[51] A.M. Ionescu, S. Cristoloveanu, A. Rusu, A. Chovet, and A. Hassein-Bey, Proceedings of the International SOI Conference, p. 144, 1993

52 M. Bucher, C. Lallement, C.C. Enz, Proceedings of the Conference on Microelectronic Test Structures, Vo. 9, p. 145, (Trento, Italy), 1996

53 B. Iñiguez, L.F. Ferreira, B. Gentinne, and D. Flandre, IEEE Transactions on Electron Devices, Vol. 43, no. 4, p. 568, 1996

54 C. Mallikarjun and K.N. Bhat, IEEE Transactions on Electron Devices, Vol. 37, no. 9, p. 2039, 1990

55 E.A. Vittoz, Proceedings of the 23rd European Solid-State Device Research Conference (ESSDERC), Ed. by. J. Borel, P. Gentil, J.P. Noblanc, A. Nouailhat and M. Verdone, Editions Frontières, p. 927, 1993

56 K. Lee, M. Shur, T.A. Fjeldly, and T. Ytterdal, *Semiconductor Device Modelling for VLSI*, Englewood Cliffs, NJ, Prentice-Hall, 1993

57 B. Iñiguez, L.F. Ferreira, B. Gentinne, and D. Flandre, IEEE Transactions on Electron Devices, Vol. 43, no. 4, p. 568, 1996

58 B. Iniguez, V. Dessard, D. Flandre, B. Gentinne, IEEE Transactions on Electron Devices, Vol. 46, no. 12, p. 2295, 1999

59 J.C. Sturm and K. Tokunaga, Electronics Letters, Vol. 25, p. 1233, 1989

60 JP Colinge, IEEE Electron Device Letters, Vol. EDL-6, p. 573, 1985

61 T. Ouisse, S. Cristoloveanu, G. Borel, Solid-State Electronics, Vol. 35, p. 141, 1992

62 F. Silveira, D. Flandre, and P.G.A. Jespers, IEEE Journal of Solid-State Circuits, Vol. 31, no. 9, p. 1314, 1996

63 M.A. Pavanello, J.A. Martino, V. Dessard, D. Flandre, Solid-State Electronics, Vol. 44, no. 7, p. 1219, 2000

64 M. Galeti, M. A. Pavanello, J.A. Martino, Electrochemical Society Proceedings, Vol. 2003-09, p. 48, 2003

65 E.A. Vittoz, Tech. Digest of Papers, ISSCC, p. 14, 1994

66 D. Flandre, L.F. Ferreira, P.G.A. Jespers, and J.P. Colinge, Solid-State Electronics, Vol. 39, no. 4, p. 455, 1996

67 H.K. Lim and J.G. Fossum, IEEE Transactions on Electron Devices, Vol. 31, p. 401, 1984

68 S.C. Sun and J.D. Plummer, IEEE Transactions on Electron Devices, Vol. 27, p. 1497, 1980

69 M. Yoshimi, H. Hazama, M. Takahashi, S. Kambayashi, T. Wada, K. Kato, and H. Tango, IEEE Transactions on Electron Devices, Vol. 36, p. 493, 1989

70 A. Yoshino, Proceedings of the Fourth International Symposium on Silicon-on-Insulator Technology and Devices, ed. by D.N. Schmidt, Vol. 90-6, The Electrochemical Society, p. 544, 1990

71 J.P. Colinge, C.A. Colinge, *Physics of Semiconductor Devices*, p. 195 and p. 199, Kluwer Academic Publishers, 2002

72 G. Ghibaudo, Electronics Letters, Vol. 24, no. 9, p. 543, 1988

73 P.K. McLarty, S. Cristoloveanu, O. Faynot. V. Misra, J.R. Hauser, J.J. Wortman, Solid-State Electronics, Vol. 38, no. 6, p. 1175, 1995

74 O. Faynot, S. Cristoloveanu, P. McLarty, C. Raynaud, J. Gautier, Proceedings of the IEEE International SOI Conference, p. 17, 1994

75 F.J.G. Sanchez, A. Ortiz-Conde, A. Cerdeira, M. Estrada, D. Flandre, J.J. Liou, IEEE Transactions on Electron Devices, Vol. 49, no. 1, p. 82, 2002

76 R.J. Van Overstraeten, G.J. Declerck, and P.A. Muls, IEEE Transactions on Electron Devices, Vol. ED-20, p. 1150, 1973

77 S.M. Sze, *Physics of Semiconductor Devices*, Wiley & Sons, p. 446, 1981

[78] J.P. Colinge, Technical Digest of International Electron Devices Meeting (IEDM), p. 817, 1989

[79] J.P. Colinge, IEEE Electron Device Letters, Vol. 7, p. 244, 1986

[80] J.P. Colinge, Ext. Abstracts of 5th Internat. Workshop on Future Electron Devices, Miyagi-Zao, Japan, p. 105, 1988

[81] K.Asada, H. Miki, M. Kumon, and T. Sugano, Ext. Abstracts of 8th International Workshop on Future Electron Devices, Kochi, Japan, p. 165, 1990

[82] H.O. Joachim, Y. Yamaguchi, K. Ishikawa, Y. Inoue, and T. Nishimura, IEEE Transactions on Electron Devices, Vol. 40, no. 10, p. 1812, 1993

[83] J.P. Colinge, C.A. Colinge, *Physics of Semiconductor Devices*, Kluwer Academic Publishers, p. 346, 2002

[84] L. Esaki, *Electronic properties of multilayers and low-dimensional semiconductor structures*, NATO ASI Series, Plenum Press, Series B: Physics Vol. 321, pp. 1-24, 1990

[85] T. Ando, A.B. Fowler, and F. Stern, Review of Modern Physics, Vol. 54, p. 437, 1982

[86] P.N. Butcher, *Physics of low-dimensional semiconductor structures*, Ed. by P.N. Butcher, N.H. March and M.P. Tosi, Plenum Press, p. 95, 1993

[87] F. Stern, *Physics of low-dimensional semiconductor structures*, Ed. by P.N. Butcher, N.H. March and M.P. Tosi, Plenum Press, p. 177, 1993

[88] Y. Omura, S. Horiguchi, M. Tabe and K. Kishi, IEEE Electron Device Letters, Vol. 14, no. 12, p. 569, 1993

[89] K. Uchida, J. Koga, R. Ohba, T. Numata, S.I. Takagi, Technical Digest of IEDM, p. 29.4.1, 2001

[90] J. Lolivier, S. Deleonibus, F. Balestra, Electrochemical Society Proceedings, Vol. 2003-05, p. 379, 2003

[91] T. Ernst, D. Munteanu, S. Cristoloveanu, T. Ouisse, N. Hefyene, S. Horiguchi, Y. Ono, Y. Takahashi, K. Murase, IEEE International SOI Conference Proceedings, p. 92, 1999

[92] T. Ernst, S. Cristoloveanu, G. Ghibaudo, T. Ouisse, S. Horiguchi, Y. Ono, Y. Takahashi, K. Murase, IEEE Transactions on Electron Devices, Vol. 50, no. 3, p. 830, 2003

[93] Y. Omura and K. Izumi, IEEE Electron Device Letters, Vol. 17, no. 6, p. 300, 1996

[94] Y. Omura, T. Ishiyama, M. Shoji, and K. Izumi, Electrochemical Society Proceedings, Vol. 96-3, p. 199, 1996

[95] F. Gámiz, J.B. Roldán, J.A. López-Villanueva, P. Cartujo-Cassinello, J.E. Carcerer, P. Cartujo, Electrochemical Society proceedings, Vol. 2001-3, p. 157, 2001

[96] A. Pirovano, A.L. Lacaita, G. Ghidini, G. Tallarida, IEEE Electron Device Letters, Vol. 21, no. 1, p.34, 2000

[97] M. Shoji and S. Horiguchi, Journal of Applied Physics, Vol. 82, no. 12, p. 6096, 1997

[98] M. Shoji and S. Horiguchi, Journal of Applied Physics, Vol. 85, no. 5, p. 2722, 1999

[99] J. Koga, S.I. Takagi, A. Toriumi, IEEE Transactions on Electron Devices, Vol. 49, no. 6, p. 1042, 2002

[100] Jin-Hyeok Choi, Young-June Park, Hong-Shick Min, IEEE Electron Device Letters, Vol.16, no 11, p.527, 1995

[101] F. Gámiz, J.B. Roldán, P. Cartujo-Cassinello, J.E. Carceller, J.A. López-Villanueva, S. Rodriguez, Journal of Applied Physics, Vol.86, no. 11, p. 6269, 1999

[102] F. Gámiz, J.B. Roldán, J.A. López-Villanueva, Journal of Applied Physics, Vol.83, no. 9, p. 4802, 1998

[103] A. Toriumi, J. Koga, H. Satake, A. Ohata, International Electron Devices Meeting. Technical Digest, p. 847, 1995

[104] F. Gámiz, J.A. López-Villanueva, J.B. Roldán, J.E. Carceller, P. Cartujo, IEEE Transactions on Electron Devices, Vol. 45, no. 5, p. 1122, 1998

[105] B. Dierickx, L. Warmerdam, E. Simoen, J. Vermeiren, and C. Claeys, IEEE Transactions on Electron Devices, Vol. 35, p. 1120, 1988

[106] J. Tihanyi and H Schlötterer, IEEE Transactions on. Electron Devices, Vol. 22, p. 1017, 1975

[107] G. Merckel, *NATO Course on Process and Device Modeling for Integrated Circuit Design*, Ed. by F. Van de Wiele, W. Engl and P. Jespers, Groningen, The Netherlands, Noordhoff, p. 725, 1977

[108] J.P. Colinge, IEEE Electron Device Lett., Vol. 9, p. 97, 1988

[109] J.G. Fossum, R. Sundaresan, and M. Matloubian, IEEE Transactions on Electron Devices, Vol. 34, p. 544, 1987

[110] M. Matloubian, C.E.D. Chen, B.Y. Mao, R. Sundaresan, G.P. Pollack, IEEE Transactions on Electron Devices, Vol. 37, no. 9, p. 1985, 1990

[111] K.M. Cham, S.Y. Oh, D. Chin, and J.L. Moll, *Computer Aided Design and VLSI Device Development*, Hingham, MA, Kluwer Academic Publishers, p. 240, 1986

[112] P.K. Ko, S. Tam, C. Hu, S.S. Wong, and C.G. Sodini, Technical Digest of International Electron Devices Meeting (IEDM), p. 88, 1984

[113] C. Hu, Technical Digest of International Electron Devices Meeting (IEDM), p. 176, 1983

[114] J.G. Fossum, J.Y. Choi, and R. Sundaresan, IEEE Transactions on Electron Devices, Vol. 37, p. 724, 1990

[115] C. Hu, S.C. Tam, F.C. Hsu, P.K. Ko, T.Y. Chan, and T.W. Terrill, IEEE Transactions on Electron Devices, Vol. 32, p. 375, 1985

[116] J.G. Fossum, J.Y. Choi, and R. Sundaresan, IEEE Transactions on Electron Devices, Vol. 37, p. 724, 1990

[117] J.P. Colinge, IEEE Transactions on Electron Devices, Vol. 34, p. 2173, 1987

[118] J.G. Fossum, J.Y. Choi, and R. Sundaresan, IEEE Transactions on Electron Devices, Vol 37, p. 724, 1990

[119] J.P. Colinge, IEEE Transactions on Electron Devices, Vol. 34, p. 2173, 1987

[120] L.T. Su, H. Fang, J.E. Chung, and D.A. Antoniadis, Technical Digest of International Electron Devices Meeting (IEDM), p. 349, 1992

[121] G. Reimbolt and A.J. Auberton-Hervé, IEEE Transactions on Electron Devices, Vol. 14, p. 364, 1993

[122] S.M. Guwaldi, A. Zaleski, D.E Ioannou, S. Cristoloveanu, G.J. Campisi, and H.L. Hughes, Proceedings of the Fifth International Symposium on Silicon-on-Insulator Technology and Devices, Ed. by W.E. Bayley, Proc. Vol. 92-13, The Electrochemical Society, p. 157, 1992

[123] B. Zhang, A. Yoshino, and T.P. Ma, Proceedings of the Fifth International Symposium on Silicon-on-Insulator Technology and Devices, Ed. by. W.E. Bayley, Proc. Vol. 92-13, The Electrochemical Society, p. 163, 1992

[124] B. Yu, Z.J. Ma, G. Zhang, and C. Hu, Solid-State Electronics, Vol. 39, no. 12, p. 1791, 1996

[125] T. Ouisse, S. Cristoloveanu, and G. Borel, Proceedings of IEEE SOS/SOI Technology Conference, p. 38, 1990

[126] D.E. Ioannou, in "Physical and Technical Problems of SOI Structures and Devices", Kluwer Academc Publishers, NATO ASI Series - High Technology, Vol. 4, Ed. by J.P. Colinge, V.S. Lysenko and A.N. Nazarov, p. 199, 1995

[127] S.P. Sinha, F.L. Duan, and D.E. Ioannou, Proceedings of the IEEE International SOI Conference, p. 18, 1996

[128] D.E. Ioannou, F.L. Duan, S.P. Sinha, A. Zaleski, IEEE Transactions on Electron Devices, Vol. 45, no. 5, p. 1147, 1998

[129] S.R. Banna, P.C.H. Chan, M. Chan, S.K.H. Fung, P.K. Ko, IEEE Transactions on Electron Devices, Vol. 45, no. 1, p. 206, 1999

[130] S.H. Renn, E. Rauly, J.L. Pelloie, F. Balestra, IEEE Transactions on Electron Devices, Vol. 5, no. 5, p. 1140, 1998

[131] P.H. Woerlee, C. Juffermans, H. Lifka, W. Manders, F. M. Oude Lansink, G.M. Paulzen, P. Sheridan, and A. Walker, Technical Digest of IEDM, p. 583, 1990

[132] Y. Omura and K. Izumi, Extended Abstracts of the Solid-State Devices and Materials Conference (SSDM), p. 496, 1992

[133] J.R. Davis, A.E. Glaccum, K. Reeson, and P.L.F. Hemment, IEEE Electron Device Letters, Vol. 7, p. 570, 1986

[134] J.G. Fossum, R. Sundaresan, and M. Matloubian, IEEE Transactions on Electron Devices, Vol. 34, p. 544, 1987

[135] M. Matloubian, C.E.D. Chen, B.Y. Mao, R. Sundaresan, G.P. Pollack, IEEE Transactions on Electron Devices, Vol. 37, no. 9, p. 1985, 1990

[136] J.Y. Choi and J.G. Fossum, Proceedings of the IEEE SOS/SOI Technology Conference, p. 21, 1990

[137] C.E.D. Chen, M. Matloubian, R. Sundaresan, B.Y. Mao, C.C. Wei, and G.P. Pollack, IEEE Electron Device Letters, Vol. 9, p. 636, 1988

[138] R. Sundaresan and C.E.D. Chen, Proceedings of the Fourth International Symposium on Silicon-on-Insulator Technology and Devices, Ed. by D.N. Schmidt, Vol. 90-6, The Electrochemical Society, p. 437, 1990

[139] A.J. Auberton-Hervé, Proceedings of the Fourth International Symposium on Silicon-on-Insulator Technology and Devices, ed. by D.N. Schmidt, Vol. 90-6, The Electrochemical Society, p. 455, 1990

[140] H.S. Sheng, S.S. Li, R.M. Fox, and W.S. Krull, IEEE Transactions on Electron Devices, Vol. 36, no. 3, p. 488, 1989

[141] A.S. Grove, *Physics and Technology of Semiconductor Devices*, J. Wiley & Sons, p. 230, 1967

[142] J.Y. Choi and J.G. Fossum, Proceedings of the IEEE SOS/SOI Technology Conference, p. 21, 1990

[143] M. Haond and J.P. Colinge, Electronics Letters, Vol. 25, p.1640 , 1989

[144] S.C. Lin, B. Kuo, IEEE Transactions on Electron Devices, Vol. 49, no. 11, p. 2016, 2002

[145] K.K. Young and J.A. Burns, IEEE Transactions on Electron Devices, Vol. 35, p. 426, 1988

[146] F.L. Duan, S.P. Sinha, D.E. Ioannou, and F.T. Brady, IEEE Transactions on Electron Devices, Vol. 44, no. 6, p. 972, 1997

[147] D. Chang, S. Veeraraghavan, M. Mendicino, M. Rashed, D. Connelly, S. Jallepalli, J. Candelaria, IEEE International SOI Conference Proceedings p. 155, 1998

[148] M. Mendicino, O. Zia, B. Min, D. Chang, X. Xu, Electrochemical Society Proceedings, Vol. 99-3, p. 335, 1999

[149] T. Ohno, M. Takahashi, Y. Kado, T. Tschuiya, IEEE Transactions on Electron Devices, Vol. 45, no. 5, p. 1071, 1998

[150] T. Shino, H. Nii, S. Kawanaka, K. Inoh, Y. Katsumata, M. Yoshimi, H. Ishiuchi, Proceedings of the IEEE International SOI Conference, p. 89, 2001

[151] P.H. Woerlee, C. Juffermans, H. Lifka, W. Manders, F.M. Oude Lansink, G.M. Paulzen, P. Sheridan, and A. Walker, Technical Digest of IEDM, p. 583, 1990

[152] F. Deng, R.A. Johnson, W.B. Dubbelday, G.A. Garcia, P.M. Asbeck, and S.S. Lau, Proceedings of the IEEE International SOI Conference, p. 78, 1996

[153] M. Yoshimi, M. Terauchi, M. Takahashi, K. Matsuzawa, N. Shigyo,and Y. Ushiku, Extended Abstracts of the International Electron Device Meeting, p. 429, 1994

[154] A. Nishiyama, O. Arisumi, and M. Yoshimi, Proceedings of the IEEE International SOI Conference, p. 68, 1996

[155] S. Krishnan, J.G. Fossum, and M.M. Pelella, Proceedings of the IEEE International SOI Conference, p. 140, 1996

[156] T. Iwamatsu, Y. Yamaguchi, Y. Inoue, T. Nishimura, IEEE Transactions on Electron Devices, Vol. 42, p. 1934, 1995

[157] Y.H. Koh, M.R. Oh, J.W. Lee, J.W. Yang, W.C. Lee, H.K. Kim, IEEE Transactions on Electron Devices, Vol. 45, no. 5, p. 1063, 1998

[158] J.G. Fossum, Proceedings of the Fourth Ynternational Symposium on Silicon-on-Insulator Technology and Devices, Ed. by D.N. Schmidt, Vol. 90-6, The Electrochemical Society, p. 491, 1990

[159] J.Y. Choi and J.G. Fossum, Proceedings of the IEEE SOS/SOI Technology Conference, p. 21, 1990

[160] M. Pelella, J.G. Fossum, D. Suh, S. Krishnan, and K.A. Jenkins, Proceedings of the IEEE International SOI Conference, p. 8, 1995

[161] A. Wei and D.A. Antoniadis, IEEE Electron Device letters, Vol. 15, no. 5, p. 193, 1996

[162] F. Assaderaghi, G. Shahidi, L. Wagner, M. Hsieh, M. Pelella, S. Chu, R.H. Dennard, and B. Davari, IEEE Electron Device Letters, Vol. 18, no. 6, p. 241, 1997

[163] J.A. Mandelman, J.E. Barth, J.K. DeBrosse, R.H. Dennard, H.L. Kalter, J. Gautier, and H.I. Hanafi, Proceedings of the IEEE International SOI Conference, p. 136, 1996

[164] M. Terauchi and M. Yoshimi, Proceedings of the IEEE International SOI Conference, p. 138, 1996

[165] J. Gautier, K.A. Jenkins, and Y.C. Sun, Technical Digest of the IEDM, p. 623, 1995

[166] T. Saraya, M. Takamiya, T.N. Duyet, T. Tanaka, H. Ishikuro, T. Hiramoto, and T. Ikoma, Proceedings of the IEEE International SOI Conference, p. 70, 1996

[167] T. Matsumoto, S. Maeda, Y. Hirano, K. Eikyu, Y. Yamaguchi, S. Maegawa, M. Inuishi, T. Nishimura, IEEE Transactions on Electron Devices, Vol. 49, no. 1, p. 55, 2002

[168] O. Faynot, J.L. Pelloie, M. Belleville, Electrochemical Society Proceedings, Vol. 2001-3, p. 211, 2001

[169] K.A. Jenkins, Y. Taur, and J.Y.C. Sun, Proceedings of the IEEE International SOI Conference, p. 72, 1996

[170] R.A. Schiebel, T.W. Houston, R. Rajgopal, K. Joyner, J.G. Fossum, D. Suh, and S. Krishnan, Proceedings of the IEEE International SOI Conference, p. 125, 1995

[171] K.A. Jenkins, J.Y.C. Sun, and J. Gautier, IEEE Electron Device Letters, Vol. 17, no. 11, p. 7, 1996

[172] F. Assaderaghi, G.G. Shahidi, M. Hargrove, K. Hathorn, H. Hovel, S. Kulkarni, W. Rausch, D. Sadana, D. Schepis, R. Schulz, D. Yee, J. Sun, R. Dennard, and B. Davari, Digest of Technical Papers, Symposium on VLSI Technology, p. 122, 1996

[173] M.R. Casu, P. Flatresse, Proceedings of the IEEE International SOI Conference, p. 62, 2002

[174] O. Faynot, T. Poiroux, J. Cluzel, M. Belleville, J. de Pontcharra, Proceedings of the IEEE International SOI Conference, p. 35, 2002

[175] R.V. Joshi, W. Hwang, S.C. Wilson, G. Shahidi, C. T Chuang, Proceedings of the IEEE International SOI Conference, p. 79, 1999

[176] A. Wei and D.A. Antoniadis, Proceedings of the IEEE International SOI Conference, p. 74, 1996

[177] H.K. Lim and J.G. Fossum, IEEE Transactions on Electron Devices, Vol. 31, no. 9, p. 1251, 1985

[178] K. Ueda, H. Morinaka, Y. Yamaguchi, T. Iwamatsu, I.J. Lim, Y. Inoue, K. Mashiko, and T. Sumi, Proceedings of the IEEE International SOI Conference, p. 142, 1996

[179] S.C. Chin, Y.C. Tseng, and J.C.S. Woo, Proceedings of the IEEE International SOI Conference, p. 144, 1996

[180] K. Verhaege, G. Groeseneken, J.P. Colinge, and H.E. Maes, IEEE Electron Device Letters, Vol. 14, no. 7, p. 326, 1993

[181] K. Verhaege, G. Groeseneken, J.P. Colinge, and H.E. Maes, Microelectronics and Reliability, Special Issue 'Reliability Physics of Advanced Electron Devices', Vol. 35, no. 256, Issue 3, p. 555, 1994

[182] J.S.T. Huang, Proceedings of the IEEE International SOI Conference, p. 122, 1993

[183] S.K.H. Fung, N. Zamdmer, I. Yang, M. Sherony, Shih-Hsieh Lo, L. Wagner, T.C. Chen, G. Shahidi, F. Assaderaghi, Proceedings of the IEEE International SOI Conference, p. 122, 2000.

[184] T. Poiroux, O. Faynot, C. Tabone, H. Tigelaar, H. Mogul, N. Bresson, S. Cristoloveanu, Proceedings of the IEEE International SOI Conference, p. 99, 2002

[185] H. Wan, S.K.H. Fung, P. Su, M. Chan, C. Hu, Proceedings of the IEEE International SOI Conference, p. 140, 2002

[186] J.W. Yang, J.G. Fossum, G.O. Workman, C.L. Huang, Solid-State Electronics, Vol. 48, no. 2, p. 259, 2004

[187] A. Mercha, J.M. Rafí, E. Simoen, E. Augendre, C. Claeys, IEEE Transactions on Electron Devices, Vol. 50, no. 7, p. 1675, 2003

[188] L. Vancaillie, V. Kilchitska, P. Delatte, L. Demeûs, H. Matsuhashi, F. Ishikawa, D. Flandre, , Proceedings of the IEEE International SOI Conference, p. 79, 2003

[189] B.W. Min, L. Kang, D. Wu, D. Caffo, J. Hayden, M.A. Mendicino, Proceedings of the IEEE International SOI Conference, p. 169, 2002

[190] M.M. Pelella, W. Maszara, S. Sundararajan, S. Sinha, A. Wei, D. Ju, W. En, S. Krishnan, D. Chan, S. Chan, P. Yeh, M. Lee, D. Wu, M. Fuselier, R. vanBentum, G. Burbach, C. Lee, G. Hill, D. Greenlaw, C. Riccobene, O. Karlsson, D. Wristers, N. Kepler, IEEE International SOI Conference Proceedings, p. 1, 2001

[191] K. Bernstein and N.J. Rohrer, *SOI Circuit Design Concepts*, Kluwer Academic Publishers, 2000

[192] A. Marshall and S. Natarajan, SOI Design: Analog, *Memory and Digital Techniques*, Kluwer Academic Publishers, 2002

[193] F. Allibert, T. Ernst, J. Pretet, N. Hefyene, C. Perret, A. Zaslavsky, S. Cristoloveanu, Solid-State Electronics, Vol. 45, p. 559, 2001

[194] J.H. Yi, Y.J. Park, H.S. Min, Proceedings of the IEEE International SOI Conference, p. 53, 2002

[195] V. Kilchytska, D. Levacq, D. Lederer, J.P. Raskin, D. Flandre, IEEE Electron Device Letters, Vol. 24, no. 6, p. 414, 2003

[196] J. Jomaah, G. Ghibaudo, F. Balestra, and J.L. Pelloie, Proceedings of the IEEE International SOI Conference, p. 114, 1995

[197] J.S. Brodsky, R.M. Fox, D.T. Zweidinger, and S. Veeraraghavan, IEEE Transactions on Electron Devices, Vol. 44, no. 6, p. 957, 1997

[198] D. Yachou and J. Gautier, Proceedings of the 24th European Solid State Device Research Conference (ESSDERC), Ed. by. C. Hill and P. Ashburn, Editions Frontières, p. 787, 1994

[199] M. Asheghi, B. Behkam, K. Yazdani, R. Joshi, K.E. Goodson, Proceedings of the IEEE International SOI Conference, p. 51, 2002

[200] L.J. Mc Daid, S. Hall, P.H. Mellor, W. Eccleston, and J.C. Alderman, Electronics Letters, Vol. 25, p. 827, 1989

[201] L.T. Su, K.E. Goodson, D.A. Antoniadis, M.I. Flik, and J.E. Chung, Technical Digest of IEDM, p. 357, 1992

[202] M. Berger and Z. Chai, IEEE Transactions on Electron Devices, Vol. 38, p. 871, 1991

[203] Y. Cheng and T.A. Fjeldly, IEEE Transactions on Electron Devices, Vol. 43, no. 8, p. 1291, 1996

[204] K.A. Jenkins, R.L Franch, Proceedings of the IEEE International SOI Conference, p. 161, 2003

[205] O. Le Neel and M. Haond, Electronics Letters, Vol. 26, p. 74, 1991

[206] D. Yachou, J. Gautier and C. Raynaud, Proceedings of the IEEE International SOI Conference, p. 148, 1993

[207] B.M. Tenbroek, M.S.L. Lee, W. Redman-White, R.J.T. Bunyan, and M.J. Uren, IEEE Transactions on Electron Devices, Vol. 43, no. 12, p. 2240, 1996

[208] B.M. Tenbroek, W. Redman-White, M.S.L. Lee, R.J.T. Bunyan, M.J. Uren, and K.M. Brunso IEEE Transactions on Electron Devices, Vol. 43, no. 12, p. 2227, 1996

[209] B.M. Tenbroek, M.S.L. Lee, W. Redman-White, C.F. Edwards, R.J.T. Bunyan, M.J. Uren, Proceedings of the IEEE International SOI Conference Proceedings, p. 156, 1997

[210] J.P. Colinge, IEEE Transactions on Electron Devices, Vol. 37, p. 718, 1990

[211] K.W. Su and J.B. Kuo, IEEE Transactions on Electron Devices, Vol. 44, no. 5, p. 832, 1997

[212] B. Iñíguez, B. Gentinne, V. Dessard, D. Flandre, Prodings of the IEEE international SOI Conference, p. 92, 1997

[213] T.W. MacElwee and D.I. Calder, Proceedings of the Second International Symposium on Ultra Large Scale Integration Science and Technology, Ed. by C.M. Osburn and J.M. Andrews, Vol. 89-9, The Electrochemical Society, p. 693, 1989

[214] N.J. Thomas and J.R. Davis, Proceedings of the IEEE SOS/SOI Technology Conference, p. 130, 1989

[215] E. Rauly, B. Iñiguez, D. Flandre, Electrochemical and Solid-State Letters, Vol. 4, no. 3, p. G28, 2001

[216] H.K. Lim and J.G. Fossum, IEEE Transactions on Electron Devices, Vol. 30, p. 1244, 1983

[217] D.J. Wouters, J.P. Colinge, and H.E. Maes, IEEE Transactions on Electron Devices, Vol. 37, p. 2022, 1990

[218] J.P. Colinge, D. Flandre and F. Van de Wiele, Solid-State Electronics, Vol. 37, No 2, p. 289, 1994

[219] F. Van de Wiele and P. Paelinck, Solid-State Electronics, Vol. 32, p. 567, 1989

[220] M. Bucher, C. Lallement, and C.C. Enz, Proceedings of the 1996 IEEE International Conference on Microelectronic Test Structures, Vol. 9, p. 145, 1996

[221] M.A. Pavanello, J.A. Martino, and J.P Colinge, Solid-State Electronics, Vol. 41, no. 1, p. 111, 1997

[222] R.N. Simons, Electronics Letters, Vol. 30, p. 654, 1994

[223] J.P. Raskin, I. Huynen, R. Gillon, D. Vanhoenacker, and J.P. Colinge, Proceedings of the IEEE International SOI Conference, p. 28, 1996

[224] J.N. Burghartz, B. Rejaei, IEEE Transactions on Electron Devices, Vol. 50, no. 3, p. 718, 2003

[225] A. Hürrich, P. Hübler, D. Eggert, H. Kück, W. Barthel, W. Budde, and M. Raab, Proceedings of the IEEE International SOI Conference, p. 130, 1996

[226] A. Viviani, J.P. Raskin, D. Flandre, J.P. Colinge, and D. Vanoenacker, Technical Digest of IEDM, p. 713, 1995

[227] J.-P. Raskin, A. Viviani, D. Flandre, and J.-P. Colinge, IEEE Transactions on Electron Devices, Vol. 44, no. 12, p. 2252, 1997

[228] Y. Fukuda, S. Ito, M. Ito, OKI Technical Review, Vol. 68, no. 4, p. 54, 2001

[229] C.H. Diaz, D.D. Tang, J.Y.C. Sun, IEEE Transactions on Electron Devices, Vol. 50, no. 3, p. 557, 2003

[230] V. Kilchytska, A. Nève, L. Vancaillie, D. Levacq, S. Adriaensen, H. van Meer, K. De Meyer, C. Raynaud, M. Dehan, J.P. Raskin, D. Flandre, IEEE Transactions on Electron Devices, Vol. 50, no. 3, p. 577, 2003

[231] R. Gillon, J.P. Raskin, D. Vanhoenacker, and J.P. Colinge, Proceedings of the European Microwave conference, Bologna, Italy, p. 543, 1995

[232] T. Ohguro. E. Morifuji, M. Saito, M. Ono, T. Yoshitomi, H.S. Momose, N. Ito, H. Iwai, Symposium on VLSI Technology Digest of Technical Papers, p. 132, 1996

[233] R. Reedy, J. Cable, D. Kelly, M. Stuber, F. Wright, G. Wu, Analog Integrated Circuits and Signal Processing, Vol. 24, p. 171, 2000

[234] J. Chen, J.P. Colinge, D. Flandre, R. Gillon, J.P. Raskin, D. Vanhoenacker, Journal of the Electrochemical Society, Vol. 144, p. 2437, 1997

[235] A. Hürrich, P. Hübler, D. Eggert, H. Kück, W. Barthel, W. Budde, M. Raab, Proceedings of the IEEE International SOI Conference, p. 130, 1996

[236] D. Eggert, P. Huebler, A. Huerrich, H. Kueck, W. Budde, and M. Vorwerk, IEEE Transactions on Electron Devices, Vol. 44, no. 11, p. 1981, 1997

[237] V. Ferlet-Cavrois, A. Bracale, F. Fel, O. Musseau, C. Raynaud, O. Faynot, J.L. Pelloie, Proceedings of the IEEE International SOI Conference, p. 24, 1999

[238] C.L. Chen, S.J. Spector, R.M. Blumgold, R.A. Neidhard, W.T. Beard, D.R. Yost, J.M. Knecht, C.K. Chen, M. Fritze, C.L. Cerny, J.A. Cook, P.W. Wyatt, C.L. Keast, IEEE Electron Device Letters, Vol. 23, p. 52, 2002

[239] A. Bracale, V. Ferlet-Cavrois, N. Fel, J.L. Gautier, J.L. Pelloie, J. du Port de Poncharra, Electrochemical Society Proceedings, Vol. 2001-3, p. 343, 2001

[240] T. Matsumoto, S. Maeda, K. Ota, Y. Hirano, K. Eikyu, H. Sayama, T. Iwamatsu, K. Yamamoto, T. Katoh, Y. Yamaguchi, T. Ipposhi, H. Oda, S. Maegawa, Y. Inoue, M. Inuishi, Technical Digest of IEDM, p. 219, 2001

[241] N. Zamdmer, A. Ray, J.O. Plouchart, L. Wagner, N. Fong, K.A. Jenkins, W. Jin, P. Smeys, I. Yang, G. Shahidi, F. Assaderghi, Digest of Technical Papers, Symposium on VLSI Technology, p. 85, 2001

[242] Y. Momiyama, T. Hirose, H. Kurata, K. Goto, Y. Watanabe, T. Sugii T, Technical Digest of IEDM, p. 451, 2000

[243] M. Vanmackelberg, C. Raynaud, O. Faynot, J.L. Pelloie, C. Tabone, A. Grouillet, F. Martin, G. Dambrine, L. Picheta, E. Mackowiak, P. Llinares, J. Stevenhans, E. Compagne, G. Fletcher, D. Flandre, V. Dessard, D. Vanhoenacker, J.P. Raskin, Solid-State Electronics, Vol. 46, p. 379, 2002

[244] A.O. Adan, S. Shitara, N. Tanba, M. Fukumi, T. Yoshimasu, Proceedings of the IEEE International SOI Conference, p. 30, 2000

[245] A.O. Adan, T. Yoshimasu, S. Shitara, N. Tanba, M. Fukumi, IEEE Transactions on Electron Devices, Vol. 49, no. 5, p. 881, 2002

[246] M. Vanmackelberg, S. Boret, D. Gloria, O. Rozeau, R. Go\woziecki, C. Raynaud, S. Lepilliet, G. Dambrinne, Proceedings of the IEEE International SOI Conference, p. 153, 2002

[247] J.P. Colinge, J. Chen, D. Flandre, J.P. Raskin, R. Gillon, and D. Vanhoenaecker, Proceedings of the IEEE International SOI Conference, p. 28, 1996

[248] D. Flandre, D. Vanhoenacker, Proceedings of the 1998 International Semiconductor Conference (CAS'98), Vol. 1, p. 115, 1998

[249] http://www.soi.tec.ufl.edu/

[250] http://www-device.eecs.berkeley.edu/~bsimsoi/

[251] http://www.iue.tuwien.ac.at/software/software.html

[252] http://legwww.epfl.ch/ekv/examples.html

[253] http://www.mentor.com/eldo/overview.html

[254] O. Faynot, J.L. Pelloie, M. Belleville, Electrochemical Society Proceedings, Vol. 2001-3, p. 211, 2001

[255] O. Faynot, T. Poiroux, J.L. Pelloie, Solid-State Electronics, Vol. 45, no. 4, p. 599, 2001

[256] http://www.micro.ecs.soton.ac.uk/activities/artic/stag/

Chapter 6

OTHER SOI DEVICES

Although the conventional single-gate MOSFET is the most widely used SOI device, the ease of processing SOI substrates, the full dielectric isolation of the devices and the possibility of using a back gate have led to large research activity in the field of novel SOI devices. Several different novel bipolar and MOS structures have been proposed, such as lateral bipolar and bipolar-MOS devices, vertical bipolar transistors with back gate-induced collector, high-voltage lateral devices of various kinds and multiple-gate MOS devices. This Chapter will review these devices, qualitatively explain their physics and explore their possible fields of application.

Figure 6-1 shows the SOI MOSFET "family tree". The SOS MOSFET is the ancestor of all SOI MOS devices. Single-gate MOSFETs are their direct descendants are analyzed in Chapter 5. If the gate electrode covers not a single, but two, three or more sides of a silicon island, multiple-gate devices can be fabricated. These will be described in section 6.1. Section 6.2 describes the physics of the MOS transistor with a connection between gate and body (MTCMOS/DTMOS). High-voltage and power SOI transistors are described in section 6.3, immediately followed by the SOI junction field-effect transistor (JFET) in section 6.4. Section 6.5 describes the properties of the Lubistor, while section 6.6 deals with SOI bipolar junction transistors. SOI photodiodes are described in section 6.7. The G^4 device is a mixed MOSFET/JFET described in section 6.8. Finally, section 6.9 deals with quantum-effect SOI devices such as single-electron transistors and quantum wires.

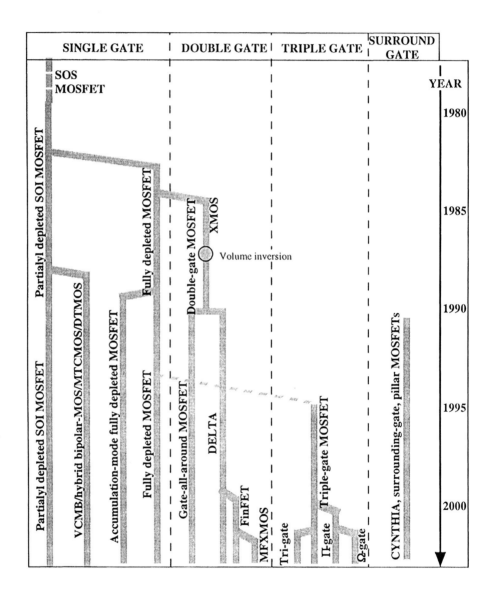

Figure 6-1. SOI MOSFET family tree as a function of time first reported (*y* axis).

6.1 MULTIPLE-GATE SOI MOSFETS

In an ever increasing need for higher current drive and better short-channel characteristics, silicon-on-insulator MOS transistors are evolving from classical, planar, single-gate devices into three-dimensional devices with multiple gates (double-, triple- or quadruple- gate devices). This section describes the evolution and the properties of such devices.

6.1.1 Multiple-gate SOI MOSFET structures

Many different forms of double-, triple- or quadruple- gate devices (multiple-gate FETS or "MuGFETs") can be found in the literature. Let us first sort them by number of gates.

6.1.1.1 Double-gate SOI MOSFETs

One of the first publication on the double-gate MOS (DGMOS) transistor concept dates back to 1984.[1] It shows that one can obtain significant reduction of short-channel effects in a device, called XMOS, where excellent control of the potential in the silicon film is achieved by using a top-and-bottom gate. The name of the device comes from its resemblance with the Greek letter Ξ. Using this configuration, a better control of the channel depletion region is obtained than in a "regular" SOI MOSFET, and, in particular, the influence of the source and drain depletion regions are kept minimal. Short-channel effects are reduced by preventing the source and drain field lines from reaching the channel region.[2] More complete modeling, including Monte-Carlo simulations, is presented in [3] in which the ultimate scaling of silicon MOSFETs is explored. According to that paper, the ultimate silicon device is a double-gate SOI MOSFET with a gate length of 30 nm, an oxide thickness of 3 nm, and a silicon film thickness of 5 to 20 nm. Such a (simulated) device shows no short-channel effects for gate lengths larger than 70 nm, and provides transconductance values up to 2300 mS/mm. The first fabricated double-gate SOI MOSFET was the "fully DEpleted Lean-channel TrAnsistor (DELTA, 1989)", where the silicon film stands vertical on its side (Figure 6-2).[4] Later implementations of vertical-channel, double-gate SOI MOSFETs include the FinFET [5], the MFXMOS [6], the triangular-wire SOI MOSFET [7] and the Δ-channel SOI MOSFET [8]. Volume inversion was discovered in 1987 [9], and the superior transconductance brought about by this phenomenon was experimentally observed in 1990 in the first practical implementation of a planar double-gate MOSFET called the "gate-all-around" (GAA) device (Figure 6-2).[10]

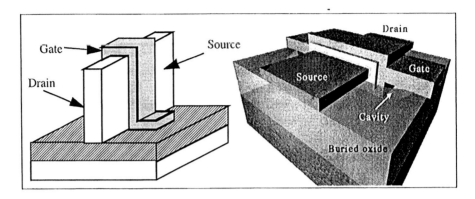

Figure 6-2. DELTA/FinFET double-gate MOS structure (left) and gate-all-around MOSFET
(GAA MOSFET, right - Courtesy X. Baie)

6.1.1.2 Triple-gate SOI MOSFETs

The triple-gate MOSFET is a thin-film, narrow silicon island with a gate on
three of its sides. Implementations include the quantum-wire SOI MOSFET
(Figure 6-3) [11] and the tri-gate MOSFET [12,13]. Improved versions feature
either a field-induced, pseudo-fourth gate such as the Π-gate device [14], the Ω-
gate device [15] and the strained-channel multi-gate device.[16] Such devices
have electrical properties between triple- and quadruple-gate devices and are
sometimes called triple-plus (3^+) gate devices or multiple-gate FETs
(MuGFETs).

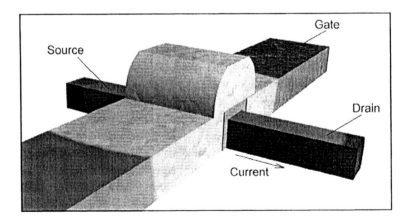

Figure 6-3. Triple-gate SOI MOSFET (Courtesy X. Baie)

6.1.1.3 Surrounding-gate SOI MOSFETs

The structure that theoretically offers the best possible control of the channel region by the gate is the surrounding-gate MOSFET. Such a device is usually fabricated using a pillar-like silicon island with a vertical channel. The structure includes the cylindrical, thin-pillar (CYNTHIA) device (circular-section device, Figure 6-4)[17] and the pillar surrounding-gate MOSFET (square-section device)[18].

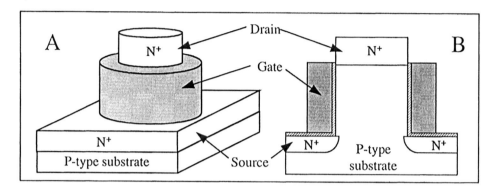

Figure 6-4. CYNTHIA device. A: 3D view, and B: cross section.

6.1.1.4 Triple-plus gate SOI MOSFETs

It is quite clear from the above considerations, that the surrounding-gate structure offers the best possible characteristics in terms of current drive and short-channel effects control. All surrounding-gate devices reported in the literature have a vertical channel current flow and non-planar nature. The source and drain are situated at different depths in the silicon film (Figure 6-4). It is, however, possible to design and fabricate quasi-surrounding-gate MOSFETs using a process similar to that used to fabricate triple-gate SOI MOSFETs. Such devices are called either Π-gate [19,20] or Ω-gate [21] MOSFETs (Figure 6-5). These devices are basically triple-gate devices with an extension of the gate electrode below the active silicon island, which increases current drive and improves short-channel effects. The gate extension can readily be formed by slightly overetching the buried oxide (BOX) during the silicon island patterning step. The gate extension forms a virtual, field-induced gate electrode underneath the device that can block drain electric field lines from encroaching on the channel region at the bottom of the active silicon. Instead the electric field lines terminate on the gate extensions. This gate structure is very effective at reducing short-channel effects. Such devices can be called 3⁺ (triple-plus)-gate devices because their characteristics lie between those of triple- and quadruple-gate devices.

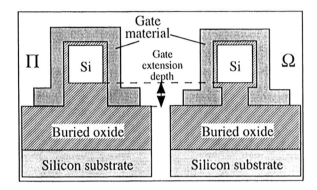

Figure 6-5. Π-gate (Pi-gate) and Ω-gate (Omega-gate) MOSFET cross-sections.

Creating gate extensions in the buried oxide may be impractical if the depth of these extensions is too great. Fortunately, this is not the case and a field-induced back-gate effect is obtained as soon as the gate extension depth reaches 5 to 10 nm, which is readily achievable through simple device processing. Efficient control of the channel region and shielding from the drain electric field lines is effective only when the width of the device is small enough.

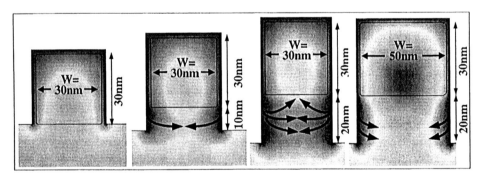

Figure 6-6. Potential distribution in Π-gate SOI MOSFETs with different gate extension depths and two different silicon island width values. $V_G=1V$, $V_{sub} = 0V$.

Figure 6-6 shows the potential distribution in Π-gate devices with $t_{si}=30$ nm and several gate extension depths. The control of the region under the silicon island by the gate potential clearly increases with the gate extension depth. In the device with an extension depth of 20 nm and $W=30$ nm, the region underneath the silicon island is at a potential close to V_G. Thus a virtual, field-induced back gate is created. In the wider device ($W=50$nm), the distance between the two gate extensions is too great for this effect to occur, and the bottom of the silicon island is not controlled by the gate potential.

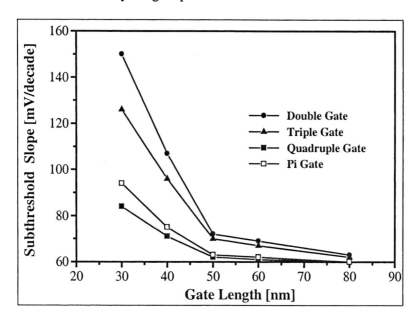

Figure 6-7. Subthreshold slope in double-, triple-, quadruple-/surrounding-gate and Π-gate MOSFETs as a function of gate length. W = t_{si} = 30 nm, t_{ox}=3nm, V_{DS}=0.1V.

Figure 6-7 compares the subthreshold swing of transistors with 2,3 and 4 gates with that of a Π-gate device. Increasing the number of gates improves the subthreshold swing since the control of the channel region by the gate(s) becomes more effective. Multiple gates also offer more shielding plates protecting the channel region from the electric field lines from the drain. It should be noted that the performances of the Π-gate structure are very close to those of a 4-gate device.

The different gate configurations for an SOI MOSFET are summarized in Figure 6-8. Practical implementations of double-gate devices may physically look either like the DELTA structure or the GAA structure (Figure 6-2).

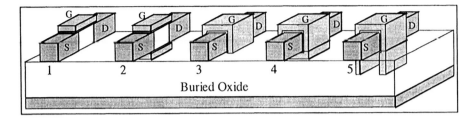

Figure 6-8. Single-gate (1), double-gate (2), triple-gate (3), quadruple-gate/surrounding-gate (4) and Π- or Ω-gate MOSFET (5)

6.1.2 Device characteristics

The main interest of multiple-gate SOI MOSFETs is two-fold: they have a high current drive per unit area of silicon real estate because of the formation of multiple channels, and they have excellent short-channel characteristics because the potential in the channel region is controlled by more than one gate.

6.1.2.1 Current drive

The current drive of multiple-gate SOI MOSFETs is essentially proportional to the total gate width. For instance, the current drive of a double-gate device is double that of a single-gate transistor with identical gate length and width. In triple-gate and vertical double-gate structures all individual devices must have the same thickness and width to have similar threshold voltage and other electrical properties. As a result the current drive is fixed to a single, discrete value, for a given gate length. For larger current drive, multi-fingered devices must be used. The current drive of a multi-fingered MOSFET is then equal to the current of an individual device multiplied by the number of fingers (also sometimes referred to as "fins" or "legs"). Considering a pitch, P, for the fingers and assuming carrier mobility is identical at all Si/SiO$_2$ interfaces, the current per unit device width is given by:

$$I_D = I_{Do} \frac{(W + 2t_{si})}{P}$$

where I_{Do} is the current of a unit-width, planar, single-gate device, and where W is the width of an individual finger, t_{si} is the silicon film thickness, and P is the finger pitch (Figure 6-9).

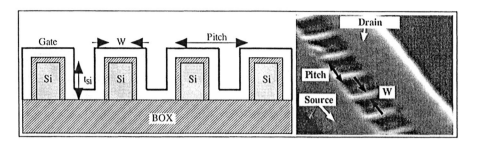

Figure 6-9. Cross section of a multi-fingered triple-gate MOSFET (left) and SEM picture of the fingers (right).[28]

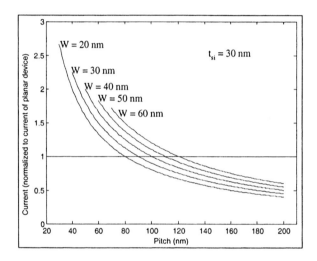

Figure 6-10. Normalized current drive in a multi-fingered, triple-gate SOI MOSFETs as a function of pitch and finger width.[28]

The FinFET device achieves high current drive through the use of a relatively thick silicon film. In that device there is no current flow at the top of the silicon island, such that $I_D=I_{Do} (2t_{si}/P)$. In triple-gate devices where $t_{si} = W$ the finger pitch needs to be smaller than $3W$ to obtain a larger current drive than in a single-gate, planar device occupying the same silicon real estate. Figure 6-10 shows the current in a multi-fingered, triple-gate MOSFET as a function of finger pitch and finger width. The silicon film thickness is 30 nm and the current is normalized to that of a planar, single-gate device occupying the same silicon real estate. One can see that the current in the triple-gate structure is, for instance, 50% higher than in a single-gate MOSFET if the finger width is 30 nm and the pitch is 60 nm.

6.1.2.2 Short-channel effects

It is possible to predict how small the silicon film thickness should be in multiple-gate devices to avoid short-channel effects (or, at least, to maintain a reasonable subthreshold swing). Subthreshold swing degradation and other short-channel effects are caused by the encroachment of electric field lines from the drain on the channel region, thereby competing for the available depletion charge, and reducing the threshold voltage. The potential distribution in the channel of a fully depleted SOI MOSFET is governed by Poisson's equation:

$$\frac{d^2\Phi(x,y,z)}{dx^2} + \frac{d^2\Phi(x,y,z)}{dy^2} + \frac{d^2\Phi(x,y,z)}{dz^2} = \frac{qN_a}{\varepsilon_{si}} \quad (6.1.1)$$

Figure 6-11 shows how the gates and the drain compete for the depletion charge. Gate control is exerted in the y- and z- directions and competes with the variation of electric field in the x-direction due to the drain voltage.

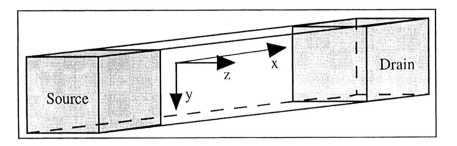

Figure 6-11. Definition of coordinate system in a multiple-gate device

In the case of a *wide* single- or double-gate device, $\dfrac{d\Phi}{dz} = 0$, and we can write:

$$\frac{d^2\Phi(x,y)}{dx^2} + \frac{d^2\Phi(x,y)}{dy^2} = \frac{qN_a}{\varepsilon_{si}} \quad (6.1.2)$$

The one-dimensional analysis of a fully depleted device yields a parabolic potential distribution in the silicon film in the y (vertical) direction. Assuming a similar distribution in the y-direction for a two-dimensional analysis we can write:[22]

$$\Phi(x, y) = c_o(x) + c_1(x) y + c_2(x) y^2 \qquad (6.1.3)$$

6.1.2.2.1 Single-gate MOSFET

Let us first examine the case of a single-gate SOI device. The boundary conditions for equation 6.1.3 are:[23]

1. $\Phi(x,0) = \Phi_f(x) = c_o(x)$ where $\Phi_f(x)$ is the front surface potential;

2. $\dfrac{d\Phi(x, y)}{dy}\bigg|_{y=0} = \dfrac{\varepsilon_{ox}}{\varepsilon_{si}} \dfrac{\Phi_f(x) - \Phi_{gs}}{t_{ox}} = c_1(x)$

   where $\Phi_{gs} = V_{gs} - V_{FBF}$ is the front gate voltage, V_{gs}, minus the front gate flat-band voltage, V_{FBF}.

3. If we assume that the buried oxide thickness is large, such that the potential difference across any finite distance in the BOX is negligible in the y direction, we can write $\dfrac{d\Phi(x, y)}{dy} \cong 0$ in the BOX. This yields:

   $\dfrac{d\Phi(x, y)}{dy}\bigg|_{y=t_{si}} = c_1(x) + 2t_{si} c_2(x) \cong 0$ and thus $c_2(x) \cong -\dfrac{c_1(x)}{2t_{si}}$.

Introducing these three boundary conditions in equation 6.1.3 we obtain:

$$\Phi(x, y) = \Phi_f(x) + \frac{\varepsilon_{ox}}{\varepsilon_{si}} \frac{\Phi_f(x) - \Phi_{gs}}{t_{ox}} y - \frac{1}{2t_{si}} \frac{\varepsilon_{ox}}{\varepsilon_{si}} \frac{\Phi_f(x) - \Phi_{gs}}{t_{ox}} y^2 \qquad (6.1.4)$$

Substituting equation 6.1.4 into equation 6.1.2, and setting $y=0$, at which depth $\Phi(x, y) = \Phi_f(x)$ we obtain:

$$\frac{d^2\Phi_f(x)}{dx^2} - \frac{\varepsilon_{ox}}{\varepsilon_{si}} \frac{\Phi_f(x) - \Phi_{gs}}{t_{si} t_{ox}} = \frac{qN_a}{\varepsilon_{si}} \qquad (6.1.5)$$

Once $\Phi_f(x)$ is determined from equation 6.1.5, $\Phi(x, y)$ can be calculated using equation 6.1.4. Equation 6.1.5, however, can be used for another purpose. If we write

$$\lambda_1 = \sqrt{\frac{\varepsilon_{si}}{\varepsilon_{ox}} t_{ox} t_{si}}$$
(6.1.6)

and

$$\varphi(x) = \Phi_f(x) - \Phi_{gs} + \frac{q N_a}{\varepsilon_{si}} \lambda_1^2$$
(6.1.7)

then equation 6.1.5 can be re-written as follows:

$$\frac{d^2 \varphi(x)}{dx^2} - \frac{\varphi(x)}{\lambda_1^2} = 0$$
(6.1.8)

This equation is a simple differential equation with a parameter, λ_1, that controls the spread of the electric potential in the x-direction. Note that $\varphi(x)$ differs from $\Phi(x)$ only by an x-independent term. Parameter λ_1 is called the "natural length" of the device and it depends on the gate oxide and silicon film thickness. Further analysis and numerical simulations show that the effective gate length of a MOS device must be larger than 5 to 10 times the natural length to prevent short-channel effects and to produce a reasonable subthreshold behaviour.[23]

6.1.2.2.2 Double-gate MOSFET

In the case of a double-gate device the boundary conditions to equation 6.1.3 are:

1. $\Phi(x,0) = \Phi(x, t_{si}) = \Phi_f(x) = c_o(x)$ where $\Phi_f(x)$ is the front surface potential;

2. $\left. \dfrac{d\Phi(x, y)}{dy} \right|_{y=0} = \dfrac{\varepsilon_{ox}}{\varepsilon_{si}} \dfrac{\Phi_f(x) - \Phi_{gs}}{t_{ox}} = c_1(x)$ where $\Phi_{gs} = V_{gs} - V_{FBF}$ is the front gate voltage, V_{gs}, minus the front gate flat-band voltage, V_{FBF}.

3. $\left. \dfrac{d\Phi(x,y)}{dy} \right|_{y=t_{si}} = -\dfrac{\varepsilon_{ox}}{\varepsilon_{si}} \dfrac{\Phi_f(x)-\Phi_{gs}}{t_{ox}} = c_1(x) + 2t_{si}\, c_2(x) = -c_1(x)$ and

thus $c_2(x) = \dfrac{-c_1(x)}{t_{si}}$.

Substituting these boundary conditions into equation 4 yields:

$$\Phi(x,y) = \Phi_f(x) + \dfrac{\varepsilon_{ox}}{\varepsilon_{si}}\dfrac{\Phi_f(x)-\Phi_{gs}}{t_{ox}}\, y - \dfrac{1}{t_{si}}\dfrac{\varepsilon_{ox}}{\varepsilon_{si}}\dfrac{\Phi_f(x)-\Phi_{gs}}{t_{ox}}\, y^2 \qquad (6.1.9)$$

The key difference between this expression and equation 6.1.4 is that the term $1/2t_{si}$ is now been replaced by $1/t_{si}$. Thus, a double-gate device with twice the thickness of a single-gate device has identical control of the short-channel effects. The natural length of the double-gate device can be derived the same way it was done for the single-gate case, which yields:

$$\lambda_2 = \sqrt{\dfrac{\varepsilon_{si}}{2\varepsilon_{ox}}t_{ox}t_{si}} \qquad (6.1.10)$$

The natural length gives a measure of the short channel effect inherent to a device structure. It represents the penetration distance of the electric field lines from the drain in the body of the device or the amount of control the drain region has on the depletion zone in the channel, as both the gate and the drain compete for that control. A small λ is desired to minimize short channel effects on the subthreshold slope. Numerical simulations establish that a device is relatively free of short-channel effects if λ has a value smaller than 5 to 10 times the gate length.[23]

6.1.2.2.3 Surrounding-gate MOSFET

The original publication of the natural length concept [23] analyzes single- and double-gate structures. It can be extended to surrounding-gate devices with a square cross section by noting that $\dfrac{d^2\Phi}{dy^2} = \dfrac{d^2\Phi}{dz^2}$ in the center of the device, where the encroachment of the electric field lines from the drain on the device body is the strongest. In that case the Poisson equation becomes:

$$\frac{d^2\Phi(x,y,z)}{dx^2} + 2\frac{d^2\Phi(x,y,z)}{dy^2} = \frac{qN_a}{\varepsilon_{si}} \qquad (6.1.11)$$

and the natural length is equal to:

$$\lambda_3 = \sqrt{\frac{\varepsilon_{si}}{4\varepsilon_{ox}}t_{ox}t_{si}} \qquad (6.1.12)$$

References [24] and [25] analyze the case of single- and double-gate devices with other boundary conditions than [23] and Reference [26] calculates the natural length in a cylindrical surrounding-gate device. The natural length corresponding to different device geometries are summarized in Table 6-1.

The natural length concept can be used to estimate the maximum silicon film thickness and device width that can be used in order to avoid short-channel effects. Figure 6-12 shows the maximum allowed silicon film thickness (and device width in a triple-gate device with $W=t_{si}$) to avoid short-channel effects. The plot is based on Equations 6.1.6, 6.1.10 and 6.1.12, and the gate oxide thickness is either equal to 1.5 nm or to a tenth of the silicon film thickness. The plot reveals that for a gate length of 50 nm, for instance, the thickness of the silicon film in a single-gate, fully depleted device needs to be approximately 3 times smaller than the gate length. If a double-gate structure is used, the silicon film thickness needs to be only half the gate length.[27]

Table 6-1. Natural length in devices with different geometries

Single gate	$\lambda_1 = \sqrt{\dfrac{\varepsilon_{si}}{\varepsilon_{ox}}t_{si}\,t_{ox}}$ [23]	$\lambda_1 = \dfrac{1}{\pi}\left(t_{si} + \dfrac{\varepsilon_{si}}{\varepsilon_{ox}}t_{ox}\right)$ [25]
Double gate	$\lambda_2 = \sqrt{\dfrac{\varepsilon_{si}}{2\varepsilon_{ox}}t_{si}\,t_{ox}}$ [23]	$\lambda_2 = \sqrt{\dfrac{\varepsilon_{si}}{2\varepsilon_{ox}}\left(1 + \dfrac{\varepsilon_{ox}t_{si}}{4\varepsilon_{si}t_{ox}}\right)t_{si}\,t_{ox}}$ [25]
Surrounding gate	$\lambda_3 \cong \sqrt{\dfrac{\varepsilon_{si}}{4\varepsilon_{ox}}t_{si}\,t_{ox}}$ [28] (square section – 6.1.12)	$\lambda_3 = \sqrt{\dfrac{2\varepsilon_{si}t_{si}^2\ln\left(1 + \dfrac{2t_{ox}}{t_{si}}\right) + \varepsilon_{ox}t_{si}^2}{16\varepsilon_{ox}}}$ [26] (circular cross section)

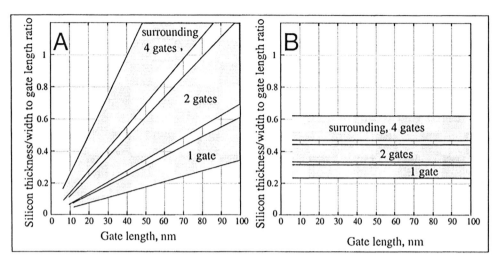

Figure 6-12. Maximum allowed silicon film thickness and device width *vs.* gate length for short-channel-free operation ($6\lambda > L > 8\lambda$). A: Gate oxide thickness is held constant ($t_{ox}=1.5$ nm), and B: $t_{ox}=t_{si}/10$.

Further relaxation is obtained using a surrounding-gate structure, where the silicon film thickness/width/diameter can be as large as the gate length. The film thickness requirements for triple-gate, Π-gate and Ω-gate devices are located between those for double-gate and surrounding-gate devices. The use of very thin silicon films is thus required for short-channel single-gate devices. However, severe mobility reduction effects have been reported in ultrathin SOI devices.[29, 30, 31] Furthermore, the use of ultra-thin films raises the issue of high source and drain resistance and tight etch selectivity tolerances. Quite clearly, the use of double-gate, and especially that of surrounding-gate structures, relaxes the requirements on film thickness.

6.1.2.3 Threshold voltage

The most common definition of the threshold voltage of a MOSFET assumes that $\Phi_S=2\Phi_F$ at threshold, *i.e.* when strong inversion is reached. This definition is inadequate for thin-film, double-gate devices, where current appears following a weak inversion mechanism. Additionally, in triple- and quadruple-gate devices, inversion may be reached in different parts of the channel region at different gate voltages.

6.1.2.3.1 Double-gate MOSFET

Francis *et al.* have developed an extensive model of the double-gate, inversion-mode SOI transistor [32,33,34,35] where the threshold voltage is defined by the transconductance change (TC) method. According to this method, the threshold voltage can be defined as the gate voltage where the derivative of the transconductance reaches a maximum, or, in mathematical terms, when $d^3 I_D/dV_G^3 = 0$.[36] Using this condition, the surface potential at threshold can be obtained:

$$\Phi_S^* = 2\Phi_F + \frac{kT}{q} \ln\left[\delta \frac{1}{1-\exp(-\alpha)}\right] \text{ where } \alpha = \frac{q}{kT}\frac{Q_D}{8C_{si}} \text{ and } \delta = \frac{C_{ox}}{4C_{si}}, \text{ all}$$

other symbols having their usual meaning. The last term of the surface potential at threshold is negative, such that Φ_S^* is smaller than $2\Phi_F$ by a value of 10 to 90 mV, which justifies the previous assumption of having a weak inversion current at threshold. The threshold voltage can be obtained analytically:

$$V_{th} = \Phi_S^* + V_{FB} + \frac{kT}{q}\frac{\alpha}{\delta}\sqrt{1+\frac{\delta}{\alpha}} \qquad (6.1.13)$$

The difference between the surface potential at inversion and $2\Phi_F$ depends on the silicon film thickness, the gate oxide thickness, and the doping concentration.

Thin-film, double-gate transistors have a low threshold voltage (around 0 volt) when the p-type channel doping is low and the top and bottom gates are made of N+ polysilicon. To obtain a higher threshold voltage it is possible to use P+ polysilicon for both gates. In that case, a threshold voltage around 1 volt is obtained, which is too high for most applications. An intermediate solution consists of using either midgap or dual-type polysilicon gate material (one gate is N+-doped and the other one is P+-doped). Figure 6-13 presents the threshold voltage in an n-channel double-gate MOSFET with low channel doping concentration ($N_a=10^{15}$ cm^{-3}), as a function of the silicon film thickness. Another solution is, of course, to use two N+-poly gates and to increase the threshold voltage by increasing the channel doping level. This solution, however, has the disadvantage of decreasing the mobility through increasing impurity scattering.

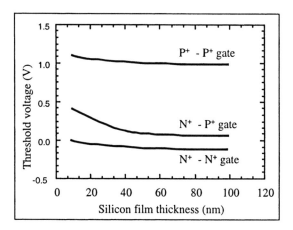

Figure 6-13. Dependence of threshold voltage in P^+-P^+, N^+-P^+ and N^+-N^+ double-gate SOI MOSFETs on silicon film thickness ($N_a=10^{15}$ cm^{-3}).

Simulations indicate that asymmetrical polysilicon gates (*i.e.*: one N^+-type gate and one P^+-type gate) yield devices with higher current drive, lower off current and better short-channel effect control than symmetrical gates (using the same gate material for both gates). This improvement is essentially due to the fact that only one channel is formed in a device with asymmetrical gates and to extended gate to gate electrical coupling and dynamic threshold voltage lowering, which enables low off current and high on current with low gate capacitance. This beneficial coupling effect is reduced in symmetrical double-gate transistors by the presence of two channels.[37,38]

Figure 6-14 shows the threshold voltage in double-gate n-channel SOI MOSFETs using either an N+ polysilicon gate and p-type channel doping, or a midgap gate material and n-type channel doping.[39] Several midgap gate materials have been reported in the literature. These include nickel and cobalt silicide [40], titanium nitride (TiN) and molybdenum. Titanium nitride and molybdenum add an extra level of flexibility to device design because their work function can be tuned using nitrogen implantation.[41,42] Threshold voltage of vertical double-gate devices such as the DELTA and FinFET device depends on the width of the silicon "fin", which plays the same role as the silicon film thickness in a planar device such as the GAA MOSFET. Ideally a silicon film should have a constant width from top to bottom. If this is not the case the relationship between the fin width and threshold voltage becomes more complicated than predicted in this section.[43]

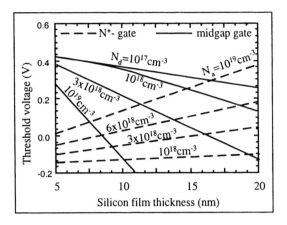

Figure 6-14. Threshold voltage in double-gate n-channel SOI MOSFETs using either an N⁺ polysilicon gate and p-type channel doping, or a midgap gate material and n-type channel doping.

6.1.2.3.2   Triple, triple-plus and quadruple gates

The 3-, 3^+, and 4-gate devices offer high current drive and reduced short-channel effects. Unlike single- or double-gate MOSFETs these devices present a non-planar silicon/gate oxide interface involving corners (four of them, usually). Inversion may form at different gate voltages in the corners and at the top or sidewall $Si-SiO_2$ interfaces. The radius of curvature of the corner has a significant impact on the device electrical characteristics and the presence or absence of a different threshold voltage at the corners than in the planar interfaces of the device. It is important to realize that in classical SOI MOSFETs, corners appearing at the edge of the device can give rise to parasitic currents which are usually undesirable, while in multiple-gate devices the corners are part of the intrinsic transistor structure. Therefore, the relationship and interaction between currents in the corner and currents in the planar surfaces of the device will be developed.

The cross section of a 3^+-gate device is shown in Figure 6-15. The thickness and width of the device are t_{si} and W, and the radii of curvature of the top and the bottom corners are noted r_{top} and r_{bot}, respectively. The gate oxide thickness is 2 nm, and $t_{si}=W=30$ nm. The depth of the gate extension in the buried oxide,

t_{ext}, is 10 nm. The gate material is N+ polysilicon with $\Phi_{MS} = -0.9V$. Because the gate material is N^+ polysilicon and the device thickness and width are small, high doping concentrations must be used to achieve useful threshold voltage values.

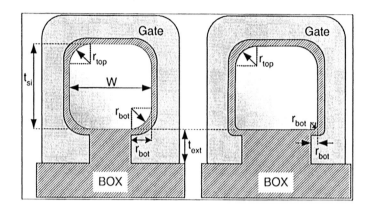

Figure 6-15. Cross section of Π/Ω-gate devices. The radii of curvature of the top and bottom corners are r_{top} and r_{bot}, respectively; $r_{top} = r_{bot}$ (left) and $r_{top} \neq r_{bot}$ (right).

Figure 6-16 presents the dg_m/dV_G characteristics of the device at $V_{DS}=0.1V$ for different doping concentrations and a top and bottom corner radius of curvature of 1 nm. The dg_m/dV_G characteristics have been used by several authors to identify the different threshold voltage(s) in accumulation-mode and double-gate SOI devices.[44,45,46,47] The maxima of the dg_m/dV_G curve correspond to the formation of channels in the device (*i.e.* they correspond to threshold voltages). The devices with the lowest doping concentrations exhibit a single maximum, indicating that both corners and edges build up channels at the same time. The devices with the more heavily doped channels have two maxima. The first of these two humps corresponds to inversion in the top corners, and the second, to top and sidewall channel formation.

The $dV_G/d(\log(I_D))$ curves of the same devices are shown in Figure 6-17. Below threshold the $dV_G/d(\log(I_D))$ values correspond to the subthreshold swing. The more lightly doped devices have a 60 mV/decade swing for most of the subthreshold current range, while the highly doped MOSFETs have higher values. Figure 6-18 presents the dg_m/dV_G characteristics in a device with a top and bottom corner radius of curvature of 5 nm. In this case a single peak is observed for all doping concentrations, which indicates that premature corner inversion has been eliminated. In that case all devices reach a subthreshold swing of 60 mV/decade over a significant range of their subthreshold current.

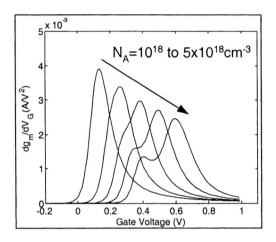

Figure 6-16. dg_m/dV_G in a 3^+-gate MOSFET for $r_{top} = r_{bot} = 1$nm. Gate is N^+ polysilicon.

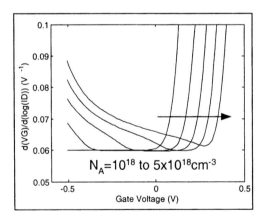

Figure 6-17. $dV_G/d(\log(I_D))$ (subthreshold swing) in a 3^+-gate MOSFET for $r_{top} = r_{bot} = 1$nm. Gate is N^+ polysilicon.

Further analysis shows that the dg_m/dV_G curves of a device with top and bottom corner radii of curvature of 5 and 1 nm, respectively, present a maximum for the highest doping concentrations. This maximum is not due to current in the top corners, but rather in the bottom corners, where inversion channels form at a lower gate voltage compared to the other device interfaces. The corner effect can thus be eliminated by using either a midgap gate material and low doping concentration in the channel, or corners with a large enough radius of curvature.[48,49]

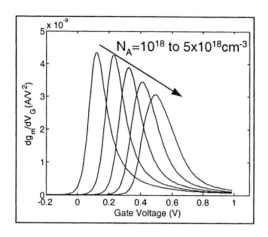

Figure 6-18. dg_m/dV_G in a 3^+-gate MOSFET for $r_{top}= r_{bot} = 5nm$. Gate is N^+ polysilicon.

6.1.2.4 Volume inversion

Volume inversion was discovered in 1987 by Balestra *et al.* [50] and was first observed in double-gate GAA MOSFETs in 1990.[51] Volume inversion is a phenomenon that appears in thin-film double-gate SOI MOSFETs in which inversion carriers are not confined near the Si/SiO2 interface, as classical device physics predicts, but rather at the center of the film. To correctly predict volume inversion one needs to solve both the Schrödinger and the Poisson equation in a self-consistent manner. Figure 6-19 shows the electron concentration in a thin-film, double-gate SOI MOSFET for gate voltages below and above threshold, predicted by classical physics (Poisson equation) or quantum mechanics (Poisson + Schrödinger).

The quantum-mechanical nature of volume inversion was first simulated and measured in 1994 [52,53] and has been explored by many research groups since. [54,55,56,57,58,59,60,61,62,63] Volume inversion is observed in triple-gate SOI MOSFETs as well.[64,65] When a double-gate MOSFET operates in the volume inversion regime, the electrons form a two-dimensional electron gas (2DEG), the thickness of which is equal to the silicon film thickness. One can, therefore, expect to observe splitting of the conduction band into subbands, since the electrons are confined in one spatial direction. The electron wave functions form standing waves in the confined dimension, x.

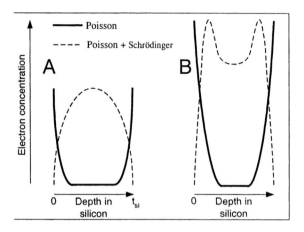

Figure 6-19. Volume inversion in a thin-film double-gate MOSFET; A: $V_G \leq V_{TH}$; B: $V_G > V_{TH}$.

The wave functions and the corresponding electron energy can be found by solving the Schrödinger equation and the Poisson equation in a self-consistent manner:

$$-\frac{\hbar^2}{2m} \frac{d^2 \Psi(x)}{dx^2} - q \Phi(x) \, \Psi(x) = E \, \Psi(x) \qquad (6.1.14)$$

and

$$\frac{d^2 \Phi(x)}{dx^2} = \frac{q(N_a + n(x))}{\varepsilon_{si}} \qquad (6.1.15)$$

The general evolution of the energy levels and wave functions is presented in Figure 6-20, for a single effective mass value. Threshold is reached when the lowest subband becomes populated. If the depletion and inversion charges are low enough (undoped device right above threshold), the potential well into which the electrons are confined can be considered as square. The position of the energy levels above the conduction band minimum, E_C, are then given by:

$$E_n = \frac{\hbar^2}{2m^*} \left(\frac{\pi n}{t_{si}} \right)^2 \qquad (6.1.16)$$

where $n = 1,2,3,...$, and the normalized wave functions are given by $\Psi(x)=sin(n\pi x/t_{si})$, where t_{si} is the silicon film thickness.

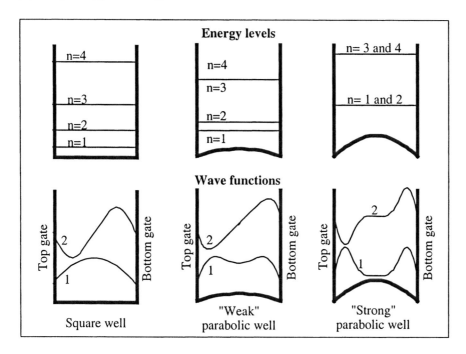

Figure 6-20. Four first energy levels (E_n) and two first wave functions (Ψ_n) in square, weak parabolic and strong parabolic potential wells for the double-gate MOSFET. The electron concentration is proportional to Ψ^2.

When the inversion charge in the film increases (right above threshold), the potential well becomes parabolic and the electrons become more attracted by the gates. The energy levels shift up and tend to pair (E_1 pairs with E_2, E_3 with E_4, etc.). Finally, for higher gate voltages the energy levels become degenerate through pairing, and the wave functions are located mostly at the interfaces, close to the gate oxides (formation of two inversion channels). Transconductance peaks are observed just above threshold, when the parabolic well is relatively weak and true volume inversion is present.

Figure 6-21 presents the evolution of the electron concentration in the 40 nm-thick device as a function of gate voltage. One can clearly see the volume inversion right above threshold and the formation of two inversion channels at higher gate voltages.

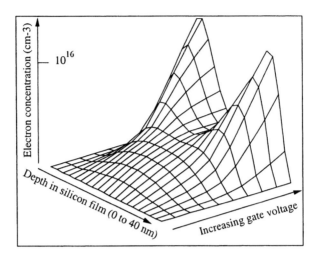

Figure 6-21. Electron concentration versus depth in the silicon film and versus $V_G - V_{TH}$ for a double-gate MOSFET (40 nm-thick device).[66]

6.1.2.5 Mobility

Volume-inversion carriers experience less interface scattering than carriers in a "regular" inversion layer since the peak electron concentration is located at some distance from the Si/SiO_2 interfaces. As a result an increase of the mobility and transconductance is observed in double-gate devices. Furthermore, the phonon scattering rate is lower in double-gate devices than in single-gate transistors.[67] The transconductance of a conventional device and that of a double-gate GAA device are compared in Figure 6-22. An additional curve labeled "SOIx2" presents the transconductance of the SOI MOSFET multiplied by two to account for both the presence of two channels in the GAA device. The gray area represents the extra drive of the double-gate device, which is attributed to volume inversion.

The dependence of mobility on film thickness in double-gate MOSFETs is illustrated in Figure 6-23. In thick films there is no interaction between the front and the back channel and there is no volume inversion. The mobility is identical to that in a bulk MOSFET and the current is twice that of a single-gate device.

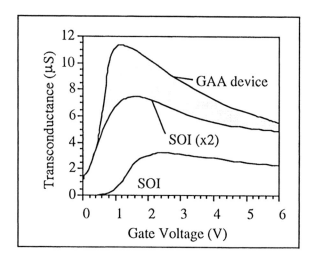

Figure 6-22. Transconductance (dI_d/dV_G) at Vds=100 mV in a conventional SOI MOSFET and a double-gate (GAA) device. [68, 69]

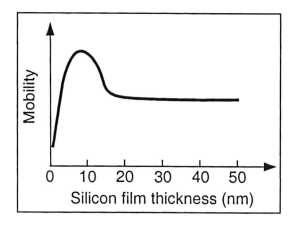

Figure 6-23. Variation of mobility with silicon film thickness in a double-gate MOSFET.[70,71]

If the film becomes thinner volume inversion appears and the mobility is increased because of reduced Si-SiO$_2$ interface scattering.[72] In thicker films the inversion carriers are concentrated near the interfaces, but in thinner films (Figure 6-24) most of the carriers are concentrated near the center of the silicon film, further away from the interface scattering centers, thereby increasing mobility. In very thin silicon films, however, the inversion carriers in the volume inversion layer do experience surface scattering because of their physical proximity to the interfaces, and mobility drops with any decrease in film thickness.[73,74]

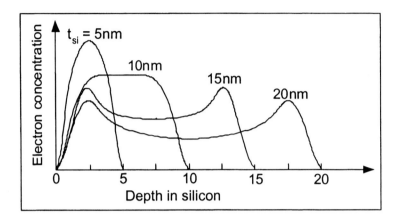

Figure 6-24. Inversion carrier concentration profile in double-gate MOSFETs of different thickness.[75]

6.2   MTCMOS/DTMOS

Interesting improvements of the electrical characteristics of an SOI MOSFET are obtained when a connection is made between the gate and the body (Figure 6-25). The simulation, fabrication and characterization of SOI transistors with gate-to-body connection were first reported in 1987, and the device was called the "voltage-controlled bipolar-MOS device (VCBM)".[76] Other research teams reproduced the device and named it the "hybrid bipolar-MOS device" [77,78], the "hybrid-mode SOI MOSFET" [79] or the "gate-controlled lateral BJT".[80] These early publications placed an emphasis on the high current drive of the device due to combined presence of both MOS and BJT currents. Later on emphasis was put on the dependence of the threshold voltage on body potential, and thus on the gate bias, through the classical body effect. The device was renamed by several teams to either the "multi-threshold CMOS (MTCMOS)" [81], the "dynamic threshold MOS (DTMOS)" [82], or the "varied-threshold MOS (VTMOS)".[83] These devices have ideal subthreshold characteristics, reduced body effect, improved current drive, and superior HF characteristics.[84] They are mostly used for very low-voltage (0.5 V) applications.[85,86]

Every enhancement-mode SOI MOSFET contains a parasitic bipolar junction transistor (BJT). One can thus attempt to add the current drive capabilities of the bipolar and the MOS transistors present in the device. This is achieved by connecting the gate, which controls the current flow in the MOS part of the device, to the floating substrate, which acts as the base of the lateral bipolar transistor (Figure 6-25). The source and the drain of the MOS transistor are also the emitter and the collector of the bipolar device, respectively.

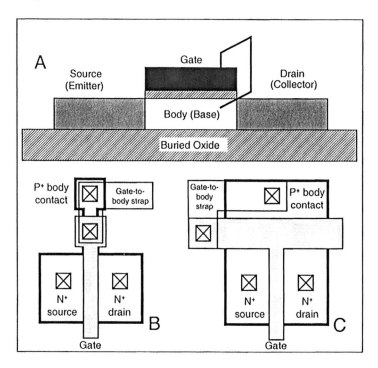

Figure 6-25. Connection between gate and body; A: schematic cross section; B: layout using conventional body contact; C: layout using T-gate body contact.

We will consider an n-channel (NPN) device, although p-channel (PNP) devices can be realized as well. When the device is OFF ($V_G=V_B=0$) the potential of the MOS substrate (the base) is low, which maximizes the value of the threshold voltage (and thereby minimizes the OFF current). When a positive gate bias is applied the potential of the MOS substrate is increased, which decreases the threshold voltage due to the body effect. This lowering of the threshold voltage increases the current drive for a given gate voltage, when compared to an MOS transistor without gate-to-body connection. Similarly, the application of a gate voltage when the device is "ON" increases the collection efficiency of the bipolar device, and reduces the effective neutral base width.[87]

Thus the presence of a gate improves the gain of the bipolar transistor, and the presence of a base contact improves both the ON and OFF characteristics of the MOS device because the threshold voltage is highest when $V_G=0$ and it is lower when $V_G>0$. Both the MOS and the bipolar transistor benefit from this symbiotic association.

If the threshold voltage of the MOS transistor is low enough, the BJT current is negligible compared to the MOS current and the device can be viewed as a MOS transistor whose threshold voltage is modulated by the gate bias. The body effect in the DTMOS/MTCMOS device is identical to that found in a bulk MOSFET, and the threshold voltage is given by:

$$V_{TH} = V_{THo} + \frac{\sqrt{2q\varepsilon_{si}N_a}}{C_{ox}}\left(\sqrt{2\Phi_F - V_{sub}} - \sqrt{2\Phi_F}\right) \qquad (6.2.1)$$

where V_{sub} is the body voltage (the source voltage is taken as a reference) and V_{THo} is the threshold voltage of the MOS device when $V_{sub}=0$. Letting $V_{sub}=V_G$ we find the expression of the threshold voltage in the MTCMOS/DTMOS device: [88,89]

$$V_{TH} = V_{THo} - \frac{\sqrt{2q\varepsilon_{si}N_a}}{C_{ox}}\left(\sqrt{2\Phi_F} - \sqrt{2\Phi_F - V_G}\right) \qquad (6.2.2)$$

This expression clearly shows that the threshold voltage decreases as the gate bias is increased. It also shows that the device is designed to operate at low supply voltage because V_G cannot exceed $2\Phi_F$. In practice V_G should not exceed 700 mV to avoid excessive body-to-source current due to forward biasing the emitter-base junction of the BJT.

Figure 6-26 shows the evolution of V_{TH} with gate voltage in a MTCMOS/DTMOS device and in the same device when the body is grounded ($V_{sub}=0$; grounded-body SOI device (GBSOI)). Because of the variation of threshold voltage with gate bias the subthreshold slope, S, of the MTCMOS/DTMOS device is very close to the ideal value of 60 mV/decade.

Figure 6-27 shows the subthreshold drain current in a MTCMOS/DTMOS device and in the same device when the body is grounded are presented. Several analytical models developed for the MTCMOS/DTMOS device can be found in the literature.[90,91] Models for the threshold voltage of the device can be found as well.[92,93]

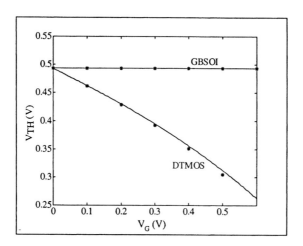

Figure 6-26. Threshold voltage *vs.* gate voltage in grounded-body SOI (GBSOI) and DTMOS devices. Solid lines represent equation 6.2.2 and the (*) symbols represent measured data; $t_{ox}=8nm$, $N_A=1.65x10^{17}cm^{-3}$.[94]

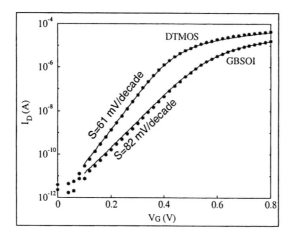

Figure 6-27. Drain current *vs.* gate voltage in GBSOI and DTMOS devices. $V_D=100mV$, $t_{ox}=8nm$, $W=10\mu m$, $L=2\mu m$.[94]

The increase of drain current due to the dependence of the threshold voltage on gate bias brings about an improvement of transconductance, compared to a regular SOI MOS device with grounded body (Figure 6-28).

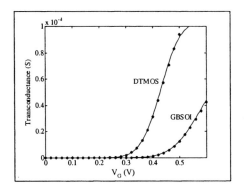

Figure 6-28. Transconductance *vs.* gate voltage in GBSOI and DTMOS devices. V_D=100mV, t_{ox}=8nm,W=10μm, L=2μm.[94]

Because of its high transconductance, the MTCMOS/DTMOS device has better RF performance than regular SOI MOS devices (higher f_T and f_{max}), especially at low supply voltage (Table 6-2).

Table 6-2. RF performance of MTCMOS/DTMOS devices.

Gate length (μm)	V_G (V)	f_T (GHz)	f_{max} (GHz)	Reference
0.25	0.6	16	33	95
0.08	0.65	140	60	96
0.25	0.7	36	-	97
0.25	0.6	18	33	98
0.25	0.7	34	-	99

MTCMOS/DTMOS devices are particularly well suited for low-voltage applications. The first circuits made using these devices were ring oscillators operating with a supply voltage as low as 700 mV. [100] Table 6-3 lists recent large scale integrated circuits (LSI) made using MTCMOS devices.[101,102,103,104,105,106,107] Low-voltage analog and RF circuits have been reported as well, [108] together with high-temperature circuits.[109]

Table 6-3. Low-voltage, low-power, digital MTCMOS circuits

Circuit	Supply voltage (V)	Operating frequency (MHz)	Operating power consumption (mW)	Standby power consumption (nW)
16-bit ALU	0.5	40	0.35	5
Communication LSI	0.5	100	1.45	-
Coding LSI	0.5	18	2	-
8-bit CPU	0.5	30	5	-
16-bit adder	0.5	50	0.16	-
Communication LSI	0.5	100	1.65	-
Wireless systems	0.5	10	1	-
54-bit adder	0.5	30	3	-

6.3 HIGH-VOLTAGE DEVICES

The full dielectric isolation provided by the SOI substrate allows one to fabricate high-voltage devices without the use of a complicated junction isolation process. Power devices can be integrated on the same substrate as CMOS logic to produce "smart power" circuits.

6.3.1 VDMOS and LDMOS

The dielectric isolation between individual transistors is an important asset of SOI technology. It allows for the integration of low-voltage logic and high-voltage transistors on a same chip. If thick silicon films are used, vertical power devices can be realized. One such device is the vertical double-diffused MOS (VDMOS), reduced surface field (RESURF) transistor.[110] Made in a 50 μm-thick silicon film, the VDMOS achieves a breakdown voltage of 480 V and an on-resistance of 0.17 Ωcm^2.[111,112]

High-voltage devices can be fabricated in thin-film SOI as well. Using the reduced surface field (RESURF) technique it is possible to design lateral power transistors with a constant electric field in a relatively long drift region, thereby avoiding a peak in electric field at PN junctions, as is the case in conventional power devices. The most successful lateral SOI power device is the RESURF lateral double-diffused MOS (LDMOS) transistor.

We will now attempt to explain the basic principle that allows thin silicon films to sustain high lateral voltage drops. In a conventional power device, the applied bias is blocked by a reverse-biased junction. Assuming uniform doping concentrations the electric field adopts a triangular profile reaching a maximum, E_{max}, at the metallurgical junction. If E_{max} rises above a given value, E_{BV}, breakdown occurs. The voltage drop across the junction space-charge region is given by: $\int_{space-ch\arg e\,region} E_x(x)\,dx$, which is represented by the area of the triangle in

Figure 6-29B and is thus equal to $\dfrac{1}{2}E_{max}W$, where W is the width of the space-charge region. In principle one could double the value of that potential drop if the electric field could be made constant and equal to E_{max} everywhere in the space-charge region. This cannot be achieved in a regular PN junction, but can be realized in a thin SOI film.

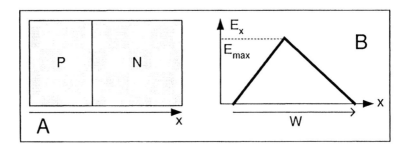

Figure 6-29. A: P-N junction; B: Lateral electric field in the junction.

Consider the structure in Figure 6-30. It consists of a P⁺-N-N⁺ junction sandwiched between two gates. The electric field in the central N-type region obeys Poisson's equation, which, assuming full depletion, can be written:

$$\frac{dE_x(x,y)}{dx} + \frac{dE_y(x,y)}{dy} = \frac{qN_d(x)}{\varepsilon_{si}} \qquad (6.3.1)$$

where the doping concentration, $N_d(x)$, can vary arbitrarily in the x-direction but is uniform in the y-direction. The voltage drop across W will be maximized if the electric field in the x-direction is constant and equal to E_{max}:

$$E_x(x,y) = E_{max} \Rightarrow \frac{dE_x(x,y)}{dx} = 0 \qquad (6.3.2)$$

$$\Downarrow$$

$$\frac{dE_y(x,y)}{dy} = \frac{q N_d(x)}{\varepsilon_{si}} \qquad (6.3.3)$$

Thus the potential drop in the x-direction can be maximized if the entirety of the depletion charge is supported by the variation of vertical electrical field induced by gates G_1 and G_2.

If we minimize the vertical electric field at the top silicon interface ($E_y(x,0) = 0$), the integration of 6.3.3 yields:

$$\frac{q}{\varepsilon_{si}} \int_0^{t_{si}} N_d(x)\,dy = \frac{q N_d(x) t_{si}}{\varepsilon_{si}} = E_y(x,t_{si}) - E_y(x,0) = E_y(x,t_{si}) \quad (6.3.4)$$

Integrating 6.3.3 we can write:

$$E_y(x,y) = \frac{q N_d(x)}{\varepsilon_{si}} y \qquad (6.3.5)$$

for $0<x<W$ and $0<y<t_{si}$. Assuming $E_x(x,t_{si})$ is independent of x and $E_y(W,y)$ is independent of y it can be shown that the potential $\Phi(x,y)$ is proportional to the xy product in the buried oxide (or bottom gate oxide).[113] Therefore, E_y is uniform in the vertical direction and E_x is uniform in the lateral direction in the BOX (Figure 6-30). Thus for a given position in the x-direction the vertical potential drop across the BOX is equal to:

$$E_{BOX}\, t_{BOX} = \frac{\varepsilon_{si}}{\varepsilon_{ox}} E_y(x,t_{si}) t_{BOX} \qquad (6.3.6)$$

Integrating 6.3.5 we find the vertical potential drop in the silicon film:

$$\Phi(x,0) - \Phi(x,t_{si}) = \int_0^{t_{si}} E_y(x,y) = \frac{q N_d(x)}{2\varepsilon_{si}} t_{si}^2 \qquad (6.3.7)$$

which can be rewritten, using 6.3.4:

$$\Phi(x,0) - \Phi(x,t_{si}) = \frac{t_{si}}{2} E_y(x,t_{si}) \qquad (6.3.8)$$

Assuming the underlying substrate is grounded ($V_{G2}=0$) the potential at the top of the silicon film is obtained by adding 6.3.6 and 6.3.8:

$$\Phi(x,0) = \left(\frac{t_{si}}{2} + \frac{\varepsilon_{si}}{\varepsilon_{ox}} t_{BOX} \right) E_y(x,t_{si}) \qquad (6.3.9)$$

If the voltage applied across the distance W is noted V, then we have $E_x(x,0)=-V/W$ and $\Phi(x,0)=Vx/W$, and therefore, equation 6.3.9 yields:

$$E_y(x,t_{si}) = \frac{V}{\left(\frac{t_{si}}{2} + \frac{\varepsilon_{si}}{\varepsilon_{ox}} t_{BOX} \right) W} \cdot x \qquad (6.3.10)$$

$E_y(x,t_{si})$ reaches a maximum value at $x=W$, where it should not exceed the breakdown field E_{BV}. Equation 6.3.9 describes the optimized electric field for achieving the highest breakdown voltage for a given lateral depletion width in the SOI device and may be regarded as the optimized RESURF condition for the SOI structure in Figure 6-30. The highest breakdown voltage can be easily calculated using 6.3.11:

$$V = \left(\frac{t_{si}}{2} + \frac{\varepsilon_{si}}{\varepsilon_{ox}} t_{BOX} \right) E_{BV} \qquad (6.3.11)$$

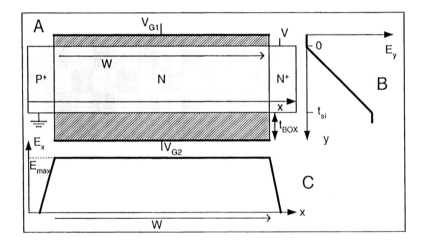

Figure 6-30. A: Drift region of a RESURF LDMOS transistor; B: Vertical electric field; C: Lateral electric field.

Thus the breakdown voltage increases with both the silicon film thickness and the BOX thickness. However, for SOI film thickness below approximately 1 μm the breakdown voltage increases with *decreasing* silicon film thickness because the ionization integrals in the *y*-direction decrease with film thickness. In other words, in thin silicon films, carriers accelerated by the vertical field do not have enough distance to gain the energy required to cause impact ionization. The theoretical breakdown voltage in SOI films is shown in Figure 6-31 for different BOX thicknesses.

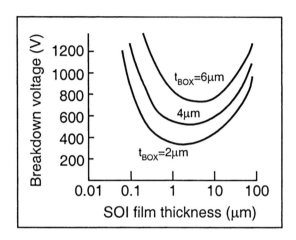

Figure 6-31. Theoretical breakdown voltage in SOI films *vs.* film thickness.

Figure 6-32 shows the cross section of an SOI RESURF LDMOS transistor. It has been shown that high breakdown voltages can be achieved in such devices. 850-V switches with an on-resistance of 13 $\Omega.mm^2$, as well as 250-V plasma display drivers [114], a 600-V power conversion system [115], and compact fluorescent lamp electronics, made using SOI LDMOS devices have been reported.[116]

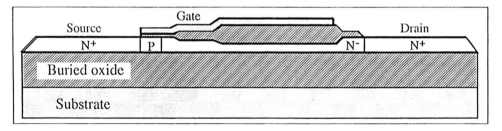

Figure 6-32. High-voltage SOI RESURF LDMOS transistor.[117]

SOI LDMOS transistors have been shown to have excellent RF characteristics, such as f_T=10.9 GHz for L=0.8 μm, 11.6 GHz for L=0.5 μm and f_T=15 GHz for L=0.6 μm.[118,119,120] LDMOS transistors have very small recovery charges, which makes them ideal devices for switching applications such as the output stage of class D power amplifiers.[121] The influence of back-gate bias on the LDMOS characteristics can be found in the literature.[122]

6.3.2 Other high-voltage devices

Other lateral high-voltage devices have been fabricated in SOI as well. These are: the lateral insulated gate bipolar transistor (LIGBT) [123,124,125,126], the IGBT mode turn-off thyristor (IGTT) [127], the lateral insulated-gate PIN transistor (LIGPT) [128], the lateral high-voltage PIN rectifier [129], the base resistance controlled thyristor (BRT) [130], and the lateral emitter-controlled thyristor (LECT).[131] SOI microthyristors might also be candidates for high-speed switching applications.[132]

Table 6-4 lists the electrical characteristics of the most important high-voltage SOI devices.

Table 6-4. Characteristics of some SOI power devices

Device type	t_{si} (μm)	V_{max} (V)	I_{max} (A)	R_{on} (Ω.mm²)	Ref.
VDMOS	6	50	-	15	133
VDMOS	15	150	2	2	134
VDMOS	-	500	10	18	135
VDMOS	-	1200	-	-	136
VDMOS	50	480	-	17	137
LDMOS	5	400	-	15	138
LDMOS	20	640	-	-	139
LDMOS	1.5	850	113mA/mm	13	140
COMFET	0.2	80	-	-	141
LIGBT	5	400	-	5	142
MOSFET	-	44	0.5	19Ω.mm	143

6.4 JUNCTION FIELD-EFFECT TRANSISTOR

Junction field-effect transistors (JFETs) can be fabricated on SOI substrates. JFETs are particularly interesting for applications where low noise, high input impedance and good radiation hardness are required. Figure 6-33 shows the cross-section of an SOI p-channel JFET [144]. The structure is fabricated as follows. Buried N$^+$ diffusions are created in the silicon overlayer of a SIMOX wafer, and p-type silicon epitaxy is carried out to increase the silicon film thickness from 0.2 to 1 μm. These buried N$^+$ diffusions will be utilized as back gates. The silicon islands are defined by means of a mesa isolation process, and ion implantation is used to form the top front-gate (N$^+$) and source and drain (P$^+$) electrodes (Figure 6-33). Modulation of the source-drain current can be achieved by applying a voltage to either the front gate or the back gate while keeping the other gate grounded, but the highest transconductance is obtained by connecting G$_1$ and G$_2$ together and using them as a single input terminal.

SOI JFETs show good radiation hardness characteristics. After a 30 krad(Si) dose irradiation (X-rays), a saturation current degradation of 20% is observed. Above 30 krad(Si) and up to 1 Mrad(Si), no further shift is observed. The current saturation degradation caused by a 1 MeV neutron fluence of 10^{14} cm^{-2} is below 10%. Such JFETs can be integrated with complementary bipolar transistors and CMOS to provide a rad-hard analog/digital SOI technology (*Durcie Mixte sur Isolant Logico-Linéaire* or DMILL, in French).[145, 146]

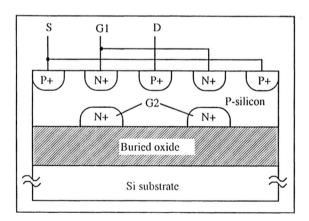

Figure 6-33. P-channel JFET. G1 is the front gate, G2 is the back gate

6.5   LUBISTOR

The lateral unidirectional bipolar-type insulated-gate transistor, or Lubistor (Figure 6-34), is an $N^+N^-P^+$ (or $N^+P^-P^+$) gated diode which exhibits triode characteristics and is able to carry current densities up to 10^5 A/cm^2. It has first been realized in SIMOX material [147], although the intrinsic properties of the device are quite independent of the SOI material being used. The operation principle of the device is the following: a positive anode to cathode voltage, V_{AK} is applied, and a positive gate to cathode voltage, V_{GK}, is used to control the current flow in the device. The device basically works as a gated PIN diode.

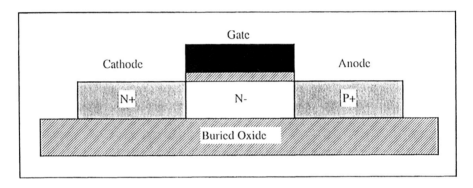

Figure 6-34. Lateral unidirectional bipolar-type insulated-gate transistor (Lubistor)

The Lubistor can be fully turned off by the gate if the silicon film thickness, t_{si}, is smaller than the Debye length, L_{DE}, defined by: $L_{DE} = \sqrt{\dfrac{2\varepsilon_{si} kT}{qN_d}}$, where N_d is the dopant concentration in the N$^-$ region. In the ON state, the device displays output characteristics in the form: $I_{AK} = A(V_{AK} - B(V_{GK} - V_{FB}))^n$, where A and B are parametric constants, V_{FB} is the flat-band voltage in the N$^-$ region, and $2 < n < 3$.

In the OFF state, the potential of the thin N$^-$ region is higher than that of both the cathode and the anode, and carrier injection does not take place even though the anode bias is higher than the built-in potential of the anode P-N junction. In the ON state, the anode potential is higher than that of the N$^-$ region, and holes are injected into the N⁻ region. Electrons are also injected in the same region from the cathode, and the device behaves like a forward-biased PIN diode. The Lubistor can be used as a neural circuit element [148,149] and as an ESD protection device.[150,151,152]

6.6 BIPOLAR JUNCTION TRANSISTORS

Different types of bipolar transistors can be realized in SOI, depending on the application and the silicon film thickness. If a pure bipolar circuit is required (e.g. ECL), a "thick" silicon layer can be employed and classical vertical transistors can be fabricated.[153,154,155,156,157] The advantages of using an SOI substrate are due in the reduction of collector-substrate capacitances, the full dielectric isolation and the reduction of sensitivity to alpha particles (reduction of the soft-error rate).[158] If the bipolar transistors have to be integrated with thin-film CMOS devices (BiCMOS and CBiCMOS), efficient lateral bipolar transistors can be realized as well. Early SOI lateral bipolar transistors suffered from high base resistance problems due to the fact that the base contact was taken laterally (MOSFET body contact).[159] More recent devices use base contacts made at the top of the device, which improves the base resistance but usually complicates the fabrication process (Figure 6-35).[160,161,162,163,164,165]

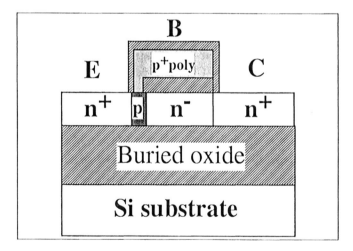

Figure 6-35. Lateral SOI bipolar transistor with top base contact.

Lateral polysilicon emitters can also be realized, at the expense of process complexity.[166] The increase of process complexity necessary to upgrade a

CMOS SOI process to (C)BiCMOS is, however, much more modest than for a bulk process, simply because of the inherent dielectric isolation of SOI devices. Table 6-5 presents the performances of some recent SOI bipolar transistors. The maximum gain of lateral devices is usually obtained for relatively low collector current values. Indeed, high current densities and high-injection phenomena are reached much more quickly in thin-film lateral devices than in vertical devices because of the small silicon thickness.

Table 6-5. Performances of SOI bipolar transistors.

Type	Base width (μm)	β_F (or h_{FE})	BV_{CEO} (V)	f_T (GHz)	Ref.
Lateral	0.2	120	2.5	4.5	167
Hybrid	0.3	10000	2.5	-	ibidem
Lateral	-	90	3	15.4	168
Lateral	-	80	>3	10	169
Lateral	-	30	2.8	20	170
Vertical	-	50	8	12.4	171
Vertical	-	110	12	4.5	172
Vertical	-	240	2.8	50	173
Lateral	0.07	88	5.3	12	174
Lateral	0.15	78	5	8	175
Vertical	-	200	35	-	176
Lateral	0.04	-	-	15	177

Because of the need for making a buried collector layer, it may seem impossible to fabricate vertical bipolar transistors in thin SOI films. An original solution to this problem has, however, been proposed, and vertical NPN transistors have been realized in 400 nm-thick SOI films (and the use of even thinner films is quite possible).[178] An ingenious solution consists of fabricating a vertical bipolar device with no buried collector diffusion. The bottom of the P- intrinsic base is directly in contact with the buried oxide. If no back gate bias is applied, very few of the electrons injected by the emitter into the base can reach the lateral collector, and the current gain of the device is extremely small. When a positive bias is applied to the back gate, however, an inversion layer is induced at the silicon-buried oxide interface, at the bottom of the base, and acts as a buried collector (Figure 6-36). Furthermore, the band bending in the vicinity of the bottom inversion layer attracts the electrons injected by the emitter into the base. As a result, collection efficiency is drastically increased, and the current gain can reach useful values. The use of a field-effect-induced buried collector makes it thus possible to obtain significant collection efficiency without the need for the formation of a diffused buried layer and an epitaxal growth step. It also solves the high-injection (high current densities) problems inherent in thin-film lateral devices. Furthermore, because the neutral base width does not depend on the collector voltage, but on the back-gate voltage, the Early effect is almost totally suppressed.[179] Since the amplification factor of a bipolar transistor is given by the Early voltage × current gain product, this feature can be extremely attractive for high-performance analog applications. Vertical SOI bipolar transistor with field-induced collector with an Early voltage over 50,000 volts and a current gain of 200 have been demonstrated. Such devices offer, thus, an Early voltage × current gain product larger than 1,000,000.[180] A model for this device can be found in the literature.[181]

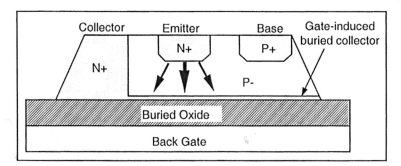

Figure 6-36. Vertical bipolar transistor with field-induced buried collector.

6.7 PHOTODIODES

PIN diodes can easily be produced using an SOI CMOS process without the addition of any mask step. The intrinsic silicon can either be protected from threshold voltage implants or be lightly doped (by threshold voltage adjust implants). The N+ cathode and the P+ anode are formed using the n-channel and p-channel device source and drain implants. PIN diodes with reverse breakdown voltages larger than 50 V can readily be formed, even in thin SOI films. PIN diodes can be useful as ESD input protection devices. Reverse-biased thin-film PIN diodes can also be used as photosensors (Figure 6-37). Electron-hole pairs are created under illumination within the large, fully depleted intrinsic region. The carriers are separated by the electric field in that region and collected by the N+ and P+ diffusions. Because the devices are made in a thin silicon film which is part of a multilayer structure (oxide passivation, silicon film, buried oxide and silicon substrate), light interferences occur and, as a result, the structure absorbs more of some wavelengths than others.[182] In general red light is not absorbed in thin silicon films, but peaks of absorption can be obtained in the blue and green regions of the visible spectrum.

Large-area, low-capacitance lateral SOI PIN photodiodes with an average external quantum efficiency of 78.6% and 68.4% at λ=430 nm and 400 nm, respectively, have been realized. The rise and fall times of the lateral PIN photodiode for light with a wavelength of 400 nm are 9.7 ns and 11.2 ns, respectively. The capacitance of the SOI PIN photodiode is 0.72 pF/mm^2. This photodiode combines a high quantum efficiency and a low capacitance with high speed.[183]

Photodiodes have been made in a SOI layer as a part of a 3-D integrated circuit. Using this technique a 10-bit linear image sensor has demonstrated. The sensor comprises a photodiode array made in an SOI layer on top of a bottom bulk Si layer containing signal processing circuits and shift registers.[184] Because of the full dielectric isolation provided by the SOI structure it is possible to generate high voltages from SOI photodiodes connected in parallel. Potentials of several thousands volts at currents of several microamperes can be obtained by photogeneration from a square centimeter photodiode array.[185]

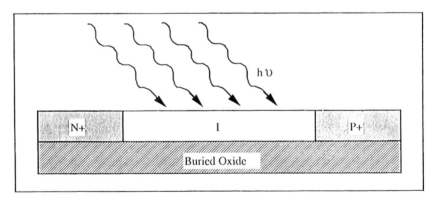

Figure 6-37. Lateral SOI PIN photodiode.[186]

6.8   G^4 FET

The 4-gate transistor (G^4 FET) has two (top and bottom) MOS gates and two (lateral) JFET gates. It is operated in accumulation mode and has the same structure as an inversion-mode partially depleted SOI MOSFET with two independent body contacts. An earlier device with similar terminal configuration was called the cross-MOS device. [187,188] In the G⁴ FET, these body contacts play the role of source and drain, whereas the former source and drain junctions now act as lateral JFET gates (Figure 6-38).

The G⁴ MOSFET has the standard front and back MOS gates plus the two lateral junctions that control the effective width of the body. The current flows from source to drain (Figure 6-38). The height and width of the conductive path are modulated by a mix of MOS and JFET effects. The conductive path can range from from a tiny quantum wire surrounded by depletion regions to a strongly accumulated body, depending on the applied gate voltages. Each gate has the ability of switching the transistor on and off. The independent action of the four gates allows for mixed-signal applications, quantum-wire effects, and quaternary logic schemes.[189,190,191] A model for the G⁴ FET has been published in the literature.[192]

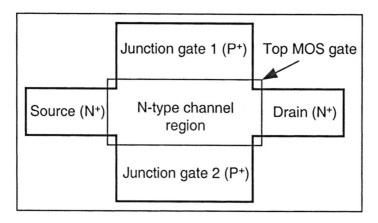

Figure 6-38. Top view of the G⁴ MOSFET. The substrate underneath the BOX is the bottom MOS gate.

6.9 QUANTUM-EFFECT DEVICES

Two-dimensionally confined injection mechanisms have been observed in insulated-gate PN junctions made in sub-10-nm-thick SOI films.

Transconductance oscillations are observed. The origin of these fluctuations is found in the splitting of the conduction and valence bands into subbands.[193]

Short quantum wire transistors have been fabricated in SOI.[194,195,196] The conductance of these devices increases in a staircase manner as a function of gate voltage. After correcting for the source and drain resistance the transconductance was found to increase in multiples of $4q^2/h$, which agrees with the Landauer formula for universal transconductance fluctuations.[197] Longer quantum wire transistors have been realized as well, and transconductance fluctuations due to subband splitting have also been observed (Figure 6-39).[198]

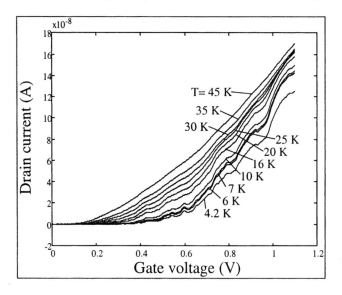

Figure 6-39. Current as a function of gate voltage in a series of 7 parallel SOI quantum wire MOSFETs at different temperatures. V_{DS}=10 mV. The device thickness and width are 86 and 100 nm, respectively.[199,200]

The fabrication of single-electron transistors (SET) on SOI has been reported as well. These device are basically a short quantum wire connected to source and drain through constrictions. The higher energy levels generated by the constrictions isolate the center of the wire from the outside world. When a gate voltage is applied current can flow through the quantum wire by a Coulomb blockade mechanism made possible by the very small capacitances (2 aF) involved in the structure. Figure 6-40 shows the drain current vs. gate and drain voltages in an SOI SET. The rhombus-shaped domains where no current flows are a signature of the Coulomb blockade effect.[201] Conductance oscillations as a function of gate voltage have been observed at temperatures as high as 300 K.[202,203] Coulomb blockade oscillations have been observed at room temperature in SOI quantum-wire MOSFETs without constrictions as well. In these devices small interconnected quantum dot domains are formed, probably due to the fluctuations of oxide charge density, interface states density or gate oxide thickness along the wire's length.[204,205] SOI single-electron transistors are potential candidates for low-power logic circuits.[206]

The fabrication and successful operation of SOI single-electron (or few electrons) memory cells with floating dot gate has also been reported. In such a device, which basically behaves as a miniature EEPROM cell, a single electron can be injected into an electrically floating dot. Because the dot has a very small capacitance the injection of such a minute charge as that of a single electron gives rise to a relatively high potential (several volts) in the dot, such that the threshold voltage of the quantum wire MOSFET is significantly modified. This MOSFET can be used as a memory element, since it will present different threshold voltage values depending on whether or not there is an electron stored in the floating dot.

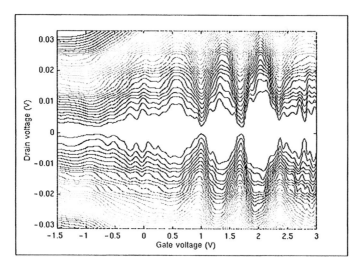

Figure 6-40. Lines of equal drain current in the V_D-V_G plane, measured in an SET. [207]

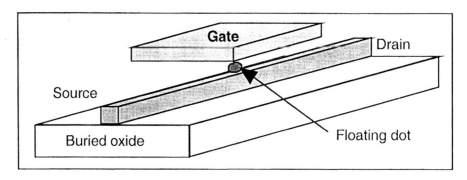

Figure 6-41. SOI single-electron memory cell MOSFET. [208,209,210,211,212]

REFERENCES

[1] T. Sekigawa and Y. Hayashi, Solid-State Electronics, Vol. 27, p. 827, 1984
[2] B. Agrawal, V.K. De, and J.D. Meindl, Proceedings of 23rd ESSDERC, Ed. by J. Borel, P. Gentil, J.P. Noblanc, A. Nouhaillat, and M. Verdone, Editions Frontières, p. 919, 1993
[3] D.J. Frank, S.E. Laux and M.V. Fischetti, Technical Digest of IEDM, p. 553, 1992
[4] D. Hisamoto, T. Kaga, Y. Kawamoto and E. Takeda, Technical Digest of IEDM, p. 833, 1989
[5] X. Huang, W.C. Lee, C. Kuo, D. Hisamoto, L. Chang, J. Kedzierski, E. Anderson, H. Takeuchi, Y.K. Choi, K. Asano, V. Subramanian, T.J. King, J. Bokor, C. Hu, Technical digest o fIEDM, p. 67, 1999
[6] Y.K. Liu, K. Ishii, T. Tsutsumi, M. Masahara, H. Takamisha and E. Suzuki, Electrochemical Society Proceedings 2003-05, p. 255, 2003 and Y. Liu, K. Ishii, T. Tsutsumi, M. Masahara, and E. Suzuki, IEEE Electron Device Letters, Vol. 24, no. 7, p. 484, 2003
[7] T. Hiramoto, IEEE International SOI Conference Proceedings, p. 8, 2001, and T. Saito, T. Saraya, T. Inukai, H. Majima, T. Nagumo, and T. Hiramoto, IEICE Transactions on Electronics, Vol. E85-C, no. 5, p. 1073, 2002
[8] Z. Jiao and A.T. Salama, Electrochem. Society Proceedings 2001-3, 403, 2001
[9] F. Balestra, S. Cristoloveanu, M. Benachir and T. Elewa, IEEE Electron Device Letters, Vol. 8, p. 410, 1987
[10] J.P. Colinge, M.H. Gao, A. Romano, H. Maes, and C. Claeys, Technical Digest of IEDM, p. 595, 1990
[11] X. Baie, J.P. Colinge, V. Bayot and E. Grivei, Proceedings of the IEEE International SOI Conference, p. 66, 1995
[12] R. Chau, B. Doyle, J. Kavalieros, D. Barlage, A. Murthy, M. Dozky, R. Arghavani and S. Datta, Extended Abstracts of the International Conference on Solid State Devices and Materials, SSDM, p. 68, 2002
[13] B.S. Doyle, S. Datta, M. Doczy, B. Jin, J. Kavalieros, T. Linton, A. Murthy, R. Rios, R. Chau, IEEE Electron Device Letters, Vol. 24, no. 4, p. 263, 2003
[14] J.T. Park, J.P. Colinge and C. H. Diaz, IEEE Electron Device Letters, 22, 405, 2001
[15] F.L. Yang, H.Y. Chen, F.C. Cheng, C.C. Huang, C.Y. Chang, H.K. Chiu, C.C. Lee, C.C. Chen H.T. Huang, C.J. Chen, H.J. Tao, Y.C. Yeo, M.S. Liang, and C. Hu, Technical Digest of IEDM, p. 255, 2002
[16] Z. Krivokapic, C. Tabery, W. Maszara, Q. Xiang, M.R. Lin, Extended Abstracts of the International Conference on Solid State Devices and Materials, SSDM, p. 760, 2003
[17] S. Miyano, M. Hirose, F. Masuoka, IEEE Transactions on Electron Devices, Vol. 39, p. 1876, 1992
[18] A. Nitayama, H. Takato, N. Okabe, K. Sunouchi, K. Hieda, F. Horiguchi, F. Masuoka, IEEE Transactions on Electron Devices, Vol. 38, p.579, 1991
[19] J.T. Park and J.P. Colinge, IEEE Transactions on Electron Devices, Vol. 49, p. 2222, 2002
[20] US patent 6,359,311
[21] F.L. Yang, H.Y. Chen, F.C. Cheng, C.C. Huang, C.Y. Chang, H.K. Chiu, C.C. Lee, C.C. Chen H.T. Huang, C.J. Chen, H.J. Tao, Y.C. Yeo, M.S. Liang, and C. Hu, Technical Digest of IEDM, p. 255, 2002
[22] K.K. Young, IEEE Transactions on Electron Devices, Vol. 37, p. 504, 1989

[23] R.H. Yan, A. Ourmazd, and K.F. Lee, IEEE Transactions on Electron Devices, Vol. 39, p. 1704, 1992

[24] S.H. Oh, D. Monroe, J.M. Hergenrother, IEEE Electron Device Letters, Vol. 21, no. 9, p. 445, 2000

[25] K. Suzuki, T. Tanaka, Y. Tosaka, H. Horie and Y. Arimoto, IEEE Transactions on Electron Devices, Vol. 40, p.2326, 1993

[26] C.P. Auth and J.D. Plummer, IEEE Electron Dev. Lett., Vol. 18, p. 74, 1997

[27] Q. Chen, E.M. Harrell, II, and J.D. Meindl, IEEE Transactions on Electron Devices, Vol. 50, no. 7, p. 1631, 2003

[28] J.P. Colinge, Electrochemical Society Proceedings, Vol. 2003-09, p. 2, 2003

[29] T. Ernst, D. Munteanu, S. Cristoloveanu, T. Ouisse, S. Horiguchi, Y. Ono, Y. Takahashi, K. Murase, Microelectronic Engineering, Vol. 48, p. 339, 1999

[30] M. Mastrapasqua, D. Esseni, G.K. Celler, F.H. Baumann, C. Fiegna,L. Selmi, E. Sangiorgi, Silicon-on-Insulator Technology and Devices X, Electrochemical Society Proceedings 2001-3, p. 97, 2001

[31] T. Ernst, D. Munteanu, S. Cristoloveanu, T. Ouisse, N. Hefyene, S. Horiguchi, Y. Ono, Y. Takahashi, K. Murase, IEEE International SOI Conference. Proceedings, p. 92, 1999

[32] P. Francis, A. Terao, D. Flandre, and F. Van de Wiele, Microelectronic Engineering, Vol. 19, p. 815, 1992

[33] P. Francis, A. Terao, D. Flandre, and F. Van de Wiele, IEEE Transactions on Electron Devices, Vol. 41-5, p. 715, 1994

[34] P. Francis, A. Terao, D. Flandre, and F. Van de Wiele, Proceedings of the 23rd ESSDERC, Editions Frontières, p. 621, 1993

[35] P. Francis, A. Terao, D. Flandre, and F. Van de Wiele, Solid-State Electronics, Vol. 38-1, p. 171, 1995

[36] A. Terao, D. Flandre, E. Lora-Tamayo, and F. Van de Wiele, IEEE Electron Device Letters, Vol. 12, p. 682, 1991

[37] K. Kim, J.G. Fossum, IEEE Transactions on Electron Devices, Vol.48, no.2, p.294, 2001

[38] K. Kim, J.G. Fossum, Proceedings of the IEEE International SOI Conference, p.98, 1999

[39] L. Chang, S. Tang, T.J. King, J. Bokor, C. Hu, Technical digest of IEDM, p. 719, 2000

[40] J. Kedzierski, M. ieong, E. Nowak, Electrochemical Society Proceedings Vol. 2003-5, p. 185, 2003

[41] H. Wakabayashi, Y. Saito, K. Takeuchi, T. Mogami, T. Kunio, Technical digest of IEDM, p. 253, 1999

[42] S. Balasubramanian, L. Chang, Y.K. Choi, D. Ha, J. Lee, P. Ranade, S. Xiong, J. Bokor, C. Hu, T.J. King, Electrochemical Society Proceedings Vol. 2003-5, p. 197, 2003

[43] X. Wu, P.C.H. Chan, M. Chan, Proceedings of the IEEE International SOI Conference, p. 151, 2003

[44] H.-S. Wong, M.H. White, T.J. Krutsck, and R.V. Booth, Solid-State Electronics, Vol. 30, no. 9, p. 953, 1987

[45] P. Francis, A. Terao, D. Flandre, and F. Van de Wiele, Solid-State Electronics, Vol. 41, no. 5, p. 715, 1994

[46] A. Terao, D. Flandre, E. Lora-Tamayo, and F. Van de Wiele, IEEE Electron Device Letters, Vol. 12, no. 12, p. 682, 1991

[47] E. Rauly, B. Iñiguez, D. Flandre and C. Raynaud, Proceedings of ESSDERC, p. 540, 2000

[48] W. Xiong, J.W. Park and J.P. Colinge, Proceedings of the IEEE International SOI Conference, p. 111, 2003

[49] J.P. Colinge, J.W. Park and W. Xiong, IEEE Electron Device Letters, Vol. 28, no. 8, p. 515, 2003

[50] F. Balestra, S. Cristoloveanu, M. Benachir, J. Brini, T. Elewa, IEEE Electron Device Letters, Vol. 8, no. 9, p. 410, 1987

[51] J.P. Colinge, M.H. Gao, A. Romano, H. Maes, and C. Claeys, Technical Digest of IEDM, 595, 1990

[52] T. Ouisse, Journal of Applied Physics, Vol. 76, p. 5979, 1994

[53] J.P. Colinge, X. Baie and V. Bayot, IEEE Electron Device Letters, Vol. 15, p. 193, 1994

[54] A. Rahman, M.S. Lundstrom, IEEE Transactions on Electron Devices, Vol. 49, no. 3, p. 481, 2002

[55] X. Baie, J.P. Colinge, Solid-State Electronics, Vol. 42, no. 4, p. 499, 1998

[56] G. Baccarani, S. Reggiani, IEEE Transactions on Electron Devices, Vol. 46, no. 8, p.1656, 1999

[57] Lixin Ge, J.G. Fossum, IEEE Transactions on Electron Devices, Vol. 49, no. 2, p.287, 2002

[58] S. Cristoloveanu, D.E. Ioannou, Superlattices & Microstructures, Vol. 8, no. 1, p. 131, 1990

[59] Y. Omura, S. Horiguchi, M. Tabe, K. Kishi, IEEE Electron Device Letters, Vol. 14, no. 12, p. 569, 1993

[60] B. Majkusiak, T. Janik, Microelectronic Engineering, Vol. 36, no. 1-4, p. 379, 1997

[61] B. Majkusiak, T. Janik, J. Walczak, IEEE Transactions on Electron Devices, Vol. 45, no. 5, p. 1127, 1998

[62] T. Ouisse, D.K. Maude, S. Horiguchi, Y. Ono, Y. Takahashi, K. Murase, S. Cristoloveanu, Physica B, Vol. 249-251, p. 731, 1998

[63] T. Xia, L.F. Register, S.K. Banerjee, IEEE Transactions on Electron Devices, Vol. 50, no. 6, p. 1511, 2003

[64] X. Baie, J.P. Colinge, V. Bayot, E. Grivei, Proceedings of the IEEE International SOI Conference, p. 66, 1995

[65] J.P. Colinge, X. Baie, V. Bayot, E. Grivei, Solid-State Electronics, Vol. 39, no. 1, p.49, 1996

[66] X. Baie, J.P. Colinge, Solid-State Electronics, Vol. 42, no. 4, p. 499, 1998

[67] F. Gámiz, J.B. Roldán, J.A. López-Villanueva, P. Cartujo-Cassinello, J.E. Carceller and P. Cartujo, Electrochemical Society Proceedings, Vol. 2001-3, p. 157, 2001

[68] J.P. Colinge, M.H. Gao, A. Romano, H. Maes, and C. Claeys, Technical Digest of IEDM, p. 595, 1990

[69] T. Ernst, S. Cristoloveanu, G. Ghibaudo, T. Ouisse, S. Horiguchi, Y. Ono, Y. Takahashi, K. .Murase, IEEE Transactions on Electron Devices, Vol. 50, no. 3, p.830, 2003

[70] B. Majkusiak, T. Janik, J. Walczak, IEEE Transactions on Electron Devices, Vol. 45, no. 5, p. 1127, 1998

[71] F. Gámiz, J.B. Roldán, J.A. López-Villanueva, P. Cartujo-Cassinello, J.E. Carceller and P. Cartujo, Electrochemical Society Proceedings, Vol. 2001-3, p. 157, 2001

[72] L. Ge, J.G. Fossum, F. Gámiz, Proceedings of the IEEE International SOI Conference, p. 153, 2003

[73] B. Majkusiak, T. Janik, J. Walczak, IEEE Transactions on Electron Devices, Vol. 45, no. 5, p. 1127, 1998

74   T. Ernst, S. Cristoloveanu, G. Ghibaudo, T. Ouisse, S. Horiguchi, Y. Ono, Y. Takahashi, K.
      Murase, IEEE Transactions on Electron Devices, Vol. 50, no. 3, p.830, 2003
75   B. Majkusiak, T. Janik, J. Walczak, IEEE Transactions on Electron Devices, Vol. 45, no. 5, p.
      1127, 1998
76   J.P. Colinge, IEEE Transactions on Electron Devices, Vol. 34, p. 845, 1987
77   S.A. Parke, C. Hu, and P.K. Ko, IEEE Electron Device Lett., Vol. 14, p. 234, 1993
78   S.S. Rofail, and Y.K. Seng, IEEE Transactions on Electron Devices, Vol. 44, p. 1473, 1997
79   M. Matloubian, IEEE International SOI Conference Proceedings, p. 106, 1993
80   Z. Yan, M.J. Deen, and D.S. Malhi, IEEE Transactions on Electron Devices, Vol. 44, p. 118,
      1997
81   T. Douseki, S. Shigematsu, J. Yamada, M. Harada, H. Inokawa, and T. Tsuchiya, IEEE
      Journal of Solid-State Circuits, Vol. 32, p. 1604, 1997
82   F. Assaderaghi, D. Sinitsky, S. A. Parke, J. Bokor, P.K. Ko, and C. Hu, Tech. Digest of
      IEDM, p. 809, 1994
83   Z. Xia, Y. Ge, Y. Zhao, Proceedings 22nd International Conference on Microelectronics
      (MIEL), p. 159, 2000
84   V. Ferlet-Cavrois, A. Bracale, N. Fel, O. Musseau, C. Raynaud, O. Faynot and J.L. Pelloie,
      Proceedings of the IEEE Intl. SOI Conference, p. 24, 1999
85   T. Douseki, F. Morisawa, S. Nakata and Y. Ohtomo, Extended Abstracts of the International
      Conference on Solid-State Devices and Materials (SSDM), p. 264, 2001
86   A. Yagishita, T. Saito, S. Inumiya, K. Matsuo, Y. Tsunashima, K. Suguro, IEEE Transactions
      on Electron Devices, Vol. 49, no. 3, p. 422, 2002
87   J.P. Colinge, IEEE Transactions on Electron Devices, Vol. 34, no. 4, p. 845, 1987
88   F. Assaderaghi, D. Sinitsky, S.A. Parke, J. Bokor, P.K.K. Ko, C. Hu, IEEE Transactions on
      Electron Devices, Vol. 44, no. 3, p. 414, 1997
89   Z. Xia, Y. Ge, Y. Zhao, Proceedings 22nd International Conference on Microelectronics
      (MIEL), p. 159, 2000
90   S.S. Rofail, Y.K. Seng, IEEE Transactions on Electron Devices, Vol. 44, no. 9, p.1473, 1997
91   R. Huang, Y. Y. Wang, and R. Han, Solid-State Electronics, Vol. 39, no. 12, p. 1816, 1996
92   M. Terauchi, Proceedings of the IEEE International SOI Conference, p.53, 2001
93   J.B. Kuo, K.H. Yuan, S.C. Lin, IEEE Transactions on Electron Devices, Vol. 49, no. 1, p.
      190, 2002
94   J.P. Colinge and J.T. Park, Journal of Semiconductor Technology and Science, Vol. 3, no. 4,
      2003
95   V. Ferlet-Cavrois, A. Bracale, N. Fel, O. Musseau, C. Raynaud, O. Faynot, J.L. Pelloie,
      Proceedings of the IEEE International SOI Conference, p. 24, 1999
96   Y. Momiyama, T. Hirose, H. Kurata, K. Goto, Y. Watanabe, T. Sugii T, Technical Digest.
      IEDM, p.451, 2000
97   M. Dehan, D. Vanhoenacker-Janvier, J.P. Raskin, Electrochemical Society Proceedings Vol.
      2003-05, p. 289, 2003
98   V. Ferlet-Cavrois, A. Bracale, C. Marcandella, O. Musseau, J.L. Pelloie, C. Raynaud, O.
      Faynot, Microelectronic Engineering, Vol. 48, p. 351, 1999
99   V. Kilchytska, A. Nève, L. Vancaillie, D. Levacq, S. Adriaensen, H. van Meer, K. De Meyer,
      C. Raynaud, M. Dehan, J.P. Raskin, D. Flandre, IEEE Transactions on Electron Devices, Vol.
      50, no. 3, p. 577, 2003

[100]   J.P. Colinge, Electronics Letters, Vol. 23, no. 19, p. 1023-1025, 1987
[101]   T. Douseki, H. Kyuragi, Electrochemical Society Proceedings, Vol. 2003-05, p. 209, 2003
[102]   Y. Kado, Y. Matsuya, T. Douseki, S. Nakata, M. Harada, J. Yamada, Electrochemical Society Proceedings, Vol. 2001-3, p. 277, 2001
[103]   T. Douseki, Y. Tanabe, M. Harada, T. Tsuchiya, Proceedings of the Eighth International Symposium on Silicon-on-Insulator Technology and Devices, Electrochem. Soc., p. 378, 1997
[104]   M. Urano, T. Douseki, T. Hatano, H. Fukuda, M. Harada, T. Tsuchiya, Proceedings Tenth Annual IEEE International ASIC Conference and Exhibit, p. 7, 1997
[105]   T. Douseki, S. Shigematsu, J. Yamada, M. Harada, H. Inokawa, T. Tsuchiya, Journal of Solid-State Circuits, Vol. 32, no. 10, p.1604, 1997
[106]   T. Douseki, S. Shigematsu, Y. Tanabe, M. Harada, H. Inokawa, T. Tsuchiya, Digest of Technical Papers of the ISSCC, p. 84, 1996
[107]   T. Douseki, H. Kyuragi, Proceedings of the IEEE International SOI Conference, p. 5, 2003
[108]   Y. Kado, Y. Matsuya, T. Douseki, S. Nakata, M. Harada, J. Yamada, Electrochemical Society Proceedings, Vol. 2001-3, p. 277, 2001
[109]   T. Douseki, M. Harada, T. Tsuchiya, Solid-State Electronics, Vol. 41, no. 4, p. 519, 1997
[110]   C. Harendt, U. Apel, T. Ifström, H.G. Graf, and B. Höfflinger, Proceedings of the Second International Symposium on "Semiconductor Wafer Bonding: Science, Technology, and Applications", Ed. by M.A. Schmidt, C.E. Hunt, T. Abe, and H. Baumgart, The Electrochemical Society, Proceedings Vol. 93-29, p. 129, 1993
[111]   U. Heinle, K. Pinardi, J. Olsson, Proceedings 32nd ESSDERC, p. 295, 2002
[112]   K. Pinardi, U. Heinle, S. Bengtsson, J. Olsson, Physica Scripta, Vol. T101, p. 38, 2002
[113]   S. Merchant, E. Arnold, H. Baumgart, S. Mukherjee, H. Pein, R. Pinker, Proceedings of the 3rd International Symposium on Power Semiconductor Devices and ICs (ISPSD '91), p. 31, 1991
[114]   R.P. Zingg, Proceedings of the International Symposium on Power Semiconductor Devices and ICs (ISPSD '2001), p. 343, 2001
[115]   T. Letavic, M. Simpson, E. Arnold, E. Peters, R. Aquino, J. Curcio, S. Herko, S. Mukherjee, Proceedings of the International Symposium on Power Semiconductor Devices and ICs (ISPSD '99), p. 325, 1999
[116]   R.P. Zingg, Microelectronic Engineering, Vol. 59, p. 461, 2001
[117]   R.P. Zingg, I. Weijland, H. van Zwol, P. Boos, T. Lavrijsen, T. Schoenmakers, Proceedings of the IEEE International SOI Conference, p. 62, 2000
[118]   E. McShane and K. Shenai, IEEE Transactions on Electron Devices, Vol. 49, no. 4, p. 643, 2002
[119]   S. Matsumoto, Y. Hiraoka, T. Sakai, T. Yachi, T. Ishiyama, T. Kosugi, H. Kamitsuna, M. Muraguchi, IEEE Transactions on Electron Devices, Vol. 48, no. 7, pp. 1448, 2001
[120]   J.G. Fiorenza and J.A. del Alamo, IEEE Transactions on Electron Devices, Vol. 49, no. 4, p. 687, 2002
[121]   M. Berkhout, IEEE. IEEE Journal of Solid-State Circuits, Vol. 38, no. 7, p. 1198, 2003
[122]   A. Vandooren, S. Cristoloveanu, M. Mojarradi, E. Kolawa, Electrochemical Society Proceedings, Vol. 2001-3, p. 151, 2001
[123]   Y.K. Leung, S.C. Kuehne, V.S.K. Huang, C.T. Nguyen, A.K. Paul, J.D. Plummer, and S.S. Wong, Proceedings of the IEEE International SOI Conference, p 132, 1996

[124] A. Nakagawa, N. Yasuhara, I. Omura, Y. Yamaguchi, T. Ogura, and T. Matsudai, Extended Abstracts of the IEDM, p. 229, 1992

[125] A. Nakagawa, Y. Yamaguchi, N. Yasuhara, K. Hirayama, and H. Funaki, technical Digest of IEDM, p. 477, 1996

[126] I.Y Park, S.H. Kim, Y.I. Choi, Proceedings of the IEEE International SOI Conference. P. 57, 1998

[127] T. Ogura and A. Nakagawa, Technical Digest of IEDM, p. 241, 1992

[128] A.Q. Huang, IEEE Electron Device Letters, Vol. 17, no. 6, p. 297, 1996

[129] S. Sridhar, Y.S. Huang, and B.J. Baliga, Extended Abstracts of the IEDM, p. 245, 1992

[130] S. Sridhar and B.J. Baliga, IEEE Electron Device Letters, Vol. 17, no. 11, p. 512, 1966

[131] Y.F. Zhao, A.Q. Huang, Y.K. Leung, S.S. Wong, Proceedings of the IEEE International SOI Conference. P. 55, 1998

[132] A. Zekry, I.M. Hafez, M.M. El-Hady, IEEE Transactions on Electron Devices, Vol. 49, no. 10, p. 1821, 2002

[133] J. Weyers, H. Vogt, M. Berger, W. Mach, B. Mütterlein, M. Raab, F. Richter, and F. Vogt, Proceedings of the 22nd ESSDERC, Microelectronic Engineering, Vol. 19, p 733, 1992

[134] C. Harendt, U. Apel, T. Ifström, H.G. Graf, and B. Höfflinger, Proceedings of the Second International Symposium on "Semiconductor Wafer Bonding: Science, Technology, and Applications", Ed. by M.A. Schmidt, C.E. Hunt, T. Abe, and H. Baumgart, The Electrochemical Society, Proceedings Vol. 93-29, p. 129, 1993

[135] F. Vogt, B. Mütterlein, and H. Vogt, Proceedings of the fifth International Symposium on Silicon-on-Insulator Technology and devices, Ed. by: K. Izumi, S. Cristoloveanu, P.L.F. Hemment, and G.W. Cullen, The Electrochemical Society Proceedings volume 92-13, p. 77, 1992

[136] W. Wondrak, E. Stein, and R. Held, in "Semiconductor Wafer Bonding: Science, Technology, and Applications", Ed. by U. Gösele, T. Abe, J. Haisma, and M.A.A Schmidt, Proceedings of the Electrochemical Society, Vol. 92-7, p. 427, 1992

[137] U. Heinle, K. Pinardi, J. Olsson, Proceedings 32nd ESSDERC, p. 295, 2002

[138] H. Pein, E. Arnold, H. Baumgart, R. Egloff, T. Letavic, S. Merchant, and S. Mukherjee, Proceedings ot the IEEE International SOI Conference, p. 146, 1992

[139] W. Wondrak, E. Stein, and R. Held, in "Semiconductor Wafer Bonding: Science, Technology, and Applications", Ed. by U. Gösele, T. Abe, J. Haisma, and M.A.A Schmidt, Proceedings of the Electrochemical Society, Vol. 92-7, p. 427, 1992

[140] R.P. Zingg, I. Weijland, H. van Zwol, P. Boos, T. Lavrijsen, T. Schoenmakers, Proceedings of the IEEE International SOI Conference, p. 62, 2000

[141] J.P. Colinge and S.Y. Chiang, IEEE Electron Device Letters, Vol. 7, p. 697, 1986

[142] H. Pein, E. Arnold, H. Baumgart, R. Egloff, T. Letavic, S. Merchant, and S. Mukherjee, Proceedings ot the IEEE International SOI Conference, p. 146, 1992

[143] W.P. Maszara, D. Boyko, A. Caviglia, G. Goetz, J.B. McKitterick, and J. O'Connor, Proceedings of the IEEE International SOI Conference, p. 131, 1995

[144] J.P. Blanc, J. Bonaime, E. Delevoye, J. Gauthier, J. de Pontcharra, R. Truche, E. Dupont-Nivet, J.L. Martin, and J. Montaron, Proceedings of the IEEE SOS/SOI Technology Conference, p. 85, 1990

[145] M. Dentan, E. Delagnes, N. Fourches, M. Rouger, M.C. Habrard, L. Blanquart, P. Delpierre, R. Potheau, R. Truche, J.P. Blanc, E. Delevoye, J. Gautier, J.L. Pelloie, J. de Pontcharra, O.

Flament, J.L. Leray, J.L. Martin, J. Montaron, and O. Musseau, IEEE Transactions on Nuclear Science, Vol. 40, no. 6, p. 1555, 1993

[146] O. Flament, J.L. Leray, and O. Musseau, in *Low-power HF microelectronics: a unified approach*, edited by G.A.S. Machado, IEE circuits and systems series 8, the Institution of Electrical Engineers, p. 185, 1996

[147] Y. Omura, Applied Physics Letters, Vol. 40, no. 6, p.528, 1982

[148] Y. Omura, IEEE International SOI Conference Proceedings, p.105, 1998

[149] Y. Omura, Journal of Semiconductor Technology and Science, Vol. 2, no. 1, p. 70, 2002

[150] Ming-Dou Ker, Kei-Kang Hung, H.T.H. Tang, S.C. Huang, S.S. Chen, M.C. Wang, Proceedings of the 2001 8th International Symposium on the Physical and Failure Analysis of Integrated Circuits (IPFA 2001), p. 91, 2001

[151] Y. Omura, S. Wakita, Electrochemical Society Proceedings, Vol. 2003-05, p. 313, 2002

[152] S. Voldman, D. Hui, L. Warriner, D. Young, R. Williams, J. Howard, V. Gross, W. Rausch, E. Leobandung, M. Sherony, N. Rohrer, C. Akrout, F. Assaderaghi, G. Shahidi, IEEE International SOI Conference Proceedings, p. 68, 1999

[153] U. Magnusson, H. Norström, W. Kaplan, S. Zhang, M. Jargelius, and D. Sigurd, Proceedings of the 23rd ESSDERC, Ed. by J. Borel, P. Gentil, J.P. Noblanc, A. Nouailhat, and M. Verdone, Editions Frontières, p 683, 1993

[154] S. Feindt, J.J.J. Hajjar, M. Smrtic, and J. Lapham, in "Semiconductor Wafer Bonding: Science, Technology, and Applications, Ed. by. M.A. Schmidt, C.E. Hunt, T. Abe, and H. Baumgart, Electrochemical Society Proceedings Vol. 93-29, p. 189, 1993

[155] S.J. Gaul, J.A. Delgado, G.V. Rouse, C.J. McLachlan, and W.A. Krull, Proceedings of the IEEE SOS/SOI Technology Conference, p. 101, 1989

[156] M.D. Church, Proceedings of the IEEE SOS/SOI Technology Conference, p. 175, 1989

[157] J.P. Blanc, J. Bonaime, E. Delevoye, J. Gauthier, J. de Pontcharra, R. Truche, E. Dupont-Nivet, J.L. Martin, and J. Montaron, Proceedings of the IEEE SOS/SOI Technology Conference, p. 85, 1990

[158] K. Watanabe, T. Hashimoto, M. Yoshida, M. Usami, Y. Sakai, and T. Ikeda, Electrochemical Society Proceedings, Vol. 92-7, p. 443, 1992

[159] J.P. Colinge, Electronics Letters, Vol. 22, p. 886, 1986

[160] W.M. Huang, K. Klein, M. Grimaldi, M. Racanelli, S. Ramaswami, J. Tsao, J. Foerstner, and B.Y. Hwang, Technical Digest of IEDM, p. 449, 1993

[161] G.G. Shahidi, D.D. Tang, B. Davari, Y. Taur, P. McFarland, K. Jenkins, D. Danner, M. Rodriguez, A. Megdanis, E. Petrillo, M. Polcari, and T.H. Ning, Technical Digest of IEDM, p. 663, 1991

[162] C.J. Patel, N.D. Jankovic and J.P. Colinge, *Physical and Technical Problems of SOI Structures and Devices*, edited by J.P. Colinge, V. S. Lysenko and A. N. Nazarov, NATO ASI Prtnership Sub-Series: 3. Vol. 4, Kluwer Academic Publishers, Dordrecht, p. 211, 1995

[163] M. Chan, S.K.H. Fung, C. Hu, and P.K. Ko, Proceedings of the IEEE International SOI Conference, p. 90, 1995

[164] B. Edholm, J. Olsson, and A. Söderbärg, IEEE Transactions on Electron Devices, Vol. 40, no. 12, p. 2359, 1993

[165] B. Edholm, J. Olsson, and A. Söderbärg, Microelectronic Engineering, Vol. 22, p. 379, 1993

[166] R. Dekker, W.T.A. v.d. Einden, and H.G.R. Maas, Technical Digest of IEDM, p. 75, 1993

[167] S.A. Parke, C. Hu, and P.K. Ko, IEEE Electron Device Letters, Vol. 14, p. 234, 1993

[168] R. Dekker, W.T.A. v.d. Einden, and H.G.R. Maas, Technical Digest of IEDM, p. 75, 1993
[169] W.M. Huang, K. Klein, M. Grimaldi, M. Racanelli, S. Ramaswami, J. Tsao, J. Foerstner, and B.Y. Hwang, Technical Digest of IEDM, p. 449, 1993
[170] G.G. Shahidi, D.D. Tang, B. Davari, Y. Taur, P. McFarland, K. Jenkins, D. Danner, M. Rodriguez, A. Megdanis, E. Petrillo, M. Polcari, and T.H. Ning, Technical Digest of IEDM, p. 663, 1991
[171] U. Magnusson, H. Norström, W. Kaplan, S. Zhang, M. Jargelius, and D. Sigurd, Proceedings of the 23rd ESSDERC, Ed. by J. Borel, P. Gentil, J.P. Noblanc, A. Nouailhat, and M. Verdone, Editions Frontières, p 683, 1993
[172] S. Feindt, J.J. Hajjar, M. Smrtic, and J. Lapham, in "Semiconductor Wafer Bonding: Science, Technology, and Applications, Ed. by. M.A. Schmidt, C.E. Hunt, T. Abe, and H. Baumgart, Electrochemical Society Proceedings Vol. 93-29, p. 189, 1993
[173] S. Nigrin, M.C. Wilson, S. Thomas, S. Connor, P.H. Osborne, Proceedings of IEEE International SOI Conference, p. 155, 2002
[174] H. Nii, T. Yamada, K. Inoh, T. Shino, S. Kawanaka, M. Yoshimi, Y. Katsumata, IEEE Transactions on Electron Devices, Vol. 47, no. 7, p. 1536, 2000
[175] M. Kumar, Y. Tan, J.K.O. Sin, IEEE Transactions on Electron Devices, Vol. 49, no. 1, p. 200, 2002
[176] M. Hattori, S. Kuromiya, F. Itoh, Proceedings of the IEEE International SOI Conference, p. 119, 1998
[177] T. Shino, S. Yoshitomi, H. Nii, S. Kawanaka, K. Inoh, T. Yamada, T. Fuse, Y. Katsumata, M. Yoshimi, S. Watanabe, J.I. Matsunaga, IEEE Transactions on Electron Devices, Vol. 49, no. 3, p. 414, 2002
[178] J.C. Sturm and J.F. Gibbons, in "Semiconductor-On-Insulator and Thin Film Transistor Technology", Chiang, Geis and Pfeiffer Eds., (North-Holland), MRS Symposium Proceedings, Vol. 53, p. 395, 1986
[179] T. Arnborg and A. Litwin, IEEE Transactions on Electron Devices, Vol. 42, no. 1, p. 172, 1995
[180] K. Yallup, S. Edwards and O. Creighton, Proceedings of the 24th ESSDERC, Ed. by C. Hill and P. Ashburn, Editions Frontières, p. 565, 1994
[181] T. Arnborg and A. Litwin, IEEE Transactions on Electron Devices, Vol. 42, no. 1, p. 172, 1995
[182] S. Kimura, K. Maio, T. Doi, T. Shimano, T. Maeda, IEEE Transactions on Electron Devices, Vol. 49, no. 6, p. 997, 2002
[183] H. Zimmermann, B. Muller, A. Hammer, K. Herzog, P. Seegebrecht, IEEE Transactions on Electron Devices, Vol. 49, no. 2, p. 334, 2002
[184] S. Hirose, T. Nishimura, K. Sugahara, S. Kusunoki, Y. Akasaka, N. Tsubouchi N, Symposium on VLSI Technology Digest of Technical Papers, p. 34, 1985
[185] W.B. Dubbelday, L.D. Flesner, G.A. Garcia, G.P. Imthurm, R.J. Hirschi, IEEE International SOI Conference Proceedings, p. 84, 1991
[186] J.P. Colinge, IEEE Transactions on Electron Devices, Vol.ED-33, no. 2, p. 203, 1986
[187] J.F. Gibbons, K.F. Lee, F.C. Wu, G.E.J. Eggermont, IEEE Electron Device Letters, Vol. EDL-3, no. 8, p. 191, 1982
[188] M.H. Gao, S.H. Wu, J.P. Colinge, C. Claeys, G. Declerck, IEEE International SOI Conference Proceedings, p. 138-9, 1991

[189] G.K. Celler, S. Cristoloveanu, Journal of Applied Physics, Vol. 93, no. 9, p.4955, 2003

[190] B.J. Blalock, S. Cristoloveanu, B.M. Dufrene, F. Allibert, M.M. Mojarradi, Journal of High Speed Electronics, Vol. 12, no. 2, p.511, 2002

[191] S. Cristoloveanu, B. Blalock, F. Allibert, B. Dufrene, M. Mojarradi, Proceedings of the 32nd European Solid-State Device Research Conference, p. 323, 2002

[192] B. Dufrene, B. Blalock, S. Cristoloveanu, M. Mojarradi, E.A. Kolawa, Electrochemical Society Proceedings, Vol. 2003-05, p. 367, 2003

[193] Y. Omura, IEEE Transactions on Electron Devices, Vol. 43, no. 3, p. 436, 1996

[194] Y. Nakajima, Y. Takahashi, S. Horiguchi, K. Iwadate, H. Namatsu, K. Kurihara, and M. Tabe, Extended Abstracts of the International Conference on Solid-State Devices and Materials, p. 538, 1994

[195] Y. Nakajima, Y. Takahashi, S. Horiguchi, K. Iwadate, H. Namatsu, K. Kurihara, and M. Tabe, Japanese Journal of Applied Physics, Vol. 34, p. 1309, 1995

[196] Y. Nakajima, Y. Takahashi, S. Horiguchi, K. Iwadate, H. Namatsu, K. Kurihara, and M. Tabe, Applied Physics Letters, Vol. 65, no. 22, p. 2833, 1994

[197] M. Büttiker, Y. Imry, R. Landauer, and S. Pinhas, Phys. Rev. B, Vol. 31, p. 6207, 1985

[198] J.P. Colinge, X. Baie, V. Bayot, E. Grivei, Solid-State Electronics, Vol. 39, pp. 49-51, 1996

[199] J.P. Colinge, Microelectronic Engineering, Vol. 28, pp. 423-430, 1995

[200] X. Baie, J.P. Colinge, V. Bayot, and E. Grivei, Proceedings of the IEEE International SOI Conference, p. 66, 1995

[201] J.P. Colinge and C.A. Colinge, *Physics of Semiconductor Devices*, pp. 353-360, Kluwer Academic Publishers, 2002

[202] Y. Takahashi, H. Namatsu, K. Kurihara, K. Iwadate, M. Nagase, and K. Murase, IEEE Transactions on Electron Devices, Vol. 43, no. 8, p. 1213, 1996

[203] K. Murase, Y. Takahashi, Y. Nakajima, H. Namatsu, M. Nagase, K. Kurihara, K. Iwadate, S. Horiguchi, M. Tabe, and K. Izumi, Microelectronic Engineering, Vol. 28, no. 1-4, p. 399, 1995

[204] H. Ishikuro, T. Fujii, T. Sayara, G. Hashiguchi, T. Iramoto, and T. Ikoma, Applied Physics Letters, Vol. 66, no. 24, p. 3585, 1966

[205] T. Hiramoto, H. Ishikuro, T. Fujii, T. Sayara, G. Hashiguchi, and T. Ikoma, Physica B: Condensed Matter, Vol. 227, p. 95, 1996

[206] K. Uchida, J. Koga, R. Ohba, A. Toriumi, IEEE Transactions on Electron Devices, Vol. 50, no. 7, p. 1623, 2003

[207] X. Tang, X. Baie, V. Bayot, F. Van de Wiele and J.P. Colinge, Proceedings of the IEEE International SOI Conference, p.46, 1999

[208] A. Nakajima, T. Futatsugi, K. Kosemura, T. Fukano, and N. Yokoyama, Technical Digest of IEDM, p. 952, 1996

[209] L. Guo, E. Leobandung, and S.Y. Chou, Technical Digest of IEDM, p. 955, 1996

[210] J.J. Welser, S. Tiwari, K.Y. Lee, and Y. Lee, IEEE Electron Device Letters, Vol. 18, no. 6, p. 278, 1997

[211] X. Tang, X. Baie, V. Bayot, F. Van de Wiele and J.P. Colinge, Proceedings of the IEEE International SOI Conference, p.100, 1999

[212] Xiaohui Tang, X. Baie, J. P. Colinge. A. Crahay, B. Katschmarsyj, V. Scheuren, D. Spôte, N. Reckinger, F. Van de Wiele, and V. Bayot, Solid State Electronics, Vol. 44, no. 12, p. 2259, 2000

Chapter 7

THE SOI MOSFET IN A HARSH ENVIRONMENT

SOI MOSFETs present several properties that allow them to operate in harsh environments where bulk devices would typically fail. These interesting properties are due to the small volume of silicon in which the devices are made, to the small area of the source-body and drain-body junctions, and to the presence of a back gate. In this Chapter, we will describe the behavior of the SOI MOSFET operating in two cases of extreme environments: the exposure to ionizing radiations and to high temperature.

7.1 IONIZING RADIATIONS

One of the major niche markets where SOI circuits and devices have been used extensively is the aerospace/military market SOI exhibits high hardness against transient radiation effects. The effect of radiation on an electronic device depends on the type of radiation (neutrons, heavy particles, electromagnetic radiation, etc.) to which the device is submitted. Unlike bipolar devices, MOSFETs are relatively insensitive to neutron irradiation (neutrons are responsible for a degradation in carrier lifetime by inducing displacement of atoms within the crystal lattice). Since MOSFETs are not minority carrier devices, this degradation in lifetime does not affect their electronic properties. MOS devices are, however, sensitive to the exposure to single-event upset (SEU), single-event latchup (SEL) single-event burnout (SEB), gamma-dot upset, and total-dose exposure than to neutrons. The effects created in bulk silicon MOSFETs by radiation exposure are well documented and can be found in the literature [1,2]. Table 7-1 lists the effects produced by different types of ionizing radiations in semiconductor devices. The following sections will compare the hardness of SOI and bulk devices to single-event phenomena and gamma-dot and total-dose irradiation conditions.

Table 7-1. Types of ionizing radiations and their effect on semiconductor devices.[3,4]

Particle	Physical effect	Electrical result	Environment
Photon	Oxide charge creation	Threshold voltage shift	Space, nuclear
	Photocurrent (high dose rate)	Device turn on	Nuclear bomb
	Creation of interface states	Subthreshold and mobility degradation	Space, nuclear
Heavy ion	Creation of e^--h^+ pairs	Single-event effects	Space
Neutron	Atom displacement	β_F degradation	Nuclear
	Recoil atoms	SEU	Avionics
Proton	Recoil atoms	SEU	Space
	Nuclear interaction	SEU	(solar flares)
	Oxide charge creation	Threshold voltage shift	(earth's
	Atom displacement	β_F degradation	radiation belts)
Electron	Oxide charge creation	Threshold voltage shift	Space
	Atom displacement	β_F degradation	(rad. belts)

7.1.1 Single-event phenomena

Single-event upset (SEU) is caused by the penetration of an energetic particle, such as an alpha particle or a heavy ion (cosmic ray) within a device. Indeed, when such a particle penetrates a reverse-biased junction and its depletion layer and the bulk silicon underneath it, a plasma track is produced along the particle path, where electron-hole pairs are generated.[5] The presence of this track temporarily collapses the depletion layer and distorts it in the vicinity of the track. The distortion of the depletion layer is called a "funnel" (Figure 7-1).

The funnel extends the depletion zone along the particle track such that the electrons created in the funnel drift towards the junction (Figure 7-1A). The holes move towards the grounded substrate, creating a substrate current. The collected electrons give rise to a current transient that can upset the logic state of a node. The duration of the collection of the electrons by the node is on the order of a fraction of a nanosecond. The *drift* current created in this process is called the "prompt current component". The length of the track in the silicon is typically on the order of ten micrometers. Afterwards, free electrons generated along the particle track underneath the funnel can diffuse towards the depletion region, where they create a second current (*diffusion* current), called the "delayed current component". This current is smaller in magnitude compared to the prompt current, but it lasts much longer (up to hundreds of nanoseconds or microseconds).[6] Multiple bit upsets (MBU) are sometimes observed as a result of the penetration of a heavy ion into an integrated circuit. [7,8] In an SOI device, the impinging particle ionizes the silicon along its track as well. However, because of the presence of a buried insulator layer between the active silicon film and the substrate, the charges generated in the substrate cannot be collected by the junctions of the SOI devices. The only electrons that can be collected are those produced within the thin silicon film, the thickness of which is typically 80-150 nm in rad-hard applications. The ratio of the lengths of the tracks along which the electrons are collected gives a first-order approximation of the advantages of SOI over bulk in terms of SEU hardness (*e.g.*: 10μm/100nm = 100 in the case of a 100 nm-thick SOI device).

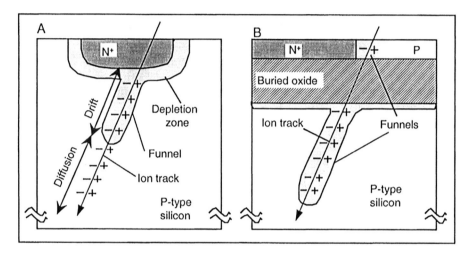

Figure 7-1. Ion strike on A: a bulk PN junction, and B: an SOI junction.

The energy imparted by a particle along the track is expressed in "linear energy transfer" (LET) units. It is defined by the following relationship:

$$LET = \frac{1}{m_v} \frac{dW}{dx} \qquad (7.1.1)$$

where x is the linear distance along the particle track, dW is the energy lost by the particle and absorbed by the silicon, and m_v is the volumic mass of silicon. The LET is usually expressed in MeV·cm²/mg. The number of electrons or holes created by a single-event upset is given by:

$$\frac{dN}{dx} = \frac{dP}{dx} = \frac{m_v}{w} LET \qquad (7.1.2)$$

where w is the energy needed to create an electron-hole pair [9]. As an illustration, a single carbon ion with an energy of 1 GeV (LET $\cong 0.24$ MeV·cm²/mg) produces 1.5×10^4 electron-hole pairs/μm in silicon. In a bulk silicon device, it produces 1.5×10^5 electron-hole pairs ($\cong 0.3$ pC) along 10 μm of track. If the electrons migrate towards the node within a time scale of

10-100 ps, an SEU current spike of 1-10 mA is then created. Because of the reduced length of the track along which the electrons are collected in SOI, the SEU current spike will be significantly smaller in an SOI device than in a bulk device. Irradiation of bulk and 0.15μm-thick SOI diodes with 11 MeV fluorine ions systematically shows a charge collected in the SOI devices 10 to 20 times smaller than in their bulk counterparts.[10] The magnitude of the impact of SEU on a circuit is measured by the upset cross section (units: cm^2/bit). The upset cross section represents the area sensitive to SEU, per bit. For example, in a memory chip it represents the area of the junctions within a single memory cell that can be upset by an SEU mechanism. The smaller this cross-section, the less the device is sensitive to particle irradiation. Light particles such as protons, electrons and neutrons have a LET that is usually too low to ionize the silicon. An LET below 1 $MeV.cm^2mg^{-1}$ generates a charge that is too small to have significant effect on a device (approximately 0.01 pC/μm). But these particles cause silicon atom recoil through a "direct hit" mechanism or induce nuclear reactions that produce energetic nuclear fragments. These recoil atoms or nucleus fragments can, in turn, act as heavy ions and cause SEU. Thus particles such as protons usually cause SEU not through direct ionization (their LET is too low), but rather through nuclear reactions resulting in recoils which lead to an indirectly generated upset.[11] The single-event upset cross-sections of several bulk and SOI circuits are presented in Figure 7-2. One can see that 150 nm-thick SOI devices are approximately 10 times less sensitive than thicker (500 nm) devices, which are in turn about 100 times harder than bulk CMOS devices. SEU events require a minimum value of LET to occur, called the "threshold LET" value. This threshold depends on both circuit layout and on the technology employed to fabricate the devices. For instance, the threshold LETs for the 500-nm and 150-nm devices in Figure 7-2 are approximately 30 and 40 $MeV.cm^2$/mg, respectively.

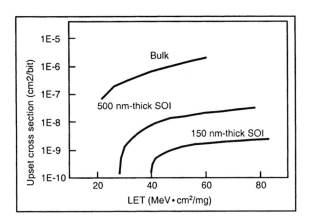

Figure 7-2. SEU cross sections for bulk and SOI circuits.[12,13]

The generation of a funnel in the substrate underneath the buried oxide of an SOI structure can also influence the characteristics of a device. Indeed, if the substrate (back-gate) bias is such that the substrate under the buried oxide is depleted, the electrons created along the track will move upwards towards the buried oxide interface (Figure 7-1B). These electrons immediately induce a positive mirror charge in the upper silicon layer. As a result, electrons are injected into the device by the external circuit to restore equilibrium, and the occurrence of a transient current is measured. This effect is not observed if the surface of the substrate underneath the buried oxide is either inverted or accumulated.[14]

The photocurrent (or "ionocurrent") generated by the impact of a particle in an SOI MOSFET can be amplified by the parasitic lateral bipolar transistor present in the device. Indeed, the hole current, I_B, created within the body of an SOI MOSFET acts as a base current for the parasitic lateral NPN bipolar transistor (n-channel device case). In response to the base current pulse induced by the particle, a collector current $I_C=\beta_F I_B$ is produced. This current adds to the current pulse caused by the SEU-induced electrons collected by the drain, such that the drain current pulse actually becomes equal to $(1+\beta_F) I_{SEU}$, where I_{SEU} is the electron current initially generated by the particle (in an n-channel device). Therefore, the presence of any bipolar transistor action, even with low gain $(\beta_F<1)$ enhances the transient current produced by SEU.[15] The problem is magnified in short-channel devices where β_F is large. In that case the bipolar gain can be large enough to make SOI transistors more sensitive to SEU than bulk devices.[16] The solutions to this problem are the use of a body tie through which part of the base current can be shunted, and/or to reduce the minority carrier lifetime in the silicon film through the use of "lifetime killer" techniques.[17] Although body ties reduce the parasitic bipolar effect, they cannot completely eliminate it. The ability of a body tie to suppress the bipolar effect depends strongly on the location of the body tie in relation to the ion strike. The farther the ion strike from the body tie, the larger the gain of the parasitic bipolar transistor, and therefore, the less efficient the body tie.[18] Circuit design solutions include the increase of RC time constant of the different circuit nodes, but this technique presents the drawback of degrading the circuit speed performance.[19] Fully depleted devices are more resistant to SEU than partially depleted devices because the bipolar gain is much smaller in FD transistors.[20]

Single-event latchup (SEL) can be caused by a heavy-ion strike in bulk CMOS integrated circuits where a parasitic n-p-n-p path exists. SEL is triggered by excess current in the base of either a parasitic npn or pnp transistor following a heavy-ion strike. Because of the regenerative feedback loop that exists between those two transistors, latchup can occur. Latchup is triggered within nanoseconds, and can cause destructive burnout within hundreds of microseconds. The holding voltage for latchup is typically on the order of 1 volt. Thus, unless the power supply voltage is interrupted, the low-resistance conduction path from the power supply to ground will be maintained.[21,22] SEL can cause permanent damage (hard error) in an integrated circuit. In SOI CMOS devices there is no p-n-p-n structure that can latchup. Therefore, SEL does not exist in SOI circuits.

Single-event burnout (SEB) can occur in power bipolar or power MOS devices. Power MOS devices contain a "parasitic" bipolar transistor structure.[23] If a heavy ion strikes the bipolar transistor the charge generated will cause current to flow in the base and raise the potential of the emitter-base junction. If the current flow is high enough, it can forward bias the emitter-base junction and the bipolar transistor turns on. After the parasitic bipolar transistor is turned on, second breakdown of the bipolar transistor can occur. This second breakdown has been referred to as current-induced avalanche.[24,25,26] Depending on the current density, the current induced in the parasitic transistor by the heavy ion will either dissipate without device degradation or will regeneratively increase until (in absence of current limiting elements) the device is destroyed. Protons have been shown to cause SEB as well.[27] There is no basic difference between bulk and thick SOI power devices, from an SEB point of view.

Single-event gate rupture (SEGR) can occur as a heavy ion passes through a gate oxide. [28,29,30] It occurs only at high oxide electric fields, such as those present during a write or clear operation in a nonvolatile E^{2}PROM cell, or in a power MOSFET. It is caused by the combination of the applied electric field and the energy imparted by the particle. As an ion passes through a gate oxide it forms a highly conducting plasma path between the silicon and the gate. If the energy is high enough, it can cause localized heating of the dielectric and potentially a thermal runaway condition. If this occurs the dielectric may locally melt or evaporate. Most likely, there is no basic difference between bulk and thick SOI power devices, from an SEGR point of view.

Single-event snapback (SES) can occur in partially depleted SOI transistors, and is enhanced by the bipolar effect. In an n-channel transistor that is turned on, the energy imparted by a particle can create enough free holes to increase the forward bias of the source-body junction, and enough electrons to increase the drain current. These conditions increase impact ionization mechanisms, which can lead to snapback (sometimes called "single-transistor latchup"). This effect has been predicted for fully depleted devices, but has not yet been observed.[31,32,33]

7.1.2 Total dose effects

Total-dose effects are caused by the cumulative exposure of insulators such as SiO_2 to ionizing radiations (X-rays or gamma rays). The dose unit is the rad(Si) or rad(SiO_2) which is defined by the deposition of 100 erg of radiation energy per gram of Si or SiO_2, respectively. Note that 1 rad(SiO_2) = 0.56 rad(Si). The rad(Si) is a CGS unit, but there exists a MKS unit for dose called "Gray" (Gy). One Gy is defined as the deposition of 1 joule of radiation energy per kilogram of matter. The equivalence between the two units is straightforward: 1 Gy = 100 rads. The number of electron-hole pairs generated is related to the energy dW absorbed by unit volume dv of the material:

$$\frac{dN}{dv} = \frac{dP}{dv} = \frac{1}{w}\frac{dW}{dv} \qquad (7.1.3)$$

where w is the effective energy needed to produce an electron-hole pair in SiO_2 (w=17 eV). The relationship between the dose, D, and the number of pairs generated is given by:[34]

$$\frac{dN}{dv} = \frac{dP}{dv} = \frac{m_v}{w} D \qquad (7.1.4)$$

It is generally given that 1 rad(Si) generates 4×10^{13} pairs per cm^3 in silicon and 7.6×10^{12} pairs per cm^3 in SiO_2. To get an idea of the magnitude of the levels of irradiation to which devices can be submitted, one can mention the following numbers: a medical or dental X-ray corresponds to less than 0.1 rad(Si). The human being becomes sick after being exposed to 100 rad(Si) and falls in an instant coma if submitted to 10 krad(Si). During their lifetime, satellites orbiting around the earth receive total doses ranging between 10 krad(Si) and 1 Mrad(Si), depending on the orbit parameters. Interplanetary spacecrafts and some electronics in nuclear reactors can be exposed to doses in excess of 10 Mrad(Si).[35] Some SOI circuits have been tested at doses up to 500 Mrad(Si).[36,37]

The main effect caused by total dose in MOS devices is the generation of charges in the oxides and the generation of interface states at the Si/SiO_2 interfaces. If the rate of energy deposition is high, a sufficient amount of electron-hole pairs can be created in the silicon resulting in photocurrents. This case, where dD/dt is high, will be considered in the next section (gamma-dot effects). Ionizing electromagnetic radiations such as X-rays and gamma rays (produced *e.g.* by a ^{60}Co source) create electron-hole pairs in silicon dioxide. Electrons are fairly mobile in SiO_2, even at room temperature, and can rapidly move out of the oxide (towards a positively biased gate electrode, in the case of a gate oxide, for instance). Holes, on the other hand, remain trapped within the oxide and contribute to the creation of a positive oxide charge, Q_{ox}. If we take the example of a gate oxide, the charge Q_{ox} is proportional to the thickness of the oxide, t_{ox}, and the resulting threshold voltage shift is, therefore, proportional to t_{ox}^2 since $\Delta V_{th} = -\frac{Q_{ox} t_{ox}}{\varepsilon_{ox}}$. The relationship between the threshold voltage shift and the dose can be written as:

$$\Delta V_{th} = -\alpha \frac{q\, m_v}{w\, \varepsilon_{ox}} t_{ox}^2\, D \qquad (7.1.5)$$

where w is the effective energy needed to produce a pair in oxide (17 eV), m_v is the volumic mass of the oxide and where the factor α accounts for the fact that only a fraction of the charges become trapped within the oxide. α is a technology-dependent parameter. Typical values of α are 0.15 for normal, unhardened oxides and 0.05 or less for rad-hard oxides. The physics of the irradiation of devices is, unfortunately, much more complex than what is described in the few equations above. Dose irradiation also creates surface states at the Si-SiO$_2$ interfaces. Contrary to oxide charges, which are always positive (which decreases the threshold voltage in n-channel devices), interface states are amphoteric and trap electrons in an n-channel device, which increases the threshold voltage. Another type of trap, called "border traps", located in the oxide very close to the silicon/oxide interface, can trap electrons as well.[38] Some oxide charges anneal out with time and even contribute to the creation of additional interface traps. This creates an effect which is called "rebound" and it is illustrated in Figure 7-3.[39,40]

The rebound effect is, of course, temperature dependent. A rebound effect can also be seen if the dose is increased without interruption above a given value (typically around 1 Mrad(Si)). In that case the generation of oxide charges saturates, while the creation of interface states does not. As a result, a rebound is seen in n-channel devices when the dose is increased.[41] The dose rate at which the device is irradiated is important as well. At low dose rates, oxide charges have time to migrate to the silicon/oxide interface and to transform into interface states. As a result, the rebound effect can occur at lower doses. This makes a difference in the radiation response between devices used in the field (*e.g.* in a spacecraft, low dose rates) and in lab conditions (higher dose rates).

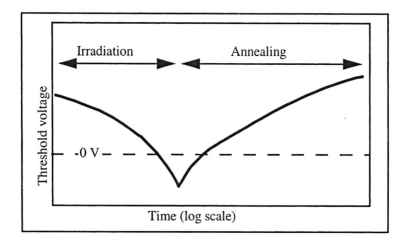

Figure 7-3. "Rebound" effect in an n-channel MOSFET exposed to dose irradiation. Annealing occurs after irradiation has stopped.

The dose-induced effects are highly dependent on the front- and back-gate bias of the device. Worst-case conditions correspond to positive gate biases (which push the holes in the oxide towards the Si-SiO$_2$ interfaces). In p-channel transistors, the generation of both positive charges and interface states cause an increase of the absolute value of the threshold voltage (*i.e.*: it becomes more negative).

The SOI MOSFET, with its many Si-SiO$_2$ interfaces (gate oxide, buried oxide and edge (field) oxide), is quite sensitive to total-dose exposure, unless special hardening techniques are used. While SOI gate oxide hardening techniques are similar to those employed in bulk (*e.g.*: low-temperature oxide growth), the techniques for prevention of back and edge leakage are specific to SOI. The classical solution for avoiding the formation of an inversion layer at the bottom of the silicon film in n-channel devices is to increase the threshold voltage of the back transistor by means of a back boron impurity implant. In some instances, use of a back-gate bias can be made to compensate for the radiation-induced generation of positive charges in the buried oxide. The creation of a peak of boron doping at the back interface implies the use of partially depleted devices. Indeed, fully depleted devices are usually too thin to allow one to realize such an implant (see Section 4.3). In addition, the presence of charges in the buried oxide and interface states at the silicon - buried oxide interface causes front threshold voltage shifts and degrades the performance of the devices if the front and back interfaces are electrically coupled, as in fully depleted devices. Partially depleted SOI circuits made in a silicon film thickness as low as 150 nm and capable of withstanding doses up to 300 Mrad(Si) have been reported.[42]

The control of edge leakage currents in n-channel devices also poses a problem since the parasitic edge transistors can be quite sensitive to dose irradiation due to the relatively thick oxides. In particular, LOCOS or oxidized mesa isolation in contact with the edges of the silicon active areas are sensitive to irradiation (see section 4.2). An example of radiation-induced edge-leakage current is presented in Figure 7-4.[43,44] Edge leakage problems can be eliminated through optimization of the sidewall doping and field oxide growth processes [45] or through the use of edgeless devices or transistors where a P+ diffusion interrupts edge leakage paths between the N+ source and drain diffusions (see section 4.6). There is usually no edge leakage problem in p-channel devices.

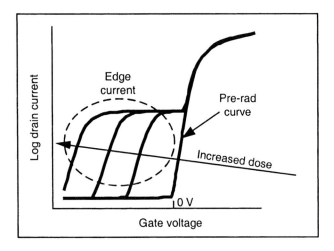

Figure 7-4. Edge leakage current caused by dose irradiation.

The positive charges generated by high irradiation levels (>500 krad(SiO^2) in the BOX of fully depleted n-channel SOI MOSFETs can invert the back-channel interface, giving rise to a leakage current, even if a negative front gate voltage is applied. This leakage current can be amplified by impact ionization at the drain junction leading to a "total-dose latch" or snapback effect. [46,47]

7.1.3 Dose-rate effects

Dose-rate (D') effects take place when a large dose of electromagnetic energy (X-rays, gamma rays) is absorbed within a short time interval by events such as nuclear explosions. Dose-rate effects are also often referred to as gamma-dot ($\dot{\Gamma}$) effects. The dose-rate unit is the rad(Si).s^{-1}, with one rad(Si) generating approximately 4×10^{13} pairs cm^{-3} in silicon. If the dose-rate is high, a significant number of electron-hole pairs can be generated in the silicon. The separation of these pairs within the depletion zones of the device gives rise to a photocurrent (prompt current). A delayed diffusion current component is observed as well. The duration of the delayed component depends on the recombination velocity of the photogenerated carriers and on other phenomena such as carrier generation by impact ionization near the reverse-biased drain junction. An example of prompt and delayed current components is shown in Figure 7-5.[48]

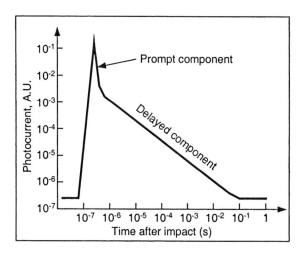

Figure 7-5. Prompt and delayed current components vs time after strike.

The generated photocurrent is expressed by:

$$I_{ph}=q \cdot V_{depl} \cdot g \cdot D'(t) \tag{7.1.6}$$

where q is the electron charge, V_{depl} is the volume of the depletion zone where charge collection occurs (also called the "sensitive volume"), and g is the carrier generation constant in silicon, which is equal to 4.2×10^{13} hole-electron pairs cm^{-3} $rad(Si)^{-1}$.[49] In the most dramatic cases, all transistors are flooded with free carriers and become so conductive that the supply voltage rail is virtually shorted to ground, and supply voltage collapse occurs. The major difference between bulk and SOI devices resides in the much smaller sensitive volune, V_{depl}, found in SOI devices (Figure 7-6). As a result, the dose-rate-induced photocurrent is significantly smaller in SOI transistors than in bulk devices.

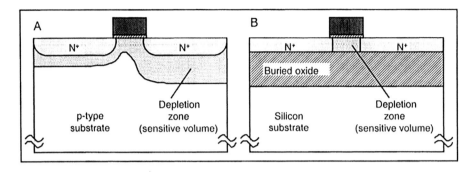

Figure 7-6. Volume of the depletion zones in bulk (A) and SOI (B) MOSFETs.

As in the case of SEU, the photocurrent generated in an SOI MOSFET can be amplified by the parasitic lateral bipolar transistor present in the device. Indeed, the hole current created within the body of an SOI MOSFET acts as a base current for the parasitic lateral NPN bipolar transistor (n-channel device case). In response to the base current pulse I_B induced by the particle, a collector current $I_C=\beta_F\ I_B$ is produced, and equation 7.1.6 becomes:

$$I_{ph}=q{\cdot}V_{depl}{\cdot}g{\cdot}\ (1+\beta_F){\cdot}D'(t) \tag{7.1.7}$$

The bipolar amplification current adds to the photocurrent caused by the gamma-dot event. Therefore, the presence of any bipolar transistor action, even with low gain ($\beta_F<1$) adds to the transient current produced by a gamma-dot event. This effect is, of course, more pronounced in short-channel devices where β_F is large. The solutions proposed to remedy this problem are the use of a body contact through which part of the base photocurrent can be shunted, and the inclusion of lifetime killers within the SOI acive region. As in the case of SEU, one must point out that the resistance of body ties is not equal to zero and that, as a result, the extraction of excess carriers through the body tie is imperfect.[50,51] Fully depleted devices are more resistant to gamma-dot effects than partially depleted devices because the bipolar gain is much smaller in FD transistors.[52]

7.2 HIGH-TEMPERATURE OPERATION

Some applications (well logging, avionics, automotive, etc.) require electronic circuits capable of operating at temperatures up to 300°C. SOI devices and circuits present three advantages in this field over their bulk counterparts: the absence of thermally-activated latch-up, reduced leakage currents and, in thin-film, fully depleted devices, a smaller variation of threshold voltage with temperature. The absence of latch-up in SOI is quite obvious and will not be discussed here, although it is an important argument for the use of SOI circuits for high-temperature applications.

7.2.1 Leakage current

The increase of junction leakage current is one of the main causes of failure in circuits operating at high temperature. The junction leakage current is, of course, proportional to the area of the junction. The strongest leakage component found in bulk CMOS operating at high temperatures originates from the "huge" well junctions.[53] These cause loss of functionality in CMOS circuits at temperatures above 180°C while the same SOI CMOS circuits can operate up to 300°C. Operation of SOI CMOS ring oscillators at temperatures even up to 500°C has been demonstrated.[54] The leakage current of a reverse-biased N^+-P junction is expressed by [55]:

$$I_{leak} = q\, A \left(\frac{D_n}{\tau_n}\right)^{1/2} \frac{n_i^2}{N_a} + q\, A\, \frac{n_i\, W}{\tau_e} \qquad (7.2.1)$$

where q is the electron charge, A is the junction area, D_n is the electron diffusion coefficient, τ_n is the electron lifetime in p-type neutral silicon, n_i is the intrinsic carrier concentration, N_a is the doping concentration in the p-type material, W is the depletion width, and $\tau_e = (\tau_n + \tau_p)/2$ is the effective lifetime related to the thermal generation process in the depletion region. The temperature dependence of n_i is given by: [56]

$$n_i = 3.9x10^{16}\ T^{3/2}\ e^{-(E_g/2kT)} \qquad (7.2.2)$$

The first term in equation 7.2.1 (diffusion component) is proportional to n_i^2, and the second term (generation component) is proportional to n_i. It is experimentally observed [57] that the leakage of SOI MOSFETs and reverse-biased diodes varies as n_i below a temperature of 100-150°C and varies as n_i^2 above those temperatures. It is worth noting that A, the area of the junction, and $A \cdot W$, the volume of the space-charge region associated with the diode, are both much smaller in SOI transistors than in bulk devices. Therefore, extremely low leakage current can be obtained in SOI devices. The leakage current at $V_{DS} = 5V$ in bulk and SOI n-channel transistors with similar geometries are compared in Figure 7-7, where the slightly higher leakage in the SOI device at room temperature is attributed to a relatively poor value of τ_e in the measured device. At high temperatures, however, where diffusion current dominates, [58,59] the leakage current is three orders of magnitude lower in the SOI devices than in the bulk device. In thin-film SOI devices it has been observed that the junction leakage current is proportional to $n_i(T)$ if the body of the device is fully depleted when the transistor is turned off. This is always the case in accumulation-mode SOI MOSFETs (Figure 7-8).

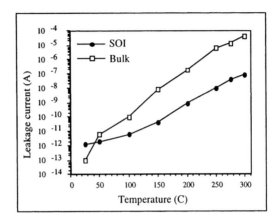

Figure 7-7. Leakage current in bulk and SOI n-channel transistors, of same geometries, as a function of temperature.

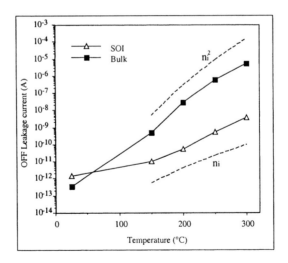

Figure 7-8. Leakage current dependence on temperature of an SOI accumulation-mode pMOS transistors for zero back gate and -3 V drain bias, and of a pMOS bulk transistor with -3 V drain and zero substrate and well bias. Evolutions proportional to n_i and n_i^2 are indicated.

The device OFF current, which is equal to the reverse-biased drain junction in enhancement-mode SOI MOSFETs, is markedly smaller in SOI than in bulk transistors, owing to the reduced junction area and to the $n_i(T)$ current increase with temperature in SOI devices. This observation is also valid for accumulation-mode devices.

Figure 7-9 presents the drain current as a function of gate voltage in an accumulation-mode p-channel SOI device, where I_{ON}/I_{OFF} ratios in excess of 100,000 and 300 are obtained at 200 and 300°C, respectively. Owing to these high I_{ON}/I_{OFF} ratios SOI logic circuits are still functional at high temperatures. For comparison, the I_{ON}/I_{OFF} ratio in a bulk transistor is only on the order of 40 at 250°C, while it is on the order of 5,000 in an SOI device.

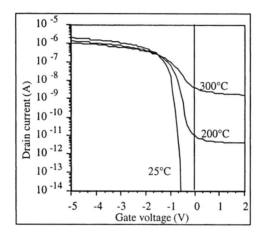

Figure 7-9. Drain current as a function of gate voltage in an accumulation-mode p-channel transistor at different temperatures.[60]

In addition to the fact that the area of source and drain junctions is much smaller in SOI than in bulk (by a factor of 15 to 100, depending on the design rules), it is also worthwhile noting that the largest of all junctions, the well junctions, are totally absent from SOI CMOS. This contributes to a drastic reduction in the overall standby current consumption of SOI circuits, compared to bulk CMOS. Figure 7-10 presents the junction leakage paths in bulk and SOI CMOS inverters. In the bulk device leakage currents flow to the substrate from the reverse-biased well junction and from the reverse-biased drain of the n-channel MOSFET (the case where the output is "high" is considered here). In the SOI inverter the only leakage current to be considered is the n-channel drain junction leakage. Consequently, the high-temperature standby power consumption is much smaller in the SOI inverter than in the bulk device.

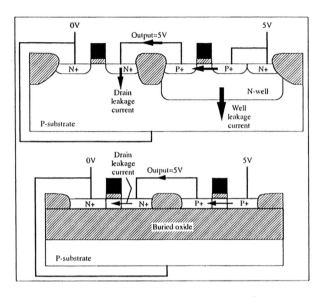

Figure 7-10. Leakage current paths in bulk (top) and SOI (bottom) CMOS inverters. No current flow to the substrate is allowed in SOI.

7.2.2 Threshold voltage

The threshold voltage of an n-channel MOSFET (bulk or SOI) is given by:

$$V_{TH} = \Phi_{MS} + 2\Phi_F - Q_{ox}/C_{ox} - Q_{depl}/C_{ox} \qquad (7.2.3)$$

where Φ_{MS}, Φ_F, Q_{ox}, Q_{depl} and C_{ox} are the metal-semiconductor work function difference, the Fermi potential, the charge density in the gate oxide, the depletion charge controlled by the gate, and the gate oxide capacitance, respectively.

In bulk and partially depleted SOI devices Q_{depl} equal to $q\,N_a\,x_{dmax}$ where x_{dmax} is the maximum depletion depth corresponding to the doping concentration, N_a, under consideration, and is equal to $\sqrt{\dfrac{4\,\varepsilon_{si}\,\Phi_F}{q\,N_a}}$. In fully depleted devices, Q_{depl} depends on back-gate bias, and has a value which can range between $q\,N_a\,t_{si}$ and $q\,N_a\,t_{si}/2$, q being the electron charge, N_a the doping concentration and t_{si} the silicon film thickness. If an N+ polysilicon gate is assumed, the value of the work function difference is given by:

$$\Phi_{MS} = -\frac{E_g}{2} - \Phi_F, \text{ with } \Phi_F = \frac{kT}{q} \ln(N_a/n_i) \qquad (7.2.4)$$

The temperature dependence of the intrinsic carrier concentration, n_i, is given by [61]:

$$n_i = 3.9x10^{16} \; T^{3/2} \; e^{-(E_g/2kT)} \qquad (7.2.5)$$

where E_g, k and T are the silicon bandgap energy, the Boltzmann constant, the temperature, respectively. The temperature dependence of the threshold voltage can be obtained from equations 7.2.3 to 7.2.5. For simplification, we will assume that Q_{ox} shows no temperature dependence over the temperature range under consideration and we will not take into account the presence of surface states at the Si-SiO$_2$ interfaces. Additionally, E_g is assumed to be independent of temperature over the temperature range under consideration.

In the case of a bulk or a partially depleted SOI device the variation of threshold voltage with temperature can be derived from the previous relationships:

$$\frac{dV_{TH}}{dT} = \frac{d\Phi_F}{dT} \; [1+\frac{q}{C_{ox}}\sqrt{\frac{\varepsilon_{si} N_a}{k \, T \ln(N_a/ni)}} \;] \qquad (7.2.6)$$

with $\quad \dfrac{d\Phi_F}{dT} = 8.63x10^{-5} \; [ln(Na) - 38.2 - \dfrac{3}{2} \; \{1+ln(T)\}] \qquad (7.2.7)$

In the case of a fully depleted, device, the depletion charge, Q_{depl}, is equal to $qN_a t_{si}/n$, where the value of n ranges between 1 and 2, depending on oxide charge and back-gate bias conditions. If one assumes that n is independent of temperature, the following dependence for the thin-film device is obtained:

$$\frac{dV_{TH}}{dT} = \frac{d\Phi_F}{dT} \qquad (7.2.8)$$

It can be seen by mere comparison of 7.2.6 and 7.2.8 that dV_{TH}/dT is smaller in fully depleted SOI devices than in bulk or partially depleted SOI devices. The ratio of threshold voltage variation in bulk devices to that in thin SOI devices is given by the bracketed term of 7.2.6, and is typically from 2 to 3, depending on the gate oxide thickness and the channel doping concentration. The right-hand term of the bracketed expression in 7.2.6 is caused by the temperature dependence of the depletion zone width, a variation which is obviously nonexistent in fully depleted devices. Typical values for dV_{TH}/dT range between -0.7 and -0.8 mV/K in fully depleted SOI MOSFETs. Much larger values of dV_{TH}/dT are observed in bulk (or partially depleted SOI) devices (see example in Table 7-2).

Table 7-2. Calculated dependence of threshold voltage on temperature, dV_{TH}/dT, (in mV/K), in bulk and SOI (t_{si}=100 nm) MOSFETs. The data are presented for two temperatures (25 and 200°C) and for the two sets of processing parameters (t_{ox} and N_a).[62]

	FDSOI T=25°C	Bulk T=25°C	FDSOI T=200°C	Bulk T=200°C
t_{ox}=19nm and N_a=1.6x10^{17} cm^{-3}	-0.74	-1.87	-0.8	-2.3
t_{ox}=32.5nm and N_a=1.2x10^{17} cm^{-3}	-0.76	-2.42	-0.82	-3.05

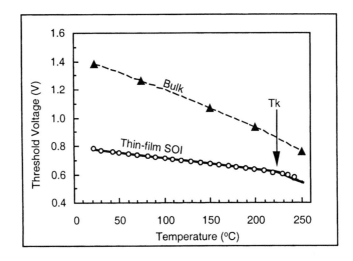

Figure 7-11. Variation of the threshold voltage of bulk and thin-film SOI n-channel MOSFETs with temperature.

When the temperature increases in thin-film devices the intrinsic carrier concentration increases, and Φ_F decreases. This gives rise to a decrease in the maximum depletion width, in such a manner that the devices are no longer fully depleted above a given critical temperature, T_k. At that temperature x_{dmax} becomes smaller than $t_{si,eff}$, where $t_{si,eff}$ is the effective electrical film thickness "seen" from the top gate, *i.e.* the silicon film thickness minus the backside depletion zone. Above T_k, the transistor behaves as a thick-film SOI MOSFET, since the film is no longer fully depleted. The temperature dependence of the threshold voltage becomes similar to that observed in bulk or PD SOI devices; dV_{TH}/dT is then governed by equation 7.2.6 (Figure 7-11). Depending on bias conditions charges can be genenerated in the BOX under high-temperature (>200°C) operation conditions, and these charges can drift in the BOX and create threshold voltage instabilities in fully depleted SOI MOSFETs.[63]

7.2.3 Output conductance

The output conductance is an important parameter in the performance of analog circuits. In partially depleted SOI devices, impact ionization-related effects (kink effect, parasitic bipolar action, etc.) tend to degrade the output conductance. This degradation is minimized, but nevertheless present, if fully depleted devices are used. It is observed that the output conductance of SOI MOSFETs improves when the temperature is increased [64], as can be seen in Figure 7-12. This is explained by several mechanisms: high temperature reduces impact ionization near the drain, excess minority carrier concentration in the device body is reduced through increased recombination, and the body potential variations are reduced owing to an increase of the saturation current of the source junction.

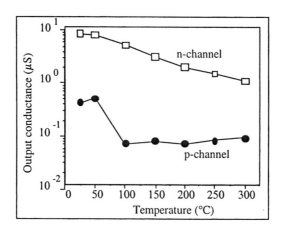

Figure 7-12. Variation of the output conductance in thin-film SOI MOSFETs (n- and p-channel) with temperature. $W/L = 20\mu m/10\mu m$ and $V_{GS} = \pm 1.5$ V.

7.2.4 Subthreshold slope

The temperature dependence of the subthreshold slope in an SOI
MOSFET is predicted by Equation 5.7.25:

$$S = \frac{kT}{q} \ln(10)\left[1 + \frac{C_b}{C_{ox1}}\right] \tag{7.2.9}$$

The capacitance C_b is the capacitance between the weak inversion layer
and the back gate. It is defined by the relationship $C_b = dQ_b/d\Phi_{s1}$, where Q_b is
the charge in the silicon film and Φ_{s1} is the front surface potential. C_b can be
calculated in a fully depleted SOI MOSFET using the depletion
approximation. The result is given by Equation 5.7.22:

$$C_b = \frac{C_{si}C_{ox2}}{C_{si} + C_{ox2}} \tag{7.2.10}$$

Since C_{si}, C_{ox1} and C_{ox2} are constant for a given device, a linear
dependence of subthreshold slope on temperature is predicted. At high
temperature, however, a nonlinear temperature dependence of the
subthreshold slope on temperature is observed (Figure 7-13). When the
temperature is higher than 150-200°C, the charge due to intrinsic free
carriers is no longer negligible compared to the depletion charge under the
gate, and the Fermi potential is significantly reduced. As a result the surface
electric field is reduced and the weak inversion layer is broadened. In
addition, the charge due to intrinsic free carriers increases capacitance
$C_b = dQ_b/d\Phi_{s1}$. All these factors contribute to increase the subthreshold slope.
Figure 7.13 shows the actual temperature dependence of the subthreshold
slope on temperature in a fully depleted SOI MOSFET.

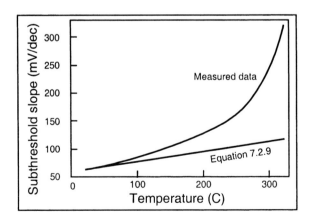

Figure 7-13. Subthreshold slope in fully depleted SOI MOSFETs *vs.* temperature.[65]

REFERENCES

[1] G.C. Messenger and M.S. Ash, *The Effects of Radiation on Electronic Systems*, Van Nostrand Reinhold Company, New York, 1986

[2] Several authors, *Commemorative Special Issue of the IEEE Transactions on Nuclear Science*, IEEE Transactions on Nuclear Science, Vol. 50, no. 3, June 2003

[3] J. R. Schwank, Short course on Silicon-on-Insulator circuits, IEEE International SOI Conference, p. 5.1, 1996

[4] J. Olsen, P.E. Becher, P.B. Fynbo, P. Raaby, and J. Schultz, IEEE Transactions on Nuclear Science, Vol. 40, no. 2, p. 74, 1993

[5] G.C. Messenger and M.S. Ash, *The Effects of Radiation on Electronic Systems*, Van Nostrand Reinhold Company, New York, 1986, p. 307

[6] J. R. Schwank, Short course on Silicon-on-Insulator circuits, IEEE International SOI Conference, p. 5.1, 1996

[7] Y. Song, K.N. Vu, J.S. Cable, A.A. Witteles, W.A. Kolasinski, R. Koga, J.H. Elder, J.V. Osborn, R.C. Martin, and N.M. Ghoniem, IEEE Transactions on Nuclear Science, Vol. 35, no. 6, p. 1673, 1988

[8] O. Musseau, F. Gardic, P. Roche, T. Corbière, R.A. Reed, S. Buchner, P. McDonald, J. Melinger, and A.B. Campbell, IEEE Transactions on Nuclear Science, Vol. 43, no. 6, p. 2879, 1996

[9] J.L. Leray, Microelectronics Engineering, Vol. 8, p. 187, 1988

[10] T. Aton, J. Seitchik, K. Joyner, T. Houston, H. Shichijo, Symposium on VLSI Technology Digest of Technical Papers, p. 98, 1996

[11] E. Normand, IEEE Nuclear and Space Radiation Effects Conference Short Course on Radiation Effects in Commercial Electronics, p. V-1, 1994

[12] G.E.Davis, L.R. Hite, T.G.W. Blake, C.E. Chen, H.W. Lam, R. DeMoyer, IEEE Transactions on Nuclear Science, Vol. 32, p. 4432, 1985

[13] J.L. Leray, E. Dupont-Nivet, O. Musseau, Y.M. Coïc, A. Umbert, P. Lalande, J.F. Péré, A.J. Auberton-Hervé, M. Bruel, C. Jaussaud, J. Margail, B. Giffard, R. Truche, and F. Martin, IEEE Transactions on Nuclear Science, Vol. 35, p. 1355, 1988

[14] J.L. Leray, E. Dupont-Nivet, O. Musseau, Y.M. Coïc, A. Umbert, P. Lalande, J.F. Péré, A.J. Auberton-Hervé, M. Bruel, C. Jaussaud, J. Margail, B. Giffard, R. Truche, and F. Martin, IEEE Transactions on Nuclear Science, Vol. 35, p. 1355, 1988

[15] L.W. Massengill, D.V. Kerns, Jr., S.E. Kerns, and M.L. Alles, IEEE Electron Device Letters, Vol. 11, p. 98, 1990

[16] Y. Tosaka, K. Suzuki, and T. Sugii, Technical Paper Digest of the Symposium on VLSI Technology, p. 29, 1995

[17] D.P. Ioannou, S. Mitra, D.E. Ioannou, S.T. Liu, W.C. Jenkins, Proceedings of the IEEE International SOI Conference, p. 83, 2003

[18] O. Musseau, V. Ferlet-Cavrois, J.L. Pelloie, S. Buchner, D. McMorrow, A.B. Campbell, IEEE Transactions on Nuclear Science, Vol. 47, p. 2196, 2000

[19] G.E. Davis, in "Silicon-On-Insulator and Buried Metals in Semiconductors", Sturm, Chen, Pfeiffer and Hemment Eds., (North-Holland), MRS Symposium Proceedings, Vol. 107, p. 317, 1988

[20] J.R. Schwank, V. Ferlet-Cavrois, M.R. Shaneyfelt, P. Paillet, P.E. Dodd, IEEE Transactions on Nuclear Science, Vol. 50, no. 3, p. 522, 2003

[21] J.R. Schwank, Short Course on Silicon-on-Insulator circuits, IEEE International SOI Conference, p. 5.1, 1996

22   P.V. Dressendorfer and A. Ochoa, IEEE Transactions on Nuclear Science, Vol. 28, p. 4288, 1981

23   M. Allenspach, C. Dachs, G.H. Johnson, R.D. Schrimpf, E. Lorfèvre, J.M. Palau, J.R. Brews, K.F. Galloway, J.L. Titus, and C.F. Wheatley, IEEE Transactions on Nuclear Science, Vol. 43, no. 6, p. 2927, 1996

24   J.R. Schwank, Short course on Silicon-on-Insulator circuits, IEEE International SOI Conference, p. 5.1, 1996

25   T.F. Wrobel, F.N. Coppage, G.L. Hash, and A.J. Smith, IEEE Transactions on Nuclear Science, Vol. 32, p. 3991, 1985

26   G.H. Johnson, R.D. Schrimpf, and K.F. Galloway, IEEE Transactions on Nuclear Science, Vol. 39, p. 1605, 1992

27   J.W. Adolphsen, J.L. Barth, and G.B. Gee, IEEE Transactions on Nuclear Science, Vol. 43, no. 6, p. 2921, 1996

28   C.F. Wheatley, J.L. Titus, D.I. Burton, and D.R. Carley, IEEE Transactions on Nuclear Science, Vol. 43, no. 6, p. 294, 1996

29   T.F. Wrobel, IEEE Transactions on Nuclear Science, Vol. 34, p. 1262, 1987

30   J.R. Schwank, Short Course on Silicon-on-Insulator circuits, IEEE International SOI Conference, p. 5.1, 1996

31   *ibidem*

32   J. Gautier and A.J. Auberton-Hervé, IEEE Electron Device Letters, Vol. 12, p. 372, 1991

33   O. Musseau, IEEE Transactions on Nuclear Science, Vol. 43, p. 603, 1996

34   J.L. Leray, Microelectronics Engineering, Vol. 8, p. 187, 1988

35   F. Wulf, D. Braunig, and A. Boden, ECFA STUDY WEEK on Instrumentation Technology for High-Luminosity Hadron Colliders, Ed. by E. Fernands and G. Jarlskog, Proc. Vol. 1, p. 109, 1989

36   J.L. Leray, E. Dupont-Nivet, J.F. Péré, Y.M. Coïc, M. Raffaelli, A.J. Auberton-Hervé, M. Bruel, B. Giffard, and J. Margail, IEEE Transactions on Nuclear Science, Vol. 37, no. 6, p. 2013, 1990

37   J.L. Leray, E. Dupont-Nivet, J.F. Péré, O. Musseau, P. Lalande, and A. Umbert, Proceedings SOS/SOI Technology Workshop, p. 114, 1989

38   D.M. Fletwood, IEEE Transactions on Nuclear Science, Vol. 39, p. 269, 1992

39   J.R. Schwank, P.S. winokur, P.J. McWorther, F.W. Sexton, P.V. Dressendorfer, D.C. Turpin, IEEE Transactions on Nuclear Science, Vol. NS-31, p. 1434, 1984

40   D.M. Fletwood, S.S. Tsao, and P.S. Winokur, IEEE Transactions on Nuclear Science, Vol. 35, p. 1361, 1988

41   G.C. Messenger and M.S. Ash, *The Effects of Radiation on Electronic Systems*, Van Nostrand Reinhold Company, New York, 1986, p. 243

42   J.L. Leray, E. Dupont-Nivet, O. Musseau, Y.M. Coïc, A. Umbert, P. Lalande, J.F. Péré, A.J. Auberton-Hervé, M. Bruel, C. Jaussaud, J. Margail, B. Giffard, R. Truche, and F. Martin, IEEE Transactions on Nuclear Science, Vol. 35, p. 1355, 1988

43   *ibidem*

44   V. Ferlet-Cavrois, O. Musseau, J.L. Leray, J.L. Pelloie, and C. Raynaud, IEEE Transactions on Electron Devices, Vol. 44, no. 6, p. 965, 1997

45   G.E. Davis, in "Silicon-On-Insulator and Buried Metals in Semiconductors", Sturm, Chen, Pfeiffer and Hemment Eds., (North-Holland), MRS Symposium Proceedings, Vol. 107, p. 317, 1988

46   V. Ferlet-Cavrois, S. Quoizola, O. Musseau, O. Flament, J.L. Leray, IEEE Transactions on Nuclear Science, Vol. 45, p. 2458, 1998

[47] J.R. Schwank, M.R. Shaneyfelt, P.E. Dodd, J.A. Burns, C.L. Keast, P.W. Wyatt, IEEE Transactions on Nuclear Science, Vol. 47, no. 3, p. 604, 2000

[48] J.R. Schwank, V. Ferlet-Cavrois, M.R. Shaneyfelt, P. Paillet, P.E. Dodd, IEEE Transactions on Nuclear Science, Vol. 50, no. 3, p. 522, 2003

[49] G.C. Messenger and M.S. Ash, *The Effects of Radiation on Electronic Systems*, Van Nostrand Reinhold Company, New York, 1986, p. 267

[50] M.L. alles, S.E. Kerns, L.W. Massengill, J.E. Clark, K.L. Jones Jr., R.E. Lowther, IEEE Transactions on Nuclear Science, Vol. 38, p. 1259, 1991

[51] V. Ferlet-Cavrois, O. Musseau, C. D'hose, C. Marcandella, G. Giraud, C. Fenouillet, J. du Port de Poncharra, IEEE Transactions on Nuclear Science, Vol. 49, p. 1456, 2002

[52] J.R. Schwank, V. Ferlet-Cavrois, M.R. Shaneyfelt, P. Paillet, P.E. Dodd, IEEE Transactions on Nuclear Science, Vol. 50, no. 3, p. 522, 2003

[53] W.A. Krull and J.C. Lee, Proceedings of the IEEE SOS/SOI Technology Workshop, p. 69, 1988

[54] W.P. Maszara, Proc. of the 4th International Symposium on Silicon-on-Insulator Technology and Devices, Ed. by D. Schmidt, the Electrochemical Society, Vol. 90-6, p. 199, 1990

[55] S.M. Sze, *Physics of Semiconductor Devices*, 2nd Edition, J. Wiley & Sons Eds, p. 91, 1981

[56] R.S. Muller and T.I. Kamins, *Device Electronics for Integrated Circuits*, 2nd Edition, J. Wiley & Sons Eds, p. 56, 1986

[57] D.P. Vu, M.J. Boden, W.R. Henderson, N.K. Cheong, P.M. Zavaracky, D.A. Adams, and M.M. Austin, Proceedings of the IEEE SOS/SOI Technology Conference, p. 165, 1989

[58] T.E. Rudenko, V.I. Kilchitskaya, and A.N. Rudenko, Microelectronic Engineering, Vol. 36, no. 1-4, p. 367, 1997

[59] T. Rudenko, A. Rudenko, V. Kilchitskaya, Electrochemical Society Proceedings, Vol. 97-23, p. 295, 1997

[60] D. Flandre, in *High-Temperature Electronics*, Ed. by. K. Fricke and V. Krozer, Materials Science and Engineering, Elsevier, Vol. B29, p. 7, 1995

[61] R.S. Muller and T.I. Kamins, *Device Electronics for Integrated Circuits*, 2nd Edition, J. Wiley & Sons Eds, p. 56, 1986

[62] G. Groeseneken, J.P. Colinge, H.E. Maes, J.C. Alderman and S. Holt, IEEE Electron Device Letters, Vol. 11, p. 329, 1990

[63] A.N. Nazarov, V.S. Lysenko, J.P. Colinge, D. Flandre, Electrochemical Society Proceedings, Vol. 2003-05, p. 455, 2003

[64] P. Francis, A. Terao, B. Gentinne, D. Flandre, JP Colinge, Technical Digest of IEDM, p. 353, 1992

[65] T. Rudenko, V. Kylchytska, J.P. Colinge, V. Dessard, D. Flandre, IEEE Electron Device Letters, Vol. 23, no. 3, p. 148, 2002

Chapter 8

SOI CIRCUITS

8.1 INTRODUCTION

SOI MOSFETs present the following advantages over bulk devices:

◊ absence of latchup in CMOS structures (see chapter 1)
◊ higher SEU and soft-error immunity (see chapter 7)
◊ reduced parasitic capacitances (see chapter 5)
◊ simplified CMOS processing (see chapter 4)

In addition, *fully depleted* devices present the following advantages:

◊ reduced body effect (see chapter 5)
◊ higher transconductance (see chapter 5)
◊ sharper subthreshold slope (see chapter 5)

while the current drive of *partially depleted* devices can be enhanced due to:

◊ kink effect (see chapter 5)
◊ drain current overshoot effects (see chapter 5)

This chapter will illustrate how these properties can be taken advantage of to produce high-performance integrated circuits. SOI technology has been employed for many years in several niche applications such as radiation-hardened and high-temperature electronics. More recently, however, it has become more common in mainstream CMOS applications such as microprocessors and mixed-mode integrated circuits.

8.2 MAINSTREAM CMOS APPLICATIONS

In the year 2000 IBM announced the use of SOI microprocessors in high-end servers, and in 2003, both IBM and Advanced Micro Devices (AMD) unveiled the first 64-bit microprocessors for personal computers, the PowerPC G5 and the Opteron, respectively. Both processors are silicon-on-insulator circuits. SOI CMOS technology is currently used by many other semiconductor manufacturers to fabricate high-speed integrated circuits. Some companies, such as IBM and AMD use partially depleted devices, while others, such as Oki, produce Bluetooth™ baseband and RF chips, microcontrollers, DRAM, SRAM and multiplexers for fiber optics using a fully depleted SOI process. These products take advantage of the reduced source and drain capacitance of SOI devices, such that a 25-35% speed advantage and a 3-fold reduction of power consumption, when compared to bulk devices, can be achieved. Table 8-1 lists some SOI VLSI circuits.

Table 8-1. Some SOI circuits

Company	Circuit	FD or PD	V_{DD}	Performance	Ref.
Samsung	Microprocessor	FD	1.5 V	600 MHz	1
Samsung	Microprocessor	FD	1.15 V	425 MHz	ibidem
IBM	Microprocessor	PD	2 V	580 MHz	2
IBM	Microprocessor	PD	1.8 V	550 MHz	3
IBM	Microprocessor	PD	1.4 V	800 MHz	4
AMD	Microprocessor	PD	1.65 V	2-3$^+$ GHz	5
Mitsubishi	DRAM	FD/PD	1 V	46 ns	6
Mitsubishi	DRAM	PD	0.9 V	-	7,8
Samsung	DRAM	PD	<1.5 V	30 ns	9
Hyundai	DRAM	PD	2.2 V	1 Gb	10
Samsung	SRAM	FD	0.9 V	20 ns	11
Samsung	SRAM	FD	0.7 V	30 ns	ibidem
Mitsubishi	Embedded DRAM	PD	<1.2 V	46 ns	12
Oki	RF and logic	FD	1-2.5 V	-	13,14

8.2.1 Digital circuits

A definite advantage of SOI technology is the increase of circuit speed and the decrease of power consumption made possible by the reduced source and drain capacitances. The increased current drive due to the reduced body coefficient further increases the speed performance in fully depleted SOI devices, while floating body effects can be used to boost the operation speed of partially depleted SOI devices. SOI CMOS ring oscillators show a steady 2× speed improvement over comparable bulk devices throughout the literature.[15,16] Larger circuits such as SRAMs and microprocessors typically show 15% to 30% improvement of access time or maximum operating frequency, even if the design is directly transferred from bulk to SOI without optimization.[17] Figure 8-1A shows power consumption vs. operating speed in FDSOI and bulk devices. The power consumption of the SOI circuits is typically 3 times smaller than that of the equivalent bulk part. Figure 8-1B compares the speed performance of FDSOI and bulk processors. In this particular example the SOI devices are at least 30% faster than their bulk counterparts, and SOI with a supply voltage of 1.5V can deliver the same speed performance as bulk with a supply voltage of 2.5V.

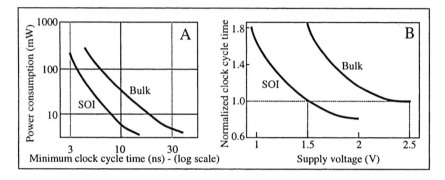

Figure 8-1. A: Relationship between power consumption and cycle time for 0.35μm design rule FD SOI CMOS and bulk CMOS [18] and B: Minimum processor operating cycle time vs. power supply voltage. The cycle time is normalized to the bulk device cycle time for $V_{DD}=2.5V$.[19]

In partially depleted SOI devices advantage can be taken from effects that enhnance current drive to improve speed performance. These effects are the kink effect (see section 5.9.1) and current overshoot effects due to the capacitive coupling between the floating body and the gate or drain (see section 5.10.3).[20,21] Table 8-2 compares the delay of CMOS inverter ring oscillators made using bulk, body-tied, partially depleted SOI (BTSOI), and floating-body, partially depleted SOI (FBSOI) MOSFETs. In these particular circuits the speed improvement due to the reduction of source and drain capacitances in SOI is modest: the body-tied SOI devices are merely 3.9 to 7.6% faster than the bulk devices. When the bodies are floating, however, the kink effect, the capacitive coupling between the body and the gate, and the associated current overshoot effects are present, and the overall speed performance is enhanced by another 21 to 24%. This speed improvement effect is routinely used to boost the speed performance in large digital circuits such as microprocessors.[22]

Table 8-2. CMOS inverter ring oscillator delay, τ_{RO}. Devices are either bulk, body-tied SOI (BTSOI) or floating-body, partially depleted SOI (FBSOI).[23]

L_{eff}=145nm and V_{DD}=1.8V	Bulk	BTSOI	FBSOI
τ_{RO}	35.7	33	25.5
SOI advantage over bulk	-	7.6%	29%
L_{eff}=70nm and V_{DD}=1.2V	Bulk	BTSOI	FBSOI
τ_{RO}	30.9	29.7	22.4
SOI advantage over bulk	-	3.9%	28%

8.2.2 Low-voltage, low-power digital circuits

SOI MOSFETs are particularly well suited for applications where low supply voltage is used. Among the different types of SOI transistors, fully depleted MOSFETs, MTCMOS/DTMOS and double/multiple-gate devices are especially well adapted to low-voltage applications because of their almost ideal subthreshold slope (near 60 mV/decade at room temperature in long-channel devices). A low value of subthreshold slope allows one to use a low threshold voltage, resulting in high current drive during the ON state, and low current in the OFF state. Figure 8-2 shows the drain current as a function of drain voltage in long-channel FD SOI transistors showing that low leakage current and low threshold voltage can be obtained simultaneously.

A simple analysis can be used to demonstrate the potential of FD SOI technology for low-voltage, low-power CMOS applications. Consider a switching CMOS gate driving a known capacitive load, C_{load}. The power consumption of the device is given by the following expression:

$$W = W_{static} + W_{dynamic} = I_{off}V_{DD} + f\,C_{load}\,V_{DD}^2 \qquad (8.3.1)$$

where I_{off} is the transistor off current, V_{DD} is the supply voltage, and f is the switching frequency. The static power consumption is due to leakage current flow through transistors that are turned off, while the dynamic power consumption is due to the charging/discharging of a load capacitance, such as a the capacitance of the transistor drain or an interconnection line.

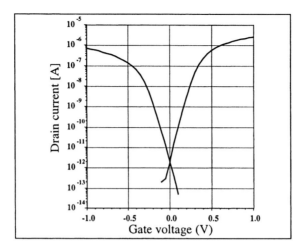

Figure 8-2. Measured current *vs.* gate voltage for an n-channel and a p-channel SOI MOSFET with W=L=3 μm at V_{DS} = ±50 mV. Because of the sharp subthreshold slope, OFF currents less than 1 pA/μm and low threshold voltage (0.4 V) can be obtained simultaneously.[24]

The switching time of a gate is proportional to the load capacitance and inversely proportional to the transistor current:

$$\tau \propto \frac{C_{load}}{I_D}$$ (8.3.2)

Reducing the supply voltage is an effective means to cut down the power consumption, according to Equation 8.3.1. Unfortunately, a reduction of supply voltage is accompanied by a reduction of the current drive, according to Equation 5.4.25 where $V_G = V_{DD}$, and therefore, a reduction of speed, according to 8.3.2. Taking the example of an n-channel device with grounded source, the maximum current drive is given by:

$$I_{dsat\_max} = \frac{1}{2n} \frac{W}{L} \mu_n C_{ox1} [V_{DD} - V_{TH}]^2$$ (8.3.3)

The reduction of current drive and circuit speed can be limited by reducing the threshold voltage. Figure 8-3 presents the ratio of saturation currents between an SOI MOSFET with a body factor n=1.05 and a threshold voltage V_{TH}=0.5V and a bulk transistor with n=1.15 and V_{TH}=0.7V. The I_{Dsat} enhancement due to the V_{TH} reduction is quite remarkable at low supply voltage.

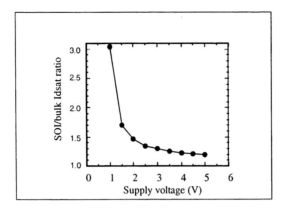

Figure 8-3. SOI/bulk I_{dsat} ratio *vs.* supply voltage. $n_{SOI} = 1.05$, $n_{bulk} = 1.15$, $V_{TH\_SOI} = 0.5$ V and $V_{TH\_bulk} = 0.7$V.

Unfortunately, any decrease of threshold voltage increases the off current of the device, defined by $I_{OFF} = I_D$ when $V_G = 0$V. The off current is directly related to the subthreshold slope. In fully depleted MOSFETs, MTCMOS/DTMOS and double/multiple-gate devices the subthreshold slope is close to 60mV/decade, which results in the lowest possible off current for any value of threshold voltage. Figure 8-4 shows the I_{ON}/I_{OFF} ratio in bulk and FDSOI MOSFETs in saturation, as a function of threshold voltage. It can be seen that, for any given threshold voltage, the SOI device has a lower OFF current (at $V_G = 0$ V) and a higher ON current (at $V_G = 1$ V) than the bulk device. The I_{ON}/I_{OFF} ratio advantage of the SOI device becomes smaller as the threshold voltage is reduced, but it still is an order of magnitude larger than in the bulk transistor for a threshold voltage of 250 mV.

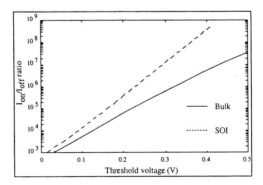

Figure 8-4. $I_{ON}(V_G=1V)/I_{OFF}(V_G=0V)$ ratio in bulk and FDSOI MOSFETs as a function of threshold voltage.[27]

The "ultimate" reduction of supply voltage can be estimated on the basis of the following observations: the dynamic power consumption of a circuit decreases with decreased supply voltage, according to equation 8.3.1. The dependence of the static power consumption on supply voltage is more complex, since it both decreases with V_{DD} (expression 8.3.1) and increases with I_{OFF}. I_{OFF} increases if V_{TH} is reduced to maintain a certain circuit speed when V_{DD} is reduced. As a result the static power consumption first increases when the supply and threshold voltage are reduced, and ultimately vanishes when $V_{DD} \to 0$. The total power consumption is dominated by the static component at low supply voltages, then decreases when the supply voltage is increased and reaches a minimum, at which point dynamic power consumption becomes the dominant component.[25] Figure 8-5 shows the power consumption of a bulk and a FDSOI CMOS gate as a function of supply voltage; both gates operate at the same clock frequency and threshold voltages vary as a function of supply voltage to allow for switching at a particular frequency. Clearly the point of minimum power dissipation is lower in the SOI device than in bulk. In general the power consumption and supply voltage reduction at the point of minimum power dissipation is reduced by over 20% by switching from bulk to SOI, and by 50% when switching from bulk to MTCMOS/DTMOS, where the subthreshold slope is virtually equal to 60 mV/decade at room temperature.[26]

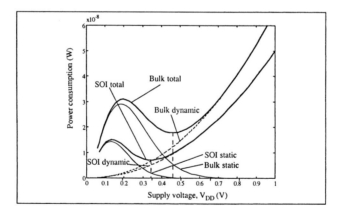

Figure 8-5. Power consumption of a bulk or FDSOI CMOS gate as a function of supply voltage.[27]

Table 8-3 lists some low-voltage, high-speed SOI circuits published in the literature. The circuits operating with a supply voltage of 0.5 V make use of transistors with a gate-to-body contact (DTMOS/MTCMOS).

Table 8-3. Low-voltage, low-power SOI circuits. MTCMOS circuits are identified by an asterisk (*).

Company	Circuit	V_{DD}	Speed	Power dissipation	Ref.
NTT	16b ALU*	0.5V	40 MHz	350 μW	28
NTT	300k gate array	1.2V	38 MHz	30 mW	29
Toshiba	16x16b multiplier	0.5V	18 ns	4 mW	30
Sharp	PLL	1.5V	1.1 GHz	-	31
Toshiba	32b ALU*	0.5V	260 MHz	2.5 mW	32
NTT	8x8 ATM switch	2V	40 Gb/s	-	33
NTT	Wireless systems*	0.5V	10 MHz	1 mW	34
Lehigh U.	Ring oscillator	0.25V	-	5 fJ/stage	35
Hiroshima U.	SRAM cell	0.5V	-	-	36
LETI	RFID chip	0.6V	13 MHz	-	37
OKI	SRAM	1 V	6.5 ns	-	38
NTT	Various circuits*	0.5 V	-	-	39
UCL	SRAM cell	0.5V	-	5 aW	40
UCL	Level keeper	0.5V	-	-	41

8.2.3 Memory circuits

The main advantages of SOI for memory chips are the following:[42,43]

- Small transistor junction capacitance, which reduces the total bit line capacitance and improves high-speed operation,
- Soft-error immunity,
- Reduced body effect, which improves source-follower transistor performance,
- Reduced leakage current, which improves the static retention time and standby current, and
- Sharp subthreshold slope of SOI transistors improves low-voltage and high-speed operation.

In the case of SRAMs, the use of SOI reduces the rate of SEU (single-event upset) and rules out SEL (single-event latchup). This explains the use of SOI for memory chips for space applications. In addition, the reduced parasitic capacitances and the reduced body effect allow for faster and lower-voltage operation.

In the case of DRAMs, the advantages of SOI are a reduction of the bit line capacitance (by 25% compared to bulk), a reduction of the access transistor leakage current, and the reduction of the soft-error sensitivity.[44,45] As a result, storage capacitance of 12 fF is sufficient for a

256 Mb, 1.5 V DRAM realized in SOI, while 34 fF is required for the equivalent bulk device. The advantage of using SOI, in terms of reducing the memory cell area, is then obvious. It is also possible to realize DRAMs that can operate with a sub-1 V power supply (which has not yet been demonstrated in bulk). In order to achieve this level of performance, a reduced C_b/C_s ratio must be achieved (C_b and C_s are the bit line and the storage capacitances, respectively). The reduction of the bit line capacitance comes "for free" with the use of SOI, while the increase of C_s is achieved through the cell design. Considering the cell retention time, it has be shown that thin-film SOI MOSFETs have a much lower leakage current than bulk or thicker SOI devices. Off-state leakage currents smaller than 1 fA/μm has been observed in fully depleted, thin-film devices.[46] Table 8-4 lists SOI memory chips published in the literature.

Table 8-4. SOI memory chips

Year	Company	Type of RAM	Ref.
1982	NTT	1kb SRAM	47
1988	Harris	High-temperature CMOS 4kb SRAM	48
1989	TI / Harris	64kb SRAM	49
1989	LETI	Thin-film 16kb SRAM	50
1990	TI	Commercial 64kb SRAM	-
1991	TI	256kb SRAM	51
1991	IBM	256kb fully depleted SRAM	52
1993	TI	1Mb SRAM, 20 ns access time @ 5V	53
1993	IBM	512 kb SRAM, 3.5 ns acc. time @ 1V	54
1993	Mitsubishi	64 kb DRAM, 1.5V operation	55
1994	Loral	256k rad-hard SRAM	56
1995	Samsung	16 Mb DRAM	57
1996	Mitsubishi	16 Mb DRAM, 0.9 V operation	58
1997	Mitsubishi	16 Mb DRAM, 46 ns @ 1V	59
1997	Hyundai	1 Gb DRAM	60
1998	Lockheed Martin	1M rad-hard SRAM	61
1999	Samsung	16 Mb DRAM, V_{DD}<1.5V	62
1999	Samsung	SRAM 20 ns, V_{DD}=0.9V	63
2000	Mitsubishi	Embedded DRAM, V_{DD}<1.2V	64
2000	NTT	64kb SRAM, 30 ns, V_{DD}=0.5V	65
2001	Micron	Embedded DRAM, SoC compatible	66
2001	Oki	SRAM, 6.5 ns, V_{DD}=1V	67
2001	NTT	128kb ROM, 30 ns, V_{DD}=0.5V	68

The DRAM architecture can also take advantage of the low body effect found in SOI devices. Bulk pass transistors suffer from body effect during the storage capacitor charging state such that the word line voltage must be boosted to $V_{WL} = V_{th} + V_{ds} + \Delta V$, where ΔV compensates for the V_{TH}

increase in the bulk pass transistor. Thus, the body effect becomes a major issue in bulk pass transistor design.[69]   The use of fully depleted SOI significantly improves the performance of pass gates because of the low body effect. In contrast to bulk, the body potential of a partially depleted transistor follows the source node potential during the charging, so that the body-to-source potential difference is kept at a constant value. For fully depleted SOI transistors there is no body potential change. Therefore, SOI transistors have better charging efficiency than bulk transistors. If the charging current is designed to be 1 $\mu A/\mu m$, then the charging efficiency, defined as the ratio of source (the storage node) voltage at 1 $\mu A/\mu m$ to the drain (the bit line) voltage, is 80% for bulk transistors, and 88% and 98% for partially depleted and fully depleted transistors, respectively (Figure 8-6). The higher charging efficiency of SOI transistors results in higher programming speed for the DRAM cell. At time *t=1ns* after being activated by the word line, a fully depleted transistor can transfer 10 times more charge than a bulk transistor. For a bulk pass transistor to achieve the same charging efficiency as an SOI transistor, the word line voltage must be increased during the charging. As shown in Figure 8-6, the gate voltage of a bulk transistor has to be 0.6 V higher than that of a fully depleted SOI MOSFET to reach a 90% charging efficiency.

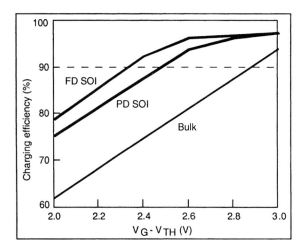

Figure 8-6. Charging efficiency as a function of gate voltage overdrive.[70]

8.2.3.1 Non volatile memory devices

Non-volatile memory (NVM) devices can be made on SOI substrates just as well as on bulk wafers. The fabrication of electrically erasable read-only memory (EEPROM) cells was first reported in 1994.[71] These devices have the same structure as bulk devices with a floating gate and a control gate. Programming is achieved by hot carrier injection in the floating gate. SOI flash EEPROM cells programmed by Fowler-Nordheim injection were have been reported as well.[72] In these devices the programming and erase steps are facilitated by the presence of a floating body (partially depleted devices are used), which allows for low-voltage program/erase operation. Flash memory devices programmed by carrier injection in a side floating gate can also be found in the literature.[73] Another type of flash memory cell employing charge injection in an opposite-side (bottom) gate has been proposed as well.[74]

Non-volatile memory operation can also be obtained by injecting electrons at the SiO_2/Si_3N_4 interface of a silicon/oxide/nitride/oxide/silicon (SONOS) structure. When used as gate dielectric and gate electrode the SONOS structure transforms a MOSFET into an NVM device.[75] The use of an SOI substrate gives more design flexibility, and it is possible to program NVM SONOS devices by injecting electrons in an oxide/nitride/oxide structure below the transistor. This approach offers the advantage of full processing compatibility with regular CMOS devices but requires the fabrication of an oxide/nitride/oxide BOX structure.[76]

8.2.3.2 Capacitorless DRAM

The traditional DRAM memory cell consists of an access transistor and a storage capacitor.[77] In partially depleted SOI devices, it is possible to store a charge in the floating body and, therefore, to realize a DRAM memory cell that does not require a capacitor. Such a capacitorless or "one-transistor" (1-T) memory cell is, of course, much smaller than a classical DRAM cell. Memory effect due to charge storage in the floating body of an SOI transistor was first reported in 1990 and called the "multistable charge-controlled memory" (MCCM) effect.[78] The study of the kink effect (section 5.9.1) shows that holes generated by impact ionization in an SOI transistor biased in saturation give the floating body a positive bias, which in turn decreases the threshold voltage. If the transistor is at low temperature (77K or 4.2K in Ref. [78]) such that generation/recombination effects are hindered, the body bias and the associated threshold voltage shift can remain for a long time. At room temperature the body bias does not last long, but it can be used to store information if it is refreshed periodically. The 1-T DRAM memory cell is based on a similar principle. A "1" binary state is stored by biasing the transistor in saturation and injecting holes in the floating body. Applying a negative bias to the bit line (device source) removes the charge from the floating body and, therefore, stores a binary "0" in the device. Additionally, other bias conditions can be used for programming the device.

[79, 80, 81, 82, 83] The retention time of the 1-T cell is on the order of a second and satisfies most DRAM requirements. Blocks of memory cells can be refreshed without the need for reading and re-writing individual cells. Using this "block refresh" scheme, information can be kept in the devices almost indefinitely.[84] It is worth noting that the 1-T memory cell works not only with PDSOI devices, but also with FDSOI transistors.

8.2.4 Analog circuits

The dc gain (A_{vo}) and the transition unit-gain frequency (f_T) of a MOSFET are given by $A_0 = -\dfrac{g_m}{I_D} V_A$ and $f_T = \dfrac{g_m}{2\pi C_L}$ where V_A is the Early voltage of the device, and C_L is its load capacitance. The g_m/I_D ratio is a measure of the efficiency to translate current (and hence power) into transconductance (and hence amplification). The larger the g_m/I_D ratio, the larger the transconductance for a given current value. The high g_m/I_D value of fully depleted SOI devices allows one to realize near-optimal micropower analog designs (g_m/I_D values of 35 V^{-1} are obtained, while g_m/I_D reaches typical values of 25 V^{-1} in bulk MOSFETs – see section 5.5.1 and Figure and 8-7).[85]

CMOS operational amplifiers (op amps) can be designed on the basis of the knowledge of g_m/I_D and the Early voltage, V_A, as a function of the scaled drain current, $I_D/(W/L)$.[86] This design technique was used to synthesize a simple 1-stage opamp with specified active device size (W/L) and output capacitance (C_L). Figure 8-8 clearly shows that the A_{vo} of the SOI amplifier is 25 to 35% larger than in bulk counterpart, while the static bias current, I_{dd}, is reduced. The simulation is based on the assumption that n is close to 1.1 and 1.5 for fully depleted SOI and bulk MOSFETs, respectively, and that V_A is similar for both device types. Taking into account the reduction of parasitic source/drain and intrinsic gate capacitances [87] found in SOI in addition to the lower n value, it can be further demonstrated that a 2-stage Miller opamp can be synthesized to achieve an A_{vo} 15 dB larger and a I_{dd} 3 times smaller in SOI than in bulk for similar area and identical bandwidth considerations (Figure 8-10). Conversely, total area and current consumption can be divided by a factor of 2-3 when all other specifications are kept constant.

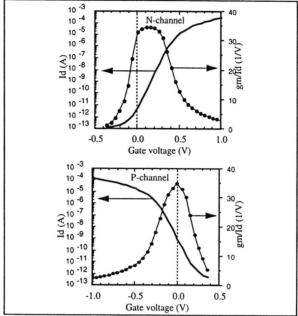

Figure 8-7. g_m/I_D and I_D in 600µm/20µm SOI transistors. V_{back}= 0V (fully depleted n-channel) and V_{back} = -1.5V (accumulation-mode p-channel). Silicon film and gate oxide thickness are 80 and 30 nm, respectively.

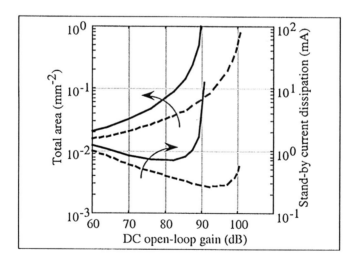

Figure 8-8. Comparison of bulk (−) and SOI (--) Miller amplifier for C_L = 10 pF and f_T = 10 MHz.

8.2.5 Mixed-mode circuits

As we have seen in the two previous sections, SOI technology is well adapted for the fabrication of low-voltage digital and analog circuits. In addition, good RF performance can be obtained from SOI transistors (see section 5.14 and table 5-5). Mixed-mode circuits combine analog, digital and RF devices. If their complexity is large they are sometimes referred to as "system on a chip" (SoC). In addition to offering superior digital, analog and RF transistor performance, SOI technology enables the reduction of crosstalk between devices and the fabrication of high-quality inductors.[88]

The word "crosstalk" describes the undesired transmission of an electrical signal from one device to another. This can happen because of capacitive coupling between metal lines, or through the substrate, which acts as a transmission medium linking different devices. The higher the resistivity of the substrate, the lower the crosstalk. The use of high-resistivity silicon as a substrate is impractical in bulk silicon technology. When SOI wafers are used the devices are separated from the silicon substrate by a dielectric material (the BOX), which reduces crosstalk,[89,90,91] and there is no restriction on the use of high-resistive silicon underneath the buried oxide, which furthers reduces crosstalk.[92,93,94] Figure 8-9 shows the crosstalk (or magnitude of electromagnetic power transmitted, S_{21}, measured using the S-parameter technique) between two junctions located 50 μm apart. The junctions are made on either bulk silicon or SOI substrates with different resistivities. One can see that the crosstalk at 1 GHz is reduced by 20 dB compared to bulk when a high-resistivity SOI substrate is used. Tthe improvement in RF performance brought about by the use of high-resistivity substrates disappears if an inversion layer is present at the substrate/BOX interface. Such a layer acts as a low-resistive path that transmits electrical signals and increases crosstalk.[95]

Inductors are critical components of RF circuits. Integrated inductors are spiral structures made using the chip metal levels. The inductance of these components depends on the number of turns. Because of the capacitive coupling between the metal and the substrate and because of dielectric losses in the substrate at high frequencies, spiral inductors made on silicon (or SOI) substrates have a poor quality factor, Q. The value of Q for inductors made on "regular" SOI wafers (resistivity=20 Ω.cm) is in the order of 4 to 6. If high-resistivity silicon is used as a substrate underneath the BOX, quality factors can be increased to values as high as 11 [96], 15 [97], or 22 [98] when single metal-layer inductors are used, and 50 [99] when 4 metal layers are used to form the inductor. Higher quality factor values can be obtained if a full dielectric substrate such as SOS is used, in which case there is no silicon underneath the inductors. Figure 8-10 shows an example of quality factor *vs.* frequency in single metal level, planar spiral inductors made on bulk, low-resistivity SOI, high-resistivity SOI and SOS substrates.

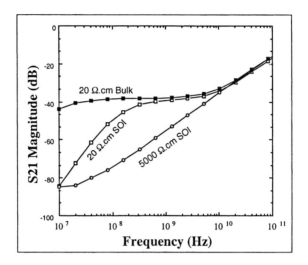

Figure 8-9. Crosstalk between nodes separated by a distance of 50 μm on different substrates.[92]

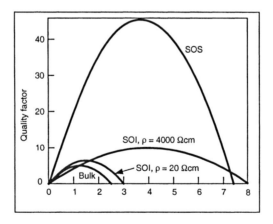

Figure 8-10. Quality factor of single metal level spiral inductors made on bulk, SOI [100, 101, 102, 103] and SOS substrates.[104]

8.3 NICHE APPLICATIONS

SOI was used in specific niche application circuits long before being adopted for mainstream CMOS applications. The most important of these niche applications are high-temperature electronics, radiation-hardened circuits, and smart-power circuits.

8.3.1 High-temperature circuits

The excellent behavior of SOI MOSFETs at high temperature renders them highly suitable for high-temperature IC applications. The major causes of failure in bulk CMOS logic at high temperature are excess power consumption due to high leakage current, and degradation of logic levels and noise margin. These effects are strongly reduced when SOI is used. Latch-up is of course totally suppressed when SOI MOSFETs are used. SOI CMOS inverters exhibit full functionality and very little change in static characteristics at temperatures up to 320°C.[105] The switching voltage remains stable if fully depleted devices are used, owing to the small variation of the threshold voltage as a function of temperature. In bulk devices the leakage current becomes very high when temperature is increased. That leakage current flows from the supply rail to the grounded substrate through all reverse-biased PN junctions.

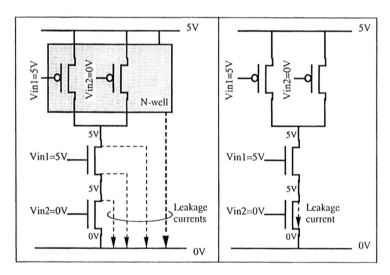

Figure 8-11. Leakage current paths in a bulk (left) and an SOI (right) CMOS NAND gate. Both gates have a high and a low input bias. All bulk junction leakage currents flow in parallel towards the substrate. The leakage of the SOI gate is limited to that of a single transistor.[106]

If we take the NAND gate of Figure 8.11 as an example we find that the total current leaking from V_{DD} to ground is equal to the *sum* of currents in three reverse-biased source/drain junctions, plus the leakage current of the n-well. In SOI circuits, the drain leakage current of each individual transistor flows towards its source, not to the substrate. Thus the leakage current is restricted actual circuit branches, unlike in bulk circuits, where all leakage currents flow into the substrate. In SOI the total leakage current is equal to the current in the least leaky device, while in bulk it is equal to the sum of all leakage currents.

While digital circuit parameters include subthreshold slope, leakage current and drive current, analog devices operation relies on other parameters. For instance, standby current consumption is fixed by the operating point of the circuit rather than by junction leakage. One important parameter is the output conductance of the transistors. Indeed, the dc voltage gain of an operational amplifier (Operational Transconductance Amplifier, OTA) is proportional to the product of the transconductance of the input transistors by the output impedance of the output transistors. As the temperature is increased, the transconductance of the input transistor decreases because of the reduction of carrier mobility with temperature. The output impedance (the inverse of the output conductance), on the other hand, increases as temperature is increased (Figure 7-12). Owing to this kind of compensation effect in the input transconductance-output impedance product, the resulting dc gain of the overall amplifier is expected to be relatively independent of temperature. The measured characteristics of the device are consistent with what can be expected from theory: as the gate transconductance decreases and the output impedance increases, the dc gain of the SOI CMOS amplifier remains quite stable.[107] Additionaly, the offset voltage of the amplifier remains between 0 and 2 millivolts for all temperatures between 20 to 300°C. Such amplifiers are suitable for A-to-D converters and switched capacitor circuits having to operate in a high-temperature environment. The operation of SOI A-to-D converters at a temperature of 350°C has been demonstrated.[108]

The excellent behavior of both analog and digital SOI CMOS circuits at high temperatures suggests the use of this technology for different applications. Classical industrial applications such as well logging can take advantage of SOI CMOS for *in-situ* signal amplification and data processing in the high-temperature environment encountered deep inside the earth's crust. Automotive and aerospace industries are also big potential consumers of SOI CMOS. Indeed, car electronics (engine and brake control systems, as well as underhood electronics) has a growing need for both analog and digital circuits which are able to withstand temperatures of 200°C or above.[109] Underhood electronics must be able to withstand relatively high temperatures when a car is parked in the hot sun (up to 200°C). Much higher temperature tolerance is expected from engine monitoring electronics, such electronic injection systems, where precision control of the injector position is controlled by electronic components located close to or within the engine block.[110] Modern anti-lock braking systems (ABS) also require electronic control systems placed close to the brakes and, therefore, submitted to high temperatures when the brakes are activated. The ability of SOI CMOS to operate at high-temperatures, combined with the possibility of integrating power devices with low-power logic on a single, dielectrically isolated substrate, renders SOI the ideal technology for such purposes.[111,112] High-temperature airplane applications include on-board electronics and, of course, engine control and surface control (wing temperature). Space applications are also expected to use SOI: thermal shielding for satellites is quite heavy, and the weight of a satellite has a direct impact on how much it costs to place it in orbit. In some special applications (Venus or Mercury probes), shielding is no longer possible and high-temperature electronics is absolutely required. Commercial nuclear applications (power plant) also require high-temperature electronics for core control.[113]

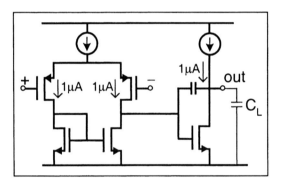

Figure 8-12. Layout of a simple low-power Miller operational amplifier. V_{dd}=3V and I_{dd}=3μA (1μA in each branch of the circuit).

In the case of analog circuits such as operational amplifiers, the increase of leakage current with temperature may result in the loss of the operating point, in bulk circuits. To illustrate this effect let us take the example of basic Miller amplifier that requires that a bias current of 1 μA flows in each of the three branches of the circuit (Figure 8.12). In a bulk design, some (or most) of this bias current flows to the substrate. When the circuit operates at high temperature and the leakage current of the junctions is high. As a result the devices are no longer properly biased and the circuit no longer operates correctly (unless the bias current is increased to compensate for the leakage losses). If SOI technology is used the leakage current can only flow through the branches of the circuit, and the circuit can operate at high temperature without losing of the bias operating point.

Figure 8-13 shows performances of an operational transconductance amplifier (OTA) whose design is based on the zero-temperature coefficient (ZTC) concept. According to ZTC there exists a bias point where the current is independent of temperature (the reduction of mobility is compensated by the reduction of threshold voltage). The open-loop dc current gain is 115 dB with 100 MHz bandwidth at room temperature. This corresponds to an amplification factor of 560,000. The dc gain is still 50 dB at a temperature of 400°C.

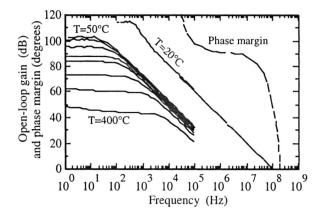

Figure 8-13. Measured Bode diagram of an operational transconductance amplifier. Measurement was carried out with a 10 pF load at 20°C and a 250 pF load at all other temperatures. The phase margin was measured at 20°C.[114]

8.3.2 Radiation-hardened circuits

Another niche market where SOI circuits and devices are used is the aerospace/military market, where radiation hardness is a key issue. High-level radiation hardness is also a prerequisite for circuits used in core instrumentation for nuclear power reactors (a containment accident may lead to more than 100 Mrad(Si) of dose exposure), nuclear well logging, control electronics in spacecraft nuclear reactors, reaction chamber instrumentation for particle colliders, and fusion reactor instrumentation. In some cases, these high-level irradiation environments may be associated with high temperature conditions.[115] When exposed to SEU or gamma-dot irradiation conditions SOI has an advantage over bulk CMOS devices. This is due to reduced active silicon volume and full dielectric isolation which gives rise to lower leakage currents in SOI than in bulk. As we have seen in Section 7.1, SOI MOSFETs are inherently harder against SEU and gamma-dot than bulk devices. SOI circuits are also free of latchup problems associed with bulk CMOS, which can be triggered by an energetic heavy ion (SEL) or gamma-dot photocurrents. Upon SEU or gamma-dot exposure SOI circuits such as 4k×1 SIMOX SRAMs show bit error rates comparable to SOS but better than those obtained in bulk CMOS.[116] The main reason for not obtaining performances superior to those of SOS is the amplification of the SEU-induced photocurrent by the parasitic lateral bipolar transistor present in SOI devices, where the minority carrier lifetime is high, compared to that of SOS material. SOI memories (64k SRAMs) offer SEU error rates of 10^{-9} errors/bit-day (worst-case geosynchronous orbit error rate) and retain functionality at dose-rate exposures up to 10^{11} rad(Si)/sec.[117] These numbers represent a hardness improvement by a factor $\cong 100$ over bulk silicon. A similar observation can be made for other types of digital circuits: Figure 8-14 shows the heavy-ion SEU cross section for the data cache bit transitions from "0" to "1" of bulk and SOI PowerPC microprocessors. The cross section of the SOI circuit is about 10 times smaller than that of the corresponding bulk part.

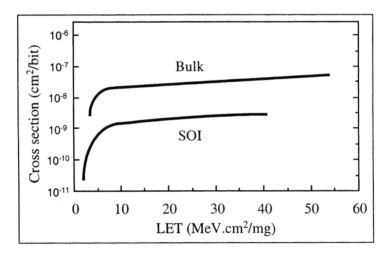

Figure 8-14. Heavy-ion SEU cross section for the data cache bit transitions from "0" to "1" of PowerPC microprocessors.[118]

Using fully-depleted technology, a 256K SOI SRAM has demonstrated under worst case SEU and prompt dose testing, an LET threshold of at least 80 MeV.cm²/mg and a prompt dose rate upset level of greater then 4×10^{10} rad(Si)/s, without design hardening. Total dose testing on transistors indicates over 100 krad(Si) capability. Together, these results represent a fully depleted SOI technology that can be used for all radiation-hardened applications except extreme total dose.[119] 1Mb SOI SRAMs fabricated on fully-depleted SOI using 0.5 μm design rules were evaluated for speed/power, prompt dose upset, SEE, and total dose hardness. As compared to the identical design fabricated on bulk CMOS, improved results were seen for performance and prompt dose hardness. On the other hand, a low n-channel snapback voltage (see section 7.1.2) was shown to degade the standby leakage, total dose hardness, and SEU hardness.[120]

Total-dose hardening of SOI circuits can be obtained using an optimized fabrication process and special transistor design.[121,122] The degradation phenomena generated by dose exposure on CMOS circuits manifests itself as a slower response time (*e.g.* an increase of the access time in SRAMs) and an increase of power consumption. These are mostly due to threshold voltage shifts and to an increase of leakage current in MOSFETs upon irradiation. Most rad-hard 16k and 64k CMOS SOI SRAMs are designed to remain functional after exposures to doses up to 1 Mrad(Si) [123,124,125], but more recent parts show 10 Mrad(SiO_2) hardness levels [126] (1 rad(SiO_2) = 0. 56 rad(Si)). The highest total-dose hardness results reported to date for large circuits mention the operation of 29101 16-bit microprocessors at doses up to 100 Mrad(SiO_2). These circuits exhibit a 30% and 100% increase of worst-case propagation delay after 10 and 30 Mrad(SiO_2) exposure, respectively. However, some parts are still functional after 100 Mrad(SiO_2) irradiation. Smaller CMOS/SIMOX circuits (frequency dividers) can show no propagation delay degradation at doses up to 1 Mrad(Si) and can even survive doses of 0.5 Grad(Si). [127,128]

The typical radiation hardness characteristics of an hardened SOI SRAM are given below (Honeywell HLX6228, 128K × 8 SRAM):

Total dose hardness	1 Mrad(SiO_2)
Neutron hardness	10^{14} cm^{-2}
Dose rate	10^{11} rad(Si)/s
SEU rate	<10^{-10} upsets/bit-day
SEL	latchup free
Read/write cycle time	14 ns (typical)
Power supply	3.3 V ±0.3 V, <12mW/MHz
Gate length	0.6 μm (L_{eff} = 0.45 μm)

Serving as another example, the specifications of the DMILL circuit family is listed below:[129,130]

Available devices: CMOS, NPN bipolar, and P-channel JFET
Input/Output analog & Digital structures available in the design kit
Hardened 4000V ESD protection included
Radiation assurance up to 10Mrad(Si) and 10^{14} neutrons cm^{-2}
Noise Monitoring
Military temperature range -55 to +125°C

8.3.3 Smart-power circuits

The full dielectric isolation provided by the SOI substrate allows one to fabricate high-voltage devices without the burden of having to use a complicated junction isolation process. Compared to bulk, the advantages of the SOI full dielectric isolation include a higher integration density, no latch-up, less leakage current, high-temperature operation, improved noise immunity, easier circuit design, and the possibility to integrate vertical DMOS transistors. Furthermore, power devices can be integrated on the same substrate as CMOS logic to produce "smart power" circuits. Relatively thick SOI films (5 to 20 μm) are usually employed, although high-voltage devices can be fabricated in thin SOI films as well.[131] Thin-film SOI RESURF LDMOS transistors are used to manufacture 250-V plasma display drivers [132], 600-V power conversion systems [133], compact fluorescent lamp electronics.[134] LDMOS transistors have very small recovery charges, which makes them ideal devices for switching applications such as the output stage of class D power amplifiers.[135] Table 6-4 in section 6.3.2 lists the electrical characteristics of the most important high-voltage SOI devices.

The flexibility of dielectric isolation allows for the fabrication of mixed chips featuring *e.g.:* CMOS + NPN bipolar + VDMOS devices (Figure 8-15) using a process which requires fewer masking steps than the equivalent bulk process.

The main drawback of dielectric isolation lies in the fact that it features a high thermal resistance between the device and the heat sink, because the thermal conductivity of SiO_2 is lower than that of silicon. As a result, the temperature in SOI power devices can rise to values higher than in bulk power devices. The difference in self-heating between SOI and bulk power devices is largest for short power transients, because the initial temperature rise in SOI devices is more rapid. As the pulse length increases, the difference in temperature rise between SOI and bulk devices converges to a constant value, which is proportional to the thickness of the buried oxide. It is worth noting that, for oxide thicknesses under 2 μm, the steady-state temperature is essentially determined by the thermal properties of the substrate and the device package, and not by the buried oxide.[136]

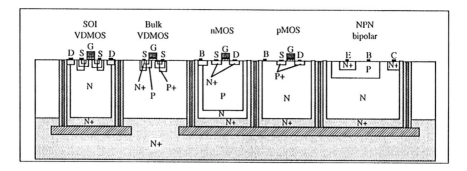

Figure 8-15. Process featuring SOI quasi-vertical DMOS, bulk vertical DMOS, CMOS and NPN devices.[137]

8.4 THREE-DIMENSIONAL INTEGRATION

The fabrication of three-dimensional (3D) integrated circuits may be considered the Holly Grail of SOI technology. In a three-dimensional integrated circuit devices are fabricated in several active layers stacked on one another (Figure 8.16).

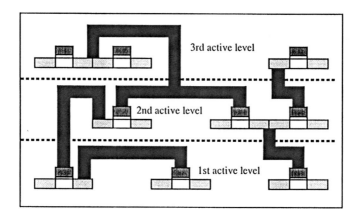

Figure 8-16. Illustration of the 3-D integrated circuit concept.

Three-dimensional integration is particularly well suited for the realization of large systems involving massive parallel processing. To illustrate this concept, consider the image-processing chip (computer vision system) described in Figure 8-17. This circuit extracts the boundaries of objects and transfers this information to a computer that recognizes the nature of the object and its orientation. The circuit operates as follows: image detection is performed by a two-dimensional photodiode array. The information contained in each pixel is digitized (we will consider only two luminosity levels, black and white, *i.e.* "1" and "0") and each pixel is compared with its eight neighbors (north, north-east, east, south-east, south, south-west, west, and north-west) to find out whether the point it represents lies inside the object, outside of it, or on its boundary. Such a comparison can be realized by software or hard-wired by means of logical XOR gates. When the edges of the object have been extracted, they can be expressed in the form of 2D vectors from which the nature and the in-plane orientation of the object can be deduced. In the case of the implementation of such a system using classical 2-D chips, the data contained in each pixel must be transferred after digitization to a computer that performs the comparison operations (on-chip digitization is possible, but at the expense of the sensor fill factor). Because of the limited amount of I/O pins of the circuits, this data transfer is carried out in a largely serial manner. If the amount of data is large (1024 x 1024 pixel array = 1 Mbit of data for each frame), the serial data transfer process quickly becomes the bottleneck of the system, due to the limited bandwidth of the data transmission bus.[138] Once in the computer, the image is reconstructed in the memory and the comparison and feature extraction operations are performed.

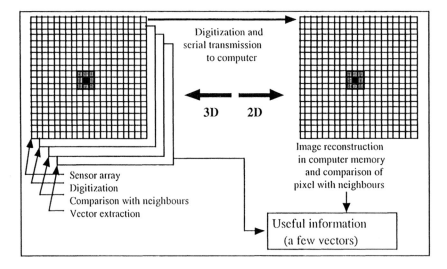

Figure 8-17. Comparison between 3-D parallel data processing and 2-D serial data processing.

In the case of a 3-D implementation of such a system, sensing, digitization, comparison and vector extraction can be carried out in four separate active device layers (Figure 8-18). All the pixels can be transferred *simultaneously* and *vertically* from the sensor array to the digitization layer. In this way, maximum sensor fill factor is obtained, which minimizes the integration time of the photodiode current. After A to D conversion, the data are transferred *simultaneously* and *vertically* to the next layer where comparison operations between pixels are performed. Each pixel is naturally surrounded by the pixels with which it has to be compared. Once this comparison operation has been performed, the data can be transferred to the bottom layer for feature extraction. The important advantage of the 3-D approach is the highly parallel inter-layer data transfer, which eliminates the data transmission bottleneck experienced in the 2D implementation. As a result, considerably higher speed performance is obtained from 3-D parallel processing systems. An increase of real-time image processing by a factor of about 1000 (compared to conventional 2D processing) has been observed in small (5x5 pixels) systems [139], and such processing speed advantage is only expected to increase as the complexity of the system increases.[140] It is also worthwhile noting that the length of the interconnections between two different active layers is typically on the order of a micrometer, while the length of interconnections between different logic blocs in 2-D circuits is generally several hundred micrometers. This also gives rise to further processing speed increase in 3-D systems. Massive parallel processing capability and reduced interconnection length offer advantage not only in image processing circuits, but also in neural networks [141], associative and common memories [142], high-density SRAMs [143,144], and particle detectors.[145]

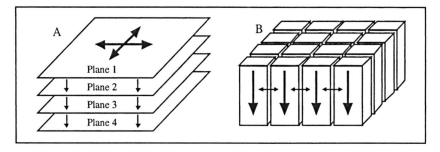

Figure 8-18. A: 3-D arrangement of 2-D circuits, B: 2-D arrangement of 3-D circuits (B).[140]

Three-dimensional ICs can be viewed as a succession of 2-D circuits stacked one upon another. Functionally, they can be considered either as a 3-D association of 2-D processing planes or a 2-D association of vertical data flow paths (Figure 8-18). In the first case, most of the data processing takes place within a horizontal plane and is then transferred to the next layer. In the case of the 2-D association of vertical circuits, the data flow from top to bottom in a highly parallel fashion, but interaction such as the comparison of a pixel with its neighbors occurs within horizontal planes. Most image processing 3-D integrated circuits belong to this category.

Three-dimensional circuits were extensively studied during the 1980's in Japan, and several remarkable circuits were successfully fabricated.[146,147] These include a moving object detector [148,149] and a character recognition system [150,151,152]. These circuits were made using laser or e-beam recrystallization, which allows for the formation of SOI layers on top of processed devices. More recently, however, other techniques, such as epitaxial lateral overgrowth (ELO – see section 2.5.1) [153], wafer bonding [154,155], and bonding/debonding [156] combined with Smart-Cut® [157] have been proposed for the fabrication of 3D circuits.

REFERENCES

[1] S.B. Park, Y.W. Kim, Y.G. Ko, K.I. Kim, I.K. Kim, H.S. Kang, J.O. Yu , K.P. Suh, IEEE Journal of Solid-State Circuits, Vol. 34, p. 1436, 1999

[2] M. Canada, C. Akroul, D. Cawlthron, J. Corr, S. Geissler, R. Houle, P. Kartschoke, D. Kramer, P. McCormick, N. Rohrer, G. Salem, L. Warriner, IEEE International Solid-State Circuits Conference, Digest of Technical Papers, p. 430, 1999

[3] A.G. Aipperspach, D.H. Allen, D.T. Cox, N.V. Phan, S.N. Storino, IEEE Journal of Solid-State Circuits, Vol. 34, p. 1430, 1999

[4] S. Geissler, D. Appenzeller, E. Cohen, S. Charlebois, P. Kartschoke, P. McCormick, N. Rohrer, G. Salem, P. Sandon, B. Singer, T. Von Reyn, J. Zimmerman, IEEE International Solid-State Circuits Conference, Digest of Technical Papers, Part Vol. 1, p.148, 2002

[5] C.N. Keltcher, K.J. McGrath, A. Ahmed, P. Conway, IEEE Micro, Vol. 23, no. 2, p. 66, 2003

[6] K. Shimomura, H. Shimano, N. Sakashita, F. Okuda, T. Oashi, Y. Yamaguchi, T. Eimori, M. Inuishi, K. Arimoto, S. Maegawa, Y. Inoue, S. Komori, K. Kyuma, IEEE Journal of Solid-State Circuits, Vol. 32, p. 1712, 1997

[7] T. Oashi, T. Eimori, F. Morishita, T. Iwamatsu, Y. Yamaguchi, F. Okuda, K. Shimomura, H. Shimano, S. Sakashita, K. Arimoto, Y. Inoue, S. Komori, M. Inuishi, T. Nishimura, H. Miyoshi, Technical Digest of the International Electron Devices Meeting, p. 609, 1996

[8] T. Eimori, T. Oashi, F. Morishita, T. Iwamatsu, Y. Yamaguchi, F. Okuda, K. Shimomura, H. Shimano, N. Sakashita, K. Arimoto, Y. Inoue, S. Komori, M. Inuishi, T. Nishimura, H. Miyoshi, IEEE Transactions on Electron Devices, Vol. 45, p. 1000, 1998

[9] J.W. Park, Y.G. Kim, I.K. Kim, K.C. Park, K.C. Lee, T.S. Jung TS, IEEE Journal of Solid-State Circuits, Vol. 34, p. 1446, 1999

[10] Y.H. Koh, M.R. Oh, J.W. Lee, J.W. Yang, W.C. Lee, C.K. Park, J.B. Park, Y.C. Heo, K.M. Rho, B.C. Lee, M.J. Chung, M. Huh, H.S. Kim, K.S. Choi, W.C. Lee, J.K. Lee, K.H. Ahn, K.W. Park, J.Y. Yang, H.K. Kim, D.H. Lee, I.S. Hwang, Technical Digest of International Electron Devices Meeting, p. 579, 1997

[11] Y.W. Kim, S.B. Park, Y.G. Ko, K.I. Kim, I.K. Kim, K.J. Bae, K.W. Lee, J.O. Y , U. Chung, K.P. Suh, Digest of Technical Papers of the IEEE International Solid-State Circuits Conference, p.432, 1999

[12] T. Yamauchi, F. Morisita, S. Maeda, K. Arimoto, K. Fujishima, H. Ozaki, T. Oshihara, IEEE Journal of Solid-State Circuits, Vol. 35, no. 8, p. 1169, 2000

[13] F. Ichikawa, Y. Nagatomo, Y. Katakura, S. Itoh, H. Matsuhashi, N. Hirashita, S. Baba, Electrochemical Society Proceedings, Vol. 2003-05, p. 123, 2003

[14] M. Itoh, Y. Kawai, S. Ito, K. Yokomizo, Y. Katakura, Y. Fukuda, F. Ichikawa, Electrochemical Society Proceedings, Vol. 2001-3, p. 331, 2003

[15] B.Y. Hwang, M. Racanelli, M. Huang, J. Foerstner, S. Wilson, T. Wetteroth, S. Wald, J. Rugg, S. Cheng, Extended Abstracts of the International Conference on Solid-State Devices and Materials, Yokohama, p. 268, 1994

[16] G.G. Shahidi, T.H. Ning, R. Dennard, B. Davari, Extended Abstracts of the International Conference on Solid-State Devices and Materials, Yokohama, p. 265, 1994

[17] M. Guerra, A. Wittkower, J. Stahman, J. Schrankler, Semiconductor international, May 1990

[18] M. Itoh, Y. Kawai, S. Ito, K. Yokomizo, Y. Katakura, Y. Fukuda, F. Ichikawa, Electrochemical Society Proceedings, Vol. 2001-3, p. 331, 2001

[19] Y. Fukuda, S. Ito, M. Ito, OKI Technical Review, Vol. 4, p. 54, 2001

[20] M.M. Pelella, W. Maszara, S. Sundararajan, S. Sinha, A. Wei, D. Ju, W. En, S. Krishnan, D. Chan, S. Chan, P. Yeh, M. Lee, D. Wu, M. Fuselier, R. van Bentum, G. Burbach, C. Lee, G. Hill, D. Greenlaw, C. Riccobene, O. Karlsson, D. Wristers, N. Kepler, Proceedings of the IEEE International SOI Conference, p. 1, 2001

[21] M.M. Pelella, J.G. Fossum, IEEE Transactions on Electron Devices, Vol. 49, no. 1, p. 96, 2002

[22] G.G. Shahidi, IBM Journal of Research & Development, Vol. 46, no. 2-3, p. 121, 2002

[23] M.A. Pelella, J.G. Fossum, IEEE Transactions on Electron Devices, Vol. 49, no. 1, pp.96, 2001

[24] D. Flandre, D. Vanhoenacker, International Semiconductor Conference (CAS'98) Proceedings, p. 115, 1998

[25] A.O. Adan, T. Naka, S. Kaneko, D. Urabe, K. Higashi, A. Kagisawa, IEEE International SOI Conference Proceedings, p. 116, 1996

[26] T. Douseki, J. Yamada, H. Kyuragi, Symposium on VLSI Circuits Digest of Technical Papers, p. 6, 2002

[27] D. Flandre, J.P. Colinge, J. Chen, D. De Ceuster, J.P. Eggermont, L. Ferreira, B. Gentinne, P.G.A. Jespers, A. Viviani, R. Gillon, J.P. Raskin, A. Vander Vorst, D. Vanhoenacker-Janvier, F. Silveira, Analog Integrated Circuits & Signal Processing, Vol. 21, no. 3, p. 213, 1999

[28] T. Douseki, S. Shigematsu, Y. Tanabe, M. Harada, H. Inokawa, T. Tsuchiya, Digest of Technical Papers of the IEEE International Solid-State Circuits Conference, p. 84, 1996

[29] M. Ino, H. Sawada, K. Nishimura, M. Urano, H. Suto, S. Date, T. Ishiara, T. Takeda, Y. Kado, H. Inokawa, T. Tsuchiya, Y. Sakakibara, Y. Arita, K. Izumi, K. Takeya, T. Sakai Digest of Technical Papers, IEEE International Solid-State Circuits Conference, p. 86, 1996

[30] T. Fuse, Y. Oowaki, M. Terauchi, S. Watanabe, M. Yoshimi, K. Ohuchi, J. Matsunaga, Digest of Technical Papers, IEEE International Solid-State Circuits Conference, p. 88, 1996

[31] A.O. Adan, T. Naka, S. Kaneko, D. Urabe, K. Higashi, A. Kasigawa, Proceedings of the IEEE International SOI Conference, p. 116, 1996

[32] T. Fuse, Y. Oowaki, Y. Yamada, M. Kamoshida, M. Ohta, T. Shino, S. Kawanaka, M. Terauchi, T. Yoshida, G. Matsubara, S. Oshioka, S. Watanabe, M. Yoshimi, K. Ohuchi, S. Manabe, Digest of Technical papers of the International Solid-State Circuits Conference, p. 286, 1997

[33] Y. Ohtomo, S. Yasuda, M. Nogawa, J. Inoue, K. Yamakoshi, H. Sawada, M. Ino, S. Hino, Y. Sato, Y. Takei, T. Watanabe, K. Takeya, Digest of Technical papers of the International Solid-State Circuits Conference, p. 154, 1997

[34] T. Douseki, H. Kyuragi, Proceedings of the IEEE International SOI Conference, p. 5, 2003

[35] H. Shang, M.H. White, D.A. Adams, Proceedings of the IEEE International SOI Conference, p. 37, 2002

[36] M. Terauchi, Proceedings of the IEEE International SOI Conference, p. 108, 2000

[37] P. Villard, J. Jomaah, B. Gomez, J. de Pontcharra, D. Save, S. Chouteau, E. Mackowiak, Electrochemical Society Proceedings, Vol. 2003-05, p. 243, 2003

[38] M. Itoh, Y. Kawai, S. Ito, K. Yokomizo, Y. Katakura, Y. Fukuda, F. Ishikawa, Electrochemical Society Proceedings, Vol. 2001-3, p. 331, 2001

[39] Y. Kado, Y. Matsuya, T. Douseki, S. Nakata, M. Harada, J. Yamada, Electrochemical Society Proceedings, Vol. 2001-3, p. 277, 200

40 D. Levacq, V. Dessard, D. Flandre, Electrochemical Society Proceedings, Vol. 2003-05, p. 249, 2003

41 D. Levacq, C. Liber, V. Dessard, D. Flandre, Proceedings of the IEEE International SOI Conference, p. 19, 2003

42 S. Natarajan, A. Marshall, IEEE International Conference on Electronics, Circuits and Systems, Part Vol. 2, p. 835, 2002

43 T. Eimori, T. Oashi, F. Morishita, T. Iwamatsu, Y. Yamaguchi, F. Okuda, K. Shimomura, H. Shimano, N. Sakashita, K. Arimoto, Y. inoue, S. Komori, M. Inuishi, T. Nishimura, H. Miyoshi, IEEE transactions on Electron Devices, Vol. 45, no. 5, p. 1000, 1998

44 T. Ikeda, S. Wakahara, Y. Tamaki, H. Higuchi, Proceedings of the IEEE International SOI Conference, p. 159, 1998

45 Y.Wada, K. Nii, H. Kuriyama, S. Maeda, K. Ueda, Y. Matsuda, Proceedings of the IEEE International SOI Conference, p. 127, 1998

46 Y. Hu, C. Teng, T.W. Houston, K. Joyner, T.J. Aton, Digest of Technical papers, Symposium on VLSI Technology, p. 128, 1996

47 K. Izumi, Y. Omura, M. Ishikawa, E. Sano, Technical Digest of the Symposium on VLSI Technology, p. 10, 1982

48 W.A. Krull, J.C. Lee, Proc. IEEE SOS/SOI Technology Workshop, p. 69, 1988

49 T. W. Houston, H. Lu, P. Mei, T.G.W. Blake, L.R. Hite, R. Sundaresan, M. Matloubian, W.E. Bailey, J. Liu, A. Peterson, G. Pollack, Proc. IEEE SOS/SOI Technology Workshop, p. 137, 1989

50 A.J. Auberton-Hervé, B. Giffard, M. Bruel, Proceedings of the IEEE SOS/SOI Technology Workshop, p. 169, 1989

51 W.E. Bayley, H. Lu, T.G.W. Blake, L.R. Hite, P. Mei, D. Hurta, T.W. Houston, G.P. Pollack, Proceedings of the IEEE International SOI Conference, p. 134, 1991

52 L.K. Wang, J. Seliskar, T. Bucelot, A. Edenfeld, N. Haddad, Technical Digest of the International Electron Device Meeting, p. 679, 1991

53 H. Lu, E. Yee, L. Hite, T. Houston, Y.D. Sheu, R. Rajgopal, C.C. CHen, J.M. Hwang, G. Pollack, Proceedings of IEEE International SOI Conference, p. 182, 1993

54 G.G. Shahidi, T.H. Ning, T.I. Chappell, J.H. Comfort, B.A. Chappell, R. Franch, C.J. Anderson, P.W. Cook, S.E. Schuster, M.G. Rosenfield, M.R. Polcari, R.H. Dennard, and B. Davari, Technical Digest of the International Electron Device Meeting, p. 813, 1993

55 T. Eimori, T. Oashi, H. Kimura, Y. Yamaguchi, T. Iwamatsu, T. Tsuruda, K. Suma, H. Hidaka, Y. Inoue, T. Nishimura, S. Satoh, and M. Miyoshi, Technical Digest of the International Electron Device Meeting, p. 45, 1993

56 F.T. Brady, T. Scott, R. Brown, J. Damato, N.F. Haddad, IEEE Transactions on Nuclear Science, Vol. 41, no. 6, p. 2304, 1994

57 H.-S. Kim, S.-B. Lee, D.-U. Choi, J.-H. Shim, K.-C. Lee, K.-P. Lee, K.-N. Kim, and J.-W. Park, Digest of Technical Papers of the Symposium on VLSI Technology, p. 143, 1995

58 T. Oashi, T. Eimori, F. Morishita, T. Iwamatsu, Y. Yamaguchi, F. Okuda, K. Shimomura, H. Shimano, S. Sakashita, K. Arimoto, Y. Inoue, S. Komori, M. Inuishi, T. Nishimura, and H. Miyoshi, Technical Digest of the International Electron Device Meeting, p. 609, 1996

59 K. Shimomura, H. Shimano, N. Sakashita, F. Okuda, T. Oashi, Y. Yamaguchi, T. Eimori, M. Inuishi, K. Arimoto, S. Maegawa, Y. Inoue, S. Komori, K. Kyuma, IEEE Journal of Solid-State Circuits, Vol. 32, no. 11, p. 1712, 1997

[60] Y.H. Koh, M.R. Oh, J.W. Lee, J.W. Yang, W.C. Lee, C.K. Park, J.B. Park, Y.C. Heo, K.M. Rho, B.C. Lee, M.J. Chung, M. Huh, H.S. Kim, K.S. Choi, W.C. Lee, J.K. Lee, K.H. Ahn, K.W. Park, J.Y. Yang, H.K. Kim, D.H. Lee, I.S. Hwang, Technical Digest of International Electron Devices Meeting, p. 579, 1997

[61] F.T. brady, R. Brown, L. Rockett, J. Vasquez, IEEE Transactions on Nuclear Science, Vol. 45, no. 6, p. 2436, 1998

[62] J.W. Park, Y.G. Kim, I.K. Kim, K.C. Park, K.C. Lee, T.S. Jung, IEEE Journal of Solid-State Circuits, Vol. 34, p. 1446, 1999

[63] Y.W. Kim, S.B. Park, Y.G. Ko, K.I. Kim, I.K. Kim, K.J. Bae, K.W. Lee, J.O. Y, U. Chung, K.P. Suh, Digest of Technical Papers of the IEEE International Solid-State Circuits Conference, p. 32, 1999

[64] T. Yamauchi, F. Morisita, S. Maeda, K. Arimoto, K. Fujishima, H. Ozaki, T. Yoshihara, IEEE Journal of Solid-State Circuits, Vol. 35, no. 8, p. 1169, 2000

[65] T. Douseki, N. Shibata, J. Yamada, Proceedings of the IEEE International SOI Conference, p. 24, 2000

[66] D. Goldman, K. DeGregorio, C.S. Kim, M. Nielson, J. Zahurak, S. Parke, 2001 IEEE International SOI Conference. Proceedings, p. 97, 2001

[67] M. Itoh, Y. Kawai, S. Ito, K. Yokomizo, Y. Katakura, Y. Fukuda, F. Ishikawa, Electrochemical Society Proceedings, Vol. 2001-3, p. 331, 2001

[68] T. Douseki, N. Shibata, J. Yamada, Proceedings of the IEEE International SOI Conference, p. 143, 2001

[69] A. Chatterjee, J. Liu, S. Aur, P.K. Mozumder, M. Rodder, I.C. Chen, Technical Digest of the International Electron Device Meeting, p. 87, 1994

[70] Y. Hu, C. Teng, T.W. Houston, K. Joyner, and T.J. Aton, Digest of Technical papers, Symposium on VLSI Technology, p. 128, 1996

[71] A. Zaleski, D.E. Ioannou, D. Flandre, J.P. Colinge, Proceedings of the IEEE International SOI Conference, p. 13, 1994

[72] M.H. Chi, A. Bergemont, Proceedings of the IEEE International SOI Conference, p. 129, 1995

[73] H. Choi, T. Tanabe, N. Kotaki, K.W. Koh, K.T. Park, H. Kurino, M. Koyanagi, Exgtended Abstracts of the International Conference on Solid-State Devices and Materials, p. 246, 2001

[74] X. Lin, M. Chan, Proceedings of the IEEE International SOI Conference, p. 71, 2002

[75] Y.K. Lee, J.S. Sim, S.K. Sung, C.J. Lee, T.H. Kim, J.D. Lee, B.G. Park, D.H. Lee, Y.W. Kim, IEEE Electron Device Letters, Vol. 23, no. 11, p. 664, 2002

[76] H. Silva, M.K. Kim, C.W. Kim, S. Tiwari, Proceedings of the IEEE International SOI Conference, p. 105, 2003

[77] J.A. Mandelman, R.H. Dennard, G.B. Bronner, J.K. DeBrosse, R. Divakaruni, Y. Li, C.J. Radens, IBM Journal of Research & Development, Vol. 46, no. 2-3, p. 187, 2002

[78] M.R. Tack, M. Gao, C.L. Claeys, G.J. Declerck, IEEE Transactions on Electron Devices, Vol. 37, no. 5, p. 1373, 1990

[79] S. Okhonin, M. Nagoga, J.M. Sallese, P. Fazan, IEEE International SOI Conference, p. 153, 2001

[80] S. Okhonin, M. Nagoga, J.M. Sallese, P. Fazan, IEEE Electron Device Letters, Vol. 23, no. 2, p. 85, 2002

[81] S. Okhonin, M. Nagoga, J.M. Sallese, P. Fazan, O. Faynot, J. de Pontcharra, S. Cristoloveanu, H. van Meer, K. De Meyer, Solid-State Electronics, Vol. 46, no. 11, p. 1709, 2002

82 P. Fazan, S. Okhonin, M. Nagoga, J.M. Salese, L. Portmann, R. Ferrant, M. Pastre, M. Blagojevic, A. Borschberg, M. Declercq, Proceedings of the IEEE International SOI Conference, p. 10, 2002

83 T. Ohsawa, K. Fujita, T. Higashi, Y. Iwata, T. Kajiyama, Y. Asao, K. Sunouchi, IEEE Journal of Solid-State Circuits, Vol. 37, no. 11, p. 1510, 2002

84 P. Fazan, S. Okhonin, M. Nagoga, Proceedings of the IEEE International SOI Conference, p. 15, 2003

85 F. Silveira, D. Flandre, and P.G.A. Jespers, IEEE Journal of Solid-State Circuits, Vol. 31, no. 9, p. 1314, 1996

86 D. Flandre, B. Gentinne, J.P. Eggermont, and P.G.A. Jespers, Proceedings of the IEEE International SOI Conference, p.99, 1994

87 Y. Omura and K. Izumi, IEEE Electron Device Letters, Vol. EDL-12, p. 655, 1991

88 E.A. McShane, K. Shenai, Proceedings of the 22nd International Conference on Microelectronics (MIEL), Vol. 1, p. 107, 2000

89 I. Rahim, I. Lim, J. Foerstner, and B.Y. Hwang, Proceedings of the IEEE International SOI Conference, p. 170, 1992

90 K. Joardar, Solid-State Electronics, Vol. 39, no. 4, p. 511, 1996

91 I. Rahim, B.Y. Hwang, and J. Foerstner, Proceedings of the IEEE International SOI Conference, p. 170, 1992

92 A. Viviani, J.P. Raskin, D. Flandre, J.P. Colinge, and D. Vanoenacker, Technical Digest of the International Electron Device Meeting, p. 713, 1995

93 J.P. Raskin, A. Viviani, D. Flandre, J.P. Colinge, IEEE Transactions on Electron Devices, Vol. 44, no. 12, p. 2252, 1997

94 D. Flandre, J.P. Colinge, J. Chen, D. De Ceuster, J.P. Eggermont, L. Ferreira, B. Gentinne, P.G.A. Jespers, A. Viviani, R. Gillon, J.P. Raskin, A. Vander Vorst, D. Vanhoenacker-Janvier, Analog Integrated Circuits and Signal Processing, Vol. 21, p. 213, 1999

95 D. Lederer, C. Desrumeaux, F. Brunier, J.P. Raskin, Proceedings of the IEEE International SOI Conference, p. 50, 2003

96 D. Eggert, P. Huebler, A. Huerrich, H. Kueck, W. Budde, M. Vorwerk, IEEE Transactions on Electron Devices, Vol. 44, no. 11, p. 1981, 1997

97 F. Ishikawa, Y. Nagatomo, Y. Katakura, S. Itoh, H. Matsuhashi, N. Hirashita, S. Baba, Electrochemical Society Proceedings, Vol. 2003-05, p. 123, 2003

98 J.O. Plouchart, Proceedings of the IEEE International SOI Conference, p. 1, 2003

99 ibidem

100 M. Vorwerk, D. Eggert, IEEE International Conference on Electronics, Circuits and Systems, Part Vol. 2, p. 31, 1998

101 D. Eggert, P. Huebler, A. Huerrich, H. Kueck, W. Budde, M. Vorwerk, IEEE Transactions on Electron Devices, Vol. 44, no. 11, p. 1981, 1997

102 Y. Fukuda, S. Ito, M. Ito, OKI Technical Review, Vol. 4, p. 54, 2001

103 F. Ichikawa, Y. Nagatomo, Y. Katakura, S. Itoh, H. Matsuhashi, N. Hirashita, S. Baba, Electrochemical Society Proceedings, Vol. 2003-05, p. 123, 2003

104 R. Reedy, J. Cable, D. Kelly, M. Stuber, F. Wright, G. Wu, Analog Integrated Circuits and Signal Processing, Vol. 25, p. 171, 2000

105 P. Francis, A. Terao, B. Gentinne, D. Flandre, J.P. Colinge, Technical Digest of the International Electron Device Meeting, p. 353, 1992

106 A.J. Auberton-Hervé, J.P. Colinge and D. Flandre, Japanese Solid State Technology, pp. 12-17, Dec. 1993 (日本語に)

[107] P. Francis, A. Terao, B. Gentinne, D. Flandre, and J.P. Colinge, Technical Digest of the International Electron Device Meeting, p. 353, 1992

[108] A. Viviani, D. Flandre, and P. Jespers, Proceedings of the IEEE SOI Conference, p. 110, 1996

[109] M. Hattori, S. Kuromiya, F. Itoh, IEEE International SOI Conference, p. 120, 1998

[110] M. Hattori, S. Kuromiya, F. Itoh, Proceedings of the IEEE International SOI Conference, p. 119, 1998

[111] J. Weyers and H. Vogt, Technical Digest of the International Electron Device Meeting, p. 225, 1992

[112] A. Nakagawa, N. Yasuhara, I. Omura, Y. Yamaguchi, T. Ogura, T. Matsudai, Technical Digest of the International Electron Device Meeting, p. 229, 1992

[113] D.B. King, "A summary of high-temperature electronics needs, research and development in the United States", presented at the EUROFORM seminar, May 1992, Darmstadt, Germany

[114] B. Gentinne, J.P. Eggermont, J.P. Colinge, Electronics Letters, Vol. 31-24, p. 2092, 1995

[115] D.M. Fletwood, F.V. Thome, S.S. Tsao, P.V. Dressendorfer, V.J. Dandini, and J.R. Schwank, IEEE Transactions on Nuclear Science, Vol. 35, p. 1099, 1988

[116] G.E.Davis, L.R. Hite, T.G.W. Blake, C.E. Chen, H.W. Lam, R. DeMoyer, IEEE Transactions on Nuclear Science, Vol. 32, p. 4432, 1985

[117] W.F. Kraus, J.C. Lee, Proceedings SOS/SOI Technology Conference, p. 173, 1989

[118] F. Irom, G.M. Swift, F.H. Farmanesh, A.H. Johnston, Proceedings of the IEEE International SOI Conference, p. 203, 2002

[119] F.T. Brady, T. Scott T, R. Brown, J. Damato, N.F. Haddad, IEEE Transactions on Nuclear Science, Vol. 41, no. 6, pt. 1, p. 2304, 1994

[120] F.T. Brady, R. Brown , L. Rockett, J. Vasquez, IEEE Transactions on Nuclear Science, Vol. 45, no. 6, pt. 1, p. 2436, 1998

[121] J.L. Leray, E. Dupont-Nivet, M Raffaelli, Y.M. Coïc, O. Musseau, J.F. Péré, P. Lalande, J. Brédy, A.J. Auberton-Hervé, M. Bruel, and B. Giffard, Annales de Physique, Colloque n°2, Vol. 14, suppl. to n°6, p. 565, 1989

[122] J.L. Leray, E. Dupont-Nivet, J.F. Péré, Y.M. Coïc, M. Rafaelli, A.J. Auberton-Hervé, M. Bruel, B. Giffard and J. Margail, IEEE Transactions on Nuclear Science, Vol. 37, no. 6, p. 2013, 1990

[123] G.T. Goeloe, G.A. Hanidu, and K.H. Lee, Tech. Prog. of IEEE Nuclear and Space Radiation Effects Conference (NSREC), paper PG-5, p. 35, 1990

[124] M.A. Guerra, Proc. of the 4th International Symposium on Silicon-on-Insulator Technology and Devices, Ed. by D. Schmidt, the Electrochemical Society, Vol. 90-6, p. 21, 1990

[125] W.F. Kraus, J.C. Lee, W.H. Newman, and J.E. Clark, Tech. Prog. of IEEE Nuclear and Space Radiation Effects Conference (NSREC), paper PG-6, p. 35, 1990

[126] L.J. Palkuti, J.J. LePage, IEEE Transactions on Nuclear Science, Vol. 29, p. 1832, 1982

[127] J.L. Leray, E. Dupont-Nivet, J.F. Péré, Y.M. Coïc, M. Rafaelli, A.J. Auberton-Hervé, M. Bruel, B. Giffard and J. Margail, IEEE Transactions on Nuclear Science, Vol. 37, no. 6, p. 2013, 1990

[128] J.L. Leray, E. Dupont-Nivet, J.F. Péré, O. Musseau, P. Lalande, and A. Umbert, Proceedings SOS/SOI Technology Conference, p. 114, 1989

[129] M. Dentan, E. Delagnes, N. Fourches, M. Rouger, M.C. Habrard, L. Blanquart, P. Delpierre, R. Potheau, R. Truche, J.P. Blanc, E. Delevoye, J. Gautier, J.L. Pelloie, J. de Pontcharra, O. Flament, J.L. Leray, J.L. Martin, J. Montaron, and O. Musseau, IEEE Transactions on Nuclear Science, Vol. 40, no. 6, p. 1555, 1993

130 O. Flament, J.L. Leray, and O. Musseau, *Low-power HF microelectronics: a unified approach*, edited by G.A.S. Machado, IEE circuits and systems series 8, the Institution of Electrical Engineers, p. 185, 1996

131 R.P. Zingg, Microelectronic Engineering, Vol. 59, p. 461, 2001

132 R.P. Zingg, Proceedings of the International Symposium on Power Semiconductor Devices and ICs (ISPSD '2001), p. 343, 2001

133 T. Letavic, M. Simpson, E. Arnold, E. Peters, R. Aquino, J. Curcio, S. Herko, S. Mukherjee, Proceedings of the International Symposium on Power Semiconductor Devices and ICs (ISPSD '99), p. 325, 1999

134 R.P. Zingg, Microelectronic Engineering, Vol. 59, p. 461, 2001

135 M. Berkhout, IEEE. IEEE Journal of Solid-State Circuits, Vol. 38, no. 7, p. 1198, 2003

136 E. Arnold, H. Pein, and S.P. Herko, Technical Digest of the International Electron Device Meeting, p. 813, 1884

137 J. Weyers, H. Vogt, M. Berger, W. Mach, B. Mütterlein, M. Raab, F. Richter, and F. Vogt, Proceedings of the 22nd ESSDERC, Microelectronic Engineering, Vol. 19, p 733, 1992

138 T. Nishimura, Y. Akasaka, Extended Abstracts of the 5th International Workshop on Future Electron Devices – Three Dimensional Integration, Miyagi-Zao, Japan, p. 1, 1988

139 T. Nishimura, Y. Inoue, K. Sugahara, S. Kusunoki, T. Kumamoto, S. Nagawa, N. Nakaya, Y. Horiba, Y. Akasaka, Technical Digest of the International Electron Device Meeting, p. 111, 1987

140 A. Terao, F. Van de Wiele, IEEE Circuits and Devices Magazine, Vol. 3, no. 6, p. 31, 1987

141 R. Aibara, Y. Mitsui, T. Ae, Extended Abstracts of the 8th International Workshop on Future Electron Devices – Three Dimensional ICs and Nanometer Functional Devices, Kochi, Japan, p. 113, 1990

142 T. Ae, Extended Abstracts of the 5th International Workshop on Future Electron Devices – Three Dimensional Integration, Miyagi-Zao, Japan, p. 55, 1988

143 Y. Inoue, T. Ipposhi, T. Wada, K. Ichinose, T. Nishimura, Y. Akasaka, Proceedings of the Symposium on VLSI Technology, p. 39, 1989

144 C.C. Liu, S. Tiwari, Proceedings of the IEEE International SOI Conference, p. 68, 2002

145 Y. Akasaka, Extended Abstracts of the 8th International Workshop on Future Electron Devices – Three Dimensional ICs and Nanometer Functional Devices, Kochi, Japan, p. 9, 1990

146 M. Nakano, Oyo Buturi, Vol. 54, no. 7, p. 652, 1985 (日本語に)

147 T. Nishimura, Y. Akasaka, Vol. 54, no. 12, p. 1274, 1985 (日本語に)

148 K. Yamazaki, Y. Itoh, A. Wada, Y. Tomita, Extended Abstracts of the 8th International Workshop on Future Electron Devices – Three Dimensional ICs and Nanometer Functional Devices, Kochi, Japan, p. 105, 1990

149 Y. Itoh, A. Wada, K. Morimoto, Y. Tomita, K. Yamazaki K, Microelectronic Engineering, Vol. 15, no. 1-4, p. 187, 1991

150 K. Kioi, S. Toyayama, M. Koba, Technical Digest of the International Electron Devices Meeting, p. 66, 1988

151 S. Toyoyama, K. Kioi, K. Shirakawa, T. Shinozaki, K. Ohtake, Extended Abstracts of the 8th International Workshop on Future Electron Devices – Three Dimensional ICs and Nanometer Functional Devices, Kochi, Japan, p. 109, 1990

152 K. Kioi, T. Shinozaki, S. Toyoyama, K. Shirakawa, K. Ohtake, S. Tsuchimoto, IEEE Journal of Solid-State Circuits, Vol. 27, no. 8, p. 1130, 1992

153 G.W. Neudeck, Electrochemical Society Proceedings, Vol. 99-3, p. 25, 1999

[154] J. Burns, L. McIlrath, J. Hopwood, C. Keast, D.P. Vu, K. Warner, P. Wyatt, Proceedings of the IEEE Onternational SOI Conference, p. 20, 2000

[155] K. Warner, J. Burns, C. Keast, R. Kunz, D. Lennon, A. Loomis, W. Mowers, D. Yost, Proceedings of the IEEE Onternational SOI Conference, p. 123, 2002

[156] C. Colinge, B. Roberds, B. Doyle, Journal of Electronic Materials, Vol. 30, p. 841, 2001

[157] H. Moriceau, O. Rayssac, B. Aspar, B. Ghyselen, Electrochemical Society Proceedings, Vol. 2003-19, p. 49, 2003

Index

Printed in the United States
112804LV00001B/12/A

9 781402 077739